Cambridge International AS and A level
Chemistry

Revision Guide

Judith Potter and Peter Cann

CAMBRIDGE
UNIVERSITY PRESS

University Printing House, Cambridge CB2 8BS, United Kingdom

One Liberty Plaza, 20th Floor, New York, NY 10006, USA

477 Williamstown Road, Port Melbourne, VIC 3207, Australia

314–321, 3rd Floor, Plot 3, Splendor Forum, Jasola District Centre, New Delhi – 110025, India

103 Penang Road, #05-06/07, Visioncrest Commercial, Singapore 238467

Cambridge University Press is part of the University of Cambridge.

It furthers the University's mission by disseminating knowledge in the pursuit of education, learning and research at the highest international levels of excellence.

www.cambridge.org
Information on this title: www.cambridge.org/9781107616653

© Cambridge University Press 2015

This publication is in copyright. Subject to statutory exception and to the provisions of relevant collective licensing agreements, no reproduction of any part may take place without the written permission of Cambridge University Press.

First published 2015

20 19 18 17 16 15 14 13 12 11 10 9 8

Printed in Great Britain by CPI Group (UK) Ltd, Croydon CR0 4YY

A catalogue record for this publication is available from the British Library

ISBN 978-1-107-61665-3 Paperback

Cambridge University Press has no responsibility for the persistence or accuracy of URLs for external or third-party internet websites referred to in this publication, and does not guarantee that any content on such websites is, or will remain, accurate or appropriate. Information regarding prices, travel timetables, and other factual information given in this work is correct at the time of first printing but Cambridge University Press does not guarantee the accuracy of such information thereafter.

NOTICE TO TEACHERS IN THE UK

It is illegal to reproduce any part of this work in material form (including photocopying and electronic storage) except under the following circumstances:

(i) where you are abiding by a licence granted to your school or institution by the Copyright Licensing Agency;

(ii) where no such licence exists, or where you wish to exceed the terms of a licence, and you have gained the written permission of Cambridge University Press;

(iii) where you are allowed to reproduce without permission under the provisions of Chapter 3 of the Copyright, Designs and Patents Act 1988, which covers, for example, the reproduction of short passages within certain types of educational anthology and reproduction for the purposes of setting examination questions.

All questions and answers provided have been written by the authors. In examinations, the way marks are awarded may be different.

Acknowledgements:

Photo credits: 12.01 Peter Cann; 13.02 Geophotos/Alamy; 13.03 Cordelia Molloy/Science Photo Library; 14.01 iStock/Thinkstock; 15.01 Paul Rapson/SPL; 16.01 NASA/SPL; 18.01, 18.02, 25.01, 27.01 Andrew Lambert/SPL; 28.01 Maxim Tupikov/Shutterstock

Table of Contents

How to use this book: a guided tour — IV

Unit 1: Moles and equations — 1
Unit 2: Atomic structure — 8
Unit 3: Electrons in atoms — 12
Unit 4: Chemical bonding — 18
Unit 5: States of matter — 26
Unit 6: Energy changes — 33
Unit 7: Redox reactions — 42
Unit 8: Equilibria — 46
Unit 9: Rates of reaction — 53
Unit 10: Periodicity — 59
Unit 11: Group 2 — 64
Unit 12: Group 17 — 68
Unit 13: Nitrogen and sulfur — 74
Unit 14: Introduction to organic chemistry — 79
Unit 15: Hydrocarbons — 88
Unit 16: Halogenoalkanes — 101
Unit 17: Alcohols, esters and carboxylic acids — 108
Unit 18: Carbonyl compounds — 117
Unit 19: Lattice energy — 123
Unit 20: Electrochemistry — 129
Unit 21: Further aspects of equilibria — 137
Unit 22: Reaction kinetics — 145
Unit 23: Entropy and Gibbs free energy — 153
Unit 24: Transition elements — 157
Unit 25: Benzene and its compounds — 166
Unit 26: Carboxylic acids and their derivatives — 174
Unit 27: Organic nitrogen compounds — 179
Unit 28: Polymerisation — 187
Unit 29: Analytical chemistry — 197
Unit 30: Organic synthesis — 209

Practical Skills 1 — 216
Practical Skills 2 — 219

Answers to Progress Check questions — 221
Answers to Exam-style questions — 240
Glossary — 255
Index — 266

How to use this book: a guided tour

Learning outcomes

You should be able to:

- define and use relative masses based on the ^{12}C scale
- define and use the mole and Avogadro's constant
- analyse simple mass spectra and use the data in relative atomic mass calculations
- define and use the terms *empirical* and *molecular formulae* and use experimental data to find them,
- write and construct equations and use them in performing mole calculations
- perform mole calculations involving solutions and gases,

Learning Outcomes – set the scene of each unit, help with navigation through the book and give a reminder of what is important about each topic

Progress check – check your own knowledge and see how well you are getting on by answering regular questions

Tip – quick suggestions to remind you about key facts and highlight important points.

Worked example – a step by step approach to answering questions, guiding you through from start to finish.

Progress check 2.01

1. a. The volume of a particular nucleus in an atom is 2.8×10^{-45} m³. The volume of the atom itself is 5.4×10^{-37} m³. Assuming that the electrons have no mass, work out the ratio:

   $$\frac{\text{average density of the nucleus}}{\text{the average density of the atom}}$$

   b. What particles make up the nucleus of the atom?

   c. From your answer to part a, what can you say about the size of the forces holding a nucleus together?

2. A stream containing a mixture of particles A and B is passed between the plates of an electric field. A is a proton. B is a proton and a neutron joined together. Draw a clearly labelled diagram of the paths of A and B when in the electric field.

2.02 Atoms and nucleons

As we saw above, the number of protons in the nucleus equals the number of electrons round the outside. Atoms of the same element have the same number of protons in the nucleus; atoms of different elements have different numbers of protons in the nucleus.

The number of protons in the nucleus is called the proton number and it is this that gives the element its position in the Periodic Table and tells us which element it is.

The number of neutrons in the nucleus increases as the proton number increases but not so regularly. Different atoms of the same element may have different numbers of neutrons in the nucleus. As the protons and neutrons account for most of the mass of the atom, the number of protons and neutrons together (the nucleon number) is also called the mass number.

Isotopes (see Unit 1) are atoms which have the same proton number but different mass numbers. They are atoms of the same element but with different numbers of neutrons.

Atoms can be described by the isotopic symbol, showing the symbol of the element, the proton number and the mass number:

$$^m_p E$$

where m = mass number and p = proton number.

Ions are formed when electrons are gained or lost. If electrons are gained, there are more electrons than protons and so a negative ion (an anion) is formed; if electrons are lost, there are fewer electrons than protons and a positive ion (a cation) is formed.

Make sure you know and can state the relative masses and charges of proton, neutron and electron. Practise writing isotopic symbols correctly.

Worked example 2.01

What are the numbers of protons, neutrons and electrons in ^4_2He, $^{51}_{23}V^{3+}$ and $^{77}_{34}Se^{2-}$?

How to get the answer:

Step 1: Use the Periodic Table to identify the elements. 'He' indicates helium, 'V' indicates vanadium and 'Se' indicates selenium.

Step 2: The lower number at the left is the proton number. ^4_2He has 2 protons, $^{51}_{23}V^{3+}$ has 23 and $^{77}_{34}Se^{2-}$ has 34.

Step 3: The upper number is the nucleon number. Remember that the neutron number equals the nucleon number minus the proton number. ^4_2He has 2 neutrons, $^{51}_{23}V^{3+}$ has 28 and $^{77}_{34}Se^{2-}$ has 43.

Step 4: The number of electrons equals the number of protons minus the charge. ^4_2He has 2 electrons, $^{51}_{23}V^{3+}$ has 20 and $^{77}_{34}Se^{2-}$ has 36.

	^4_2He	$^{51}_{23}V^{3+}$	$^{77}_{34}Se^{2-}$
protons	2	23	34
neutrons	2	28	43
electrons	2	20	36

Sample answer

Question:

a What is the empirical formula of a hydrocarbon containing 83.7% carbon? [4]

b If the $M_r = 86$, what is the molecular formula of the hydrocarbon? [2]

Answer: a

Element	%	Divide by A_r	Divide by the smallest	Convert to whole number by ×3
C	83.7	$\frac{83.7}{12} = 6.975$	$\frac{6.975}{6.975} = 1$	3
H	16.3	$\frac{16.3}{1} = 16.3$	$\frac{16.3}{6.975} = 2.34$	$2.34 \times 3 = 7$
		[1 mark]	[1 mark]	[1 mark]

empirical formula = C_3H_7 [1 mark]

b $M_r[C_3H_7] = 43$

$\frac{86}{43} = 2$ [1 mark]

Two empirical formula units form the molecular formula, C_6H_{14}. [1 mark]

Sample answer – an example of a question with an excellent answer that would score high marks

Revision checklist

Check that you know the following:

- describe and sketch the shapes of s and p orbitals
- explain and use ionisation energy
- explain why the ionisation energy changes across a period and down a group
- how to draw a diagram of the energy levels in an atom and describe the number and type of orbitals associated with them
- Deduce which orbitals are filled and how to represent this by electron configurations

Revision checklist – at the end of each unit so you can check off the topics as you revise them

Exam-style questions

1 Figure 3.09 represents the energy levels in a nitrogen atom.

Figure 3.09

a i Label the energy levels with the principal quantum number and the type of orbital it contains. [1]

   ii Show, by using half arrows to represent electrons, the electron distribution in a nitrogen atom. [2]

   iii Draw the shapes of two different types of orbital occupied by electrons in a nitrogen atom. [2]

b i Give the electronic configurations of Cl and Cl^-. [2]

   ii Which atom has the same electronic configuration as the electronic configuration of Cl^-? [1]

Total: 8

2 a i Use a Data Book to state the values of the first ionisation energies of Be and B. [1]

      ii State how and explain why the first ionisation energies of Be and B differ. [2]

   b Write an equation for the second ionisation energy of an element, M. [1]

   c The first eight ionisation energies in kJ mol^{-1} for element X are:

   1140 2100 3500 4560 5760
   8549 9939 18600

   To which group in the Periodic Table does the element belong? [1]

   d i Write equations for the first ionisation energies of F^- and of F. [2]

      ii Suggest which of the two values is the larger and explain your answer. [2]

Total: 9

Exam-style questions – prepare for examinations by completing the exam-style questions and checking your answers which are provided at the back of the book

Unit 1

Moles and equations

Learning outcomes

You should be able to:

- define and use relative masses based on the ^{12}C scale
- define and use the mole and Avogadro's constant
- analyse simple mass spectra and use the data in relative atomic mass calculations
- define and use the terms *empirical* and *molecular formulae* and use experimental data to find them,
- write and construct equations and use them in performing mole calculations
- perform mole calculations involving solutions and gases,

This unit gathers together the basic chemistry calculations. You may need to refer to it as you revise other topics in later units.

1.01 Relative masses

An atom is the smallest part of an element that can exist and still retain the identity of the element. A molecule is a group of atoms bonded together and the smallest portion of an element or compound that can exist alone. A simple ion is an atom which has lost or gained one or more electrons. Atoms, molecules and ions are so small that their masses are far too small to measure directly on a balance. For most purposes, chemists use relative masses, which are the number of times heavier a particle is than (1/12) of the mass of one atom of carbon-12. Carbon-12 is the standard.

relative atomic mass,

$$A_r = \frac{\text{average mass of one atom of an element}}{(1/12) \text{ of the mass of one atom of } ^{12}C}$$

relative molecular mass,

$$M_r = \frac{\text{average mass of one molecule}}{(1/12) \text{ of the mass of one atom of } ^{12}C}$$

The relative molecular mass is the sum of all the relative atomic masses of the atoms making up the molecule.

If compounds are ionic, then '**relative formula mass**' is used instead of 'relative molecular mass'.

Atoms of an element always have the same number of protons in the nucleus but may have different numbers of neutrons. These atoms are called **isotopes** and have the same atomic number but a different mass number.

relative isotopic mass =

$$\frac{\text{mass of one atom of the isotope}}{(1/12) \text{ of the mass of one atom of } ^{12}C}$$

Relative masses are a ratio of two masses and have no units.

1.02 The mole and the Avogadro constant

To give an easy measurement of mass, a number of atoms is chosen so that their combined mass is equal to the relative atomic mass in grams. This number of atoms is called the **mole**.

> **TIP**
> Learn the relationships between mass and the mole.
> mass of one mole = M
> mass of n moles = m
> $n = \frac{m}{M}$
> $M = A_r$ in grams

The relative atomic mass is the weighted average of relative isotopic masses and may not be a whole number; relative isotopic masses are whole numbers.

a 12.0 g C = 1 mol of carbon;

$$3.0 \text{ g C} = \frac{3.0}{12.0} \text{ mol}$$
$$= 0.25 \text{ mol}$$

b 24.3 g Mg = 1 mol of magnesium;

$$6.0 \text{ g Mg} = \frac{6.0}{24.3} \text{ mol}$$

$$= 0.247 \text{ mol}$$

The number of atoms in 1 mole of carbon atoms equals the number of atoms in 1 mole of any other type of particle. The number of atoms in 1 mole is called the **Avogadro constant**.

The Avogadro constant, L, = 6.022×10^{23} mol^{-1}.

1 mol of carbon weighs 12.0 g and contains 6.022×10^{23} atoms

0.65 mol of carbon weighs 12.0×0.65 g and contains $0.65 \times 6.022 \times 10^{23}$ atoms

> **TIP**
> In mass calculations, make sure you use A_r values and not the atomic number.

1.03 Mass spectra and relative atomic mass calculations

A **mass spectrometer** is used to find the masses of atoms of individual isotopes; it produces a trace like the one for boron shown in Figure 1.01.

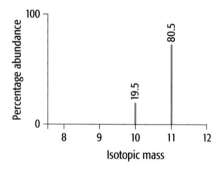

Figure 1.01 The mass spectrum for boron. The sum of the two percentage abundances = 100%.

Figure 1.01 shows two boron isotopes with relative isotopic masses of 10 and 11. The relative atomic mass for boron is the average of these values, taking account of the percentage abundances of each.

relative mass of 100 atoms of boron

$= \Sigma$ (relative isotopic mass $\times$ percentage abundance)

$$= (19.5 \times 10) + (80.5 \times 11)$$
$$= 1080.5$$
$$A_r[\text{boron}] = \frac{1080.5}{100}$$
$$= 10.81 \text{ (4 sig figs)}$$

> ### Progress check 1.01
> Be = 9.0; Mg = 24.3; Cr = 52.0; V = 50.9; Sc = 45.0; O = 16.0; Fe = 55.8
>
> 1 How many atoms are in the following:
>
>   a 3.0 g of beryllium atoms?
>
>   b 3.47 g of magnesium atoms?
>
>   c 3.71 g of chromium atoms?
>
> 2 What is the mass of the following:
>
>   a 6.022×10^{22} atoms of vanadium?
>
>   b 3.011×10^{23} atoms of scandium?
>
>   c 0.35 mol of O atoms?
>
>   d 0.018 mol of iron?
>
> 3 The four isotopes of chromium and their abundancies are given in the table.
>
> | Mass numbers | 50 | 52 | 53 | 54 |
> |---|---|---|---|---|
> | % abundances | 4.31 | 83.76 | 9.55 | 2.38 |
>
>   a Sketch the mass spectrum for chromium and label the axes.
>
>   b Work out A_r [chromium] to 3 sig figs from these values.
>
> 4 The mass spectrum for palladium is shown in Figure 1.02.
>
>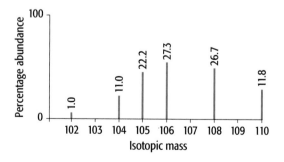
>
> Figure 1.02

a Label the axes of the spectrum.

b How many isotopes does palladium have?

c Work out the relative atomic mass of palladium using the values on the spectrum.

1.04 Empirical and molecular formulae

The molecular formula shows the actual number and types of atoms bonded together. Experimentally, you usually find the empirical formula, which is the simplest whole number ratio of all types of atoms bonded together. For example, the molecular formula of ethanoic acid is $C_2H_4O_2$ but its empirical formula is CH_2O. You can work out the empirical formula from the mass or the percentage mass of the elements combined together.

Worked example 1.01

0.50 g of compound **X** contains 0.30 g carbon, 0.133 g oxygen and the rest is hydrogen. What is the empirical formula of X?

How to get the answer:

Step 1: Find the mass of hydrogen combined = $(0.50 - 0.30 - 0.133)\,g = 0.067\,g$

Step 2: Put the elements and their masses into a table.

Step 3: Divide the masses by the respective A_r values and insert in the table.

Step 4: Divide all values by the smallest of the values in Step 3 and put into the table.

Elements	Masses combined / g	Divide by A_r	Divide by the smallest value
C	0.30	$\frac{0.30}{12} = 0.025$	$\frac{0.025}{8.31 \times 10^{-3}} = 3$
O	0.133	$\frac{0.133}{16} = 8.31 \times 10^{-3}$	$\frac{8.31 \times 10^{-3}}{8.31 \times 10^{-3}} = 1$
H	0.067	$\frac{0.067}{1} = 0.067$	$\frac{0.067}{8.31 \times 10^{-3}} = 8$

Step 5: Write the empirical formula. $X = C_3H_8O$

Progress check 1.02

1 0.350 g of compound **Y** contains 0.142 g carbon, 0.0297 g hydrogen, 0.0831 g nitrogen and 0.0949 g oxygen. Find the empirical formula of Y.

2 What is the empirical formula of an oxide of iron which contains 72.3% Fe?

The molecular formula is an integral number of empirical formula units; the molecular formula can be found if you know both the empirical formula and the relative molecular mass.

Sample answer

Question:

a What is the empirical formula of a hydrocarbon containing 83.7% carbon? [4]

b If the $M_r = 86$, what is the molecular formula of the hydrocarbon? [2]

Answer: a

Element	%	Divide by A_r	Divide by the smallest	Convert to whole number by ×3
C	83.7	$\frac{83.7}{12} = 6.975$	$\frac{6.975}{6.975} = 1$	3
H	16.3	$\frac{16.3}{1} = 16.3$	$\frac{16.3}{6.975} = 2.34$	2.34 × 3 = 7
		[1 mark]	[1 mark]	[1 mark]

empirical formula = C_3H_7 [1 mark]

b $M_r[C_3H_7] = 43$

$\frac{86}{43} = 2$ [1 mark]

Two empirical formula units form the molecular formula, C_6H_{14}. [1 mark]

1.05 Constructing equations

An equation is a shorthand way of describing a reaction using correct formulae for reactants and products.

For example, methane burns completely in oxygen to produce carbon dioxide and water. The equation becomes:

$$CH_4 + O_2 \longrightarrow CO_2 + H_2O$$

The equation needs to be balanced. Without changing the formulae, the numbers of molecules need to be adjusted so that the same number of atoms of each element are on the left-hand and right-hand sides of the arrow. In the above example, the carbon atoms balance but the hydrogen atoms need to be balanced:

$$CH_4 + O_2 \longrightarrow CO_2 + \mathbf{2}H_2O$$

and then the oxygen atoms need to be balanced:

$$CH_4 + \mathbf{2}O_2 \longrightarrow CO_2 + 2H_2O$$

Lastly, information on the physical state of reactants and products, using (g), (l), (s) or (aq), is inserted and any further information on conditions, such as temperature or pressure, is shown over the arrow:

$$CH_4(g) + 2O_2(g) \xrightarrow{burn} CO_2(g) + 2H_2O(l)$$

Progress check 1.03

Write correctly balanced equations for the following reactions.

Use formulae from this list:

$Cu(NO_3)_2$ HCl HNO_3 H_2O
Li_2CO_3 Li_2O NO NO_2 P_4
P_2O_5 O_2 $SiCl_4$ SiO_2

1. phosphorus burns in oxygen to produce phosphorus(V) oxide
2. silicon tetrachloride reacts with water to produce silica and hydrochloric acid
3. heating lithium carbonate produces carbon dioxide and lithium oxide
4. with concentrated nitric acid, copper produces nitrogen dioxide (nitrogen(IV) oxide) and water as well as copper nitrate
5. with dilute nitric acid, copper produces nitrogen monoxide (nitrogen(II) oxide) and water as well as copper nitrate

Balancing redox reactions using oxidation numbers is covered in Unit 7, section 7.

1.06 Mole calculations

A balanced equation shows the number of molecules and also the number of moles reacting and being produced.

a **Reacting masses**

As the number of moles is related to mass, a balanced equation can be used to calculate the masses reacting together and being produced in a reaction.

Worked example 1.02

What is the mass of magnesium oxide left when 1.00 g of magnesium carbonate is heated until there is no further reaction?

How to get the answer:

Step 1: Write the balanced equation:

$$MgCO_3(s) \longrightarrow MgO(s) + CO_2(g)$$

Step 2: Find the mole relationship:

1 mol $MgCO_3$ gives 1 mol MgO

Step 3: Work out $M_r[MgCO_3]$:

$[24.3 + 12.0 + (3 \times 16.0)] = 84.3$

Step 4: Work out how many moles you start with:

$$1.00 \text{ g } MgCO_3 = \frac{1.00}{84.3} \text{ mol } MgCO_3$$
$$= 0.0119 \text{ mol}$$

Step 5: Use the mole relationship:

0.0119 mol $MgCO_3$ gives 0.0119 mol MgO

Step 6: Work out $M_r[MgO]$:

$24.3 + 16.0 = 40.3$

Step 7: Find the mass:

0.0119 mol MgO weighs 0.0119×40.3 g
$= 0.480$ g

b **Reacting gas volumes**

At constant pressure and temperature, the volume of a gas is proportional to the number of moles (see Unit 5, section 5). This means that 1.0 mol of every gas occupies the same volume. At room temperature and pressure, this volume is 24 dm³.

For the reaction

$$H_2(g) + Cl_2(g) \longrightarrow 2HCl(g)$$

the equation shows that equal numbers of moles and thus equal volumes of hydrogen and chlorine react together.

The volume of carbon dioxide formed when 0.50 g ethanol is completely burnt is found by using a balanced equation to get the mole ratio:

- $CH_3CH_2OH + 3O_2 \longrightarrow 2CO_2 + 3H_2O$
- 1 mol ethanol $\longrightarrow$ 2 mol CO_2
- M_r[ethanol] = 46, so number of moles
  $= \dfrac{0.50}{46} \longrightarrow 2 \times \dfrac{0.50}{46}$ mol CO_2
- volume $CO_2 = 2 \times \dfrac{0.50}{46} \times 24$ dm³
  $= 0.522$ dm³ $= 522$ cm³

c **Reacting solutions**

The concentration of a solution is measured in mol dm⁻³.

$$\text{concentration in mol dm}^{-3} = \dfrac{\text{number of moles}}{\text{volume in dm}^3}$$

number of moles
= concentration in mol dm⁻³ × volume of solution in dm³

For example, identify M when 0.281 g of the Group 1 hydroxide, MOH, dissolves to make 100 cm³ of solution and a 10.0 cm³ portion reacts exactly with 6.68 cm³ of 0.075 mol dm⁻³ HCl:

- $MOH(aq) + HCl(aq) \longrightarrow MCl(aq) + H_2O(l)$
- number of moles HCl
  = concentration × volume
  = 0.075 mol dm⁻³ × 6.68 × 10⁻³ dm³
  = 5.01 × 10⁻⁴
- number of moles MOH in 10 cm³ of solution
  = 5.01 × 10⁻⁴
- number of moles MOH in 100 cm³ of solution
  = 5.01 × 10⁻³

- 0.281 g MOH contain 5.01 × 10⁻³ mol
- 1 mol MOH = $\dfrac{0.281}{5.01} \times 10^{-3}$ g = 56.1 g
- A_r[M] = 56.1 − 17 = 39.1
- M = potassium

d **Combustion data calculations**

An accurately known mass of an organic compound is burnt completely, to form CO_2 and H_2O. The volume or mass of carbon dioxide and the mass of water are accurately measured and the masses of C and H calculated. The empirical formula can then be worked out.

For example, Z contains carbon, hydrogen and oxygen. When 0.325 g Z is burnt completely, 265 cm³ CO_2 and 0.149 g H_2O are produced. What is the empirical formula of Z?

- 265 cm³ $CO_2 = \dfrac{265}{24\,000}$ mol = 0.0110 mol CO_2
- this contains 0.0110 mol C
- this weighs (0.0110 × 12) g = **0.132 g C**
- 0.149 g $H_2O = \dfrac{0.149}{18}$ mol H_2O
  = 8.28 × 10⁻³ mol H_2O
- this contains 0.0166 mol H = **0.0166 g H**
- mass O = (0.325 − 0.132 − 0.0166) g
  = **0.1765 g O**

Element	Mass / g	Divide by A_r	Divide by the smallest	×2
O	0.1765	$\dfrac{0.1765}{16} = 0.0110$	1	2
C	0.132	$\dfrac{0.132}{12} = 0.011$	1	2
H	0.0165	$\dfrac{0.0165}{1} = 0.0165$	1.5	3

- empirical formula = $C_2H_3O_2$

> **TIP**
> When working out reacting masses or gas volumes, always start with a correctly balanced equation.

Progress check 1.04

1. What mass of $FeCl_3$ is formed if 1.12 g $FeCl_2$ is reacted with excess chlorine?

2. Magnesium nitride, Mg_3N_2, is decomposed by water to produce magnesium oxide and ammonia (NH_3). What mass of magnesium oxide is produced by decomposing 1.00 g of the nitride?

3. A sample of sodium sulfate, Na_2SO_4, is contaminated with sodium nitrate; 5.00 g of the contaminated sample is dissolved in water and then reacted with excess aqueous barium chloride; 5.74 g of solid barium sulfate, $BaSO_4$, is formed. What mass of sodium sulfate was in the contaminated sample?

4. Compound T contains carbon, hydrogen and oxygen only. On complete combustion, 0.450 g of T produces 360 cm³ CO_2 and 0.27 g H_2O. Find the empirical formula of T.

5. Compound S contains carbon, hydrogen, oxygen and nitrogen. On complete combustion, 0.350 g of S produces 286 cm³ CO_2, 71 cm³ N_2 and 0.267 g H_2O. Find the empirical formula of S.

6. Aqueous barium chloride, $BaCl_2$, reacts with aqueous sulfuric acid, H_2SO_4, to produce a precipitate of barium sulfate and water. 10.0 cm³ of a barium chloride solution requires 15.0 cm³ of 0.0500 mol dm⁻³ H_2SO_4 for complete reaction. What is the concentration of the barium chloride solution in g dm⁻³?

7. A solution of ethanoic acid, CH_3COOH, is contaminated by oxalic acid, $(COOH)_2$. Ethanoic acid is monobasic but oxalic acid is dibasic. 10.0 cm³ of 0.0140 mol dm⁻³ NaOH requires 12.45 cm³ of the contaminated ethanoic acid solution for complete reaction. The concentration of the ethanoic acid = 0.0100 mol dm⁻³; what is the concentration of the oxalic acid?

Revision checklist

Check that you know the following:

- [ ] how to define and use 'relative atomic, isotopic, molecular and formula masses',
- [ ] how to use the mole and understand and use the concept of the Avogadro constant,
- [ ] how to analyse mass spectra in terms of isotopic abundances and calculate the relative atomic mass
- [ ] the definition, calculation and use of 'empirical' and 'molecular' formulae,
- [ ] how to construct balanced equations and use them to calculate reacting masses or gas volumes,
- [ ] how to work out the concentration of solutions and perform titration calculations

Exam-style questions

1. Sodium carbonate crystals, $Na_2CO_3 \cdot xH_2O$, react with aqueous hydrochloric acid to form sodium chloride, carbon dioxide and water. 1.350 g of $Na_2CO_3 \cdot xH_2O$ crystals react completely with 25.0 cm³ of 0.390 mol dm⁻³ HCl.

   a i Write a balanced equation for the reaction. [2]

      ii How many moles of HCl are used in the reaction? [1]

      iii How many moles of $Na_2CO_3 \cdot xH_2O$ are used in the reaction? [1]

      iv Work out x in the reaction. [3]

   b What mass of NaCl is formed in the reaction? [1]

   c i What is the maximum volume of CO_2 (measured at room temperature and pressure) which can be produced by the reaction? [1]

      ii Suggest why the actual volume of CO_2 would be less than your value in part c i. [1]

   Total: 10

2. W is a dibasic acid which contains carbon, hydrogen and oxygen only.

   a 1.50 g of W undergoes combustion analysis to form 1.73 dm³ of CO_2 and 0.488 g H_2O. Work out the empirical formula of W. [6]

   b 1.50 g of W was dissolved in water to make 1.00 dm³ of solution. 10.0 cm³ of this solution reacted exactly with 18.10 cm³ of 0.0100 mol dm⁻³ NaOH.

      i How many moles of NaOH were used? [1]

      ii How many moles of W react with the NaOH in part b i? [1]

      iii How many moles of W are contained in 1.50 g? [1]

      iv What is the M_r of W? [1]

   c Using your answers to parts a and b, what is the molecular formula of W? [2]

   Total: 12

Atomic structure

Learning outcomes

You should be able to:

- identify and describe the charge and mass of protons, neutrons and electrons
- deduce the behaviour of protons, neutrons and electrons in electric fields
- describe the mass and charge within an atom
- deduce the numbers of protons, neutrons and electrons in atoms and ions
- recognise and use isotopic symbols
- understand and use proton number and nucleon number to distinguish isotopes

2.01 Protons, neutrons and electrons

Atoms have a very small central **nucleus**; the diameter of an atom is about 10^5 times larger than the diameter of the nucleus.

There are three particles in atoms: **protons**, **neutrons** and **electrons**. The relative charges, masses and where the particles are found are shown in Table 2.01.

Particle	Relative charge	Relative mass	Where found
proton	+1	1	in nucleus
neutron	0	1	in nucleus
electron	−1	1/1840	outside nucleus

Table 2.01 The relative charges and masses of the particles in an atom.

> **TIP**
> The actual charges and masses of the protons, electrons and neutrons are very small and for most purposes we only need to know their relative values.

From Table 2.01, you can see that all the positive charges are in the tiny nucleus and so the forces holding these particles together must be very large. The negatively charged particles are spread around the outside of the nucleus. As the charge on a proton is equal and opposite to the charge on an electron, the number of protons in the nucleus must equal the number of electrons outside the nucleus in an uncharged atom.

Again looking at Table 2.01, you can see that the mass of the proton is very nearly the same as the mass of the neutron, but the mass of the electron is much smaller. Often it is sufficient to take the mass of an electron as zero as it is so much smaller than the mass of the neutron or proton. The two heavier particles are in the nucleus.

The nucleus is small, positively charged and heavy; it is surrounded by very light negatively charged particles.

The relative charges and masses of the three particles can be shown by the behaviour of beams of the particles in an electric field (see Figure 2.01).

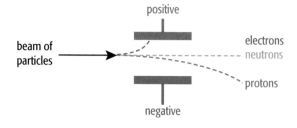

Figure 2.01 The behaviour of neutrons, electrons and protons in an electric field

The uncharged neutrons pass through the electric field undeflected; the negative electrons are attracted to the positive plate; the positive protons are attracted to the negative plate but as the mass of the proton is much greater, the deflection is less than for the electrons.

Progress check 2.01

1. a. The volume of a particular nucleus in an atom is 2.8×10^{-45} m^3. The volume of the atom itself is 5.4×10^{-37} m^3. Assuming that the electrons have no mass, work out the ratio:

    $$\frac{\text{average density of the nucleus}}{\text{the average density of the atom}}$$

   b. What particles make up the nucleus of the atom?

   c. From your answer to part a, what can you say about the size of the forces holding a nucleus together?

2. A stream containing a mixture of particles A and B is passed between the plates of an electric field. A is a proton. B is a proton and a neutron joined together. Draw a clearly labelled diagram of the paths of A and B when in the electric field.

2.02 Atoms and nucleons

As we saw above, the number of protons in the nucleus equals the number of electrons round the outside. Atoms of the same element have the same number of protons in the nucleus; atoms of different elements have different numbers of protons in the nucleus.

The number of protons in the nucleus is called the **proton number** and it is this that gives the element its position in the Periodic Table and tells us which element it is.

The number of neutrons in the nucleus increases as the proton number increases but not so regularly. Different atoms of the same element may have different numbers of neutrons in the nucleus. As the protons and neutrons account for most of the mass of the atom, the number of protons and neutrons together (the **nucleon number**) is also called the **mass number**.

Isotopes (see Unit 1) are atoms which have the same proton number but different mass numbers. They are atoms of the same element but with different numbers of neutrons.

Atoms can be described by the **isotopic symbol**, showing the symbol of the element, the proton number and the mass number:

$$^{m}_{p}E$$

where m = mass number and p = proton number.

Ions are formed when electrons are gained or lost. If electrons are gained, there are more electrons than protons and so a negative ion (an anion) is formed; if electrons are lost, there are fewer electrons than protons and a positive ion (a cation) is formed.

> **TIP** Make sure you know and can state the relative masses and charges of proton, neutron and electron. Practise writing isotopic symbols correctly.

Worked example 2.01

What are the numbers of protons, neutrons and electrons in $^{4}_{2}$He, $^{51}_{23}$V^{3+} and $^{77}_{34}$Se^{2-}?

How to get the answer:

Step 1: Use the Periodic Table to identify the elements. 'He' indicates helium, 'V' indicates vanadium and 'Se' indicates selenium.

Step 2: The lower number at the left is the proton number. $^{4}_{2}$He has 2 protons, $^{51}_{23}$V^{3+} has 23 and $^{77}_{34}$Se^{2-} has 34.

Step 3: The upper number is the nucleon number. Remember that the neutron number equals the nucleon number minus the proton number. $^{4}_{2}$He has 2 neutrons, $^{51}_{23}$V^{3+} has 28 and $^{77}_{34}$Se^{2-} has 43.

Step 4: The number of electrons equals the number of protons minus the charge. $^{4}_{2}$He has 2 electrons, $^{51}_{23}$V^{3+} has 20 and $^{77}_{34}$Se^{2-} has 36.

	$^{4}_{2}$He	$^{51}_{23}$V^{3+}	$^{77}_{34}$Se^{2-}
protons	2	23	34
neutrons	2	28	43
electrons	2	20	36

Sample answer

Question:

What particles would need to be lost from the nuclei of Th and Ra in steps 1 and 2 of the decay chain shown below?

$^{229}_{90}$Th → $^{225}_{88}$Ra → $^{225}_{89}$Ac? [5]

Answer:

$^{229}_{90}$Th → $^{225}_{88}$Ra has lost 2 protons [1 mark] and 4 nucleons so has lost 2p + 2n / He nucleus [1 mark].

$^{225}_{88}$Ra → $^{225}_{89}$Ac has gained 1 proton [1 mark] but has the same number of nucleons so has lost a neutron [1 mark]. The nucleus must have lost 1 electron for this to happen [1 mark].

Progress check 2.02

1. Use a Periodic Table to fill in the blanks in the following table.

Symbol	Species	Number of protons	Number of neutrons	Number of electrons
		56	82	54
	chloride ion		20	
$^{132}_{54}$Xe				
	strontium(II) ion		48	

2. Chromium exists as four isotopes with mass numbers 50, 52, 53 and 54. Use a Periodic Table to find which of these isotopes has the same number of neutrons in its atom as the atom of the isotope of titanium with mass number 48.

Revision checklist

Check that you know the following:

- [ ] the relative charges and masses of protons, neutrons and electrons and describe where these particles are found,
- [ ] deduce how protons, neutrons and electrons behave in an electric field,
- [ ] recognise and use isotopic symbols,
- [ ] how positive ions are formed by loss of electrons and negative ions by gain of electrons,
- [ ] deduce the numbers of protons, neutrons and electrons in any species from its isotopic symbol.

Exam-style questions

1. a i State the relative charges of the particles in the atom. [1]

   ii What is an isotope? [2]

   b i Complete the following table to show the numbers of the different particles in the atoms and ions:

Symbol	Number of neutrons	Number of protons	Number of electrons	
$^{23}_{11}A$				[1]
$^{22}_{10}B$				[1]
$^{27}_{13}C$				[1]
$^{25}_{12}D^+$				[1]
$^{26}_{12}E^{2+}$				[1]

   ii Which two particles in the table are isotopes? [1]

   iii Which particle in the table has the most mass in its nucleus? [1]

   Total: 10

2. a One of the stages in the radioactive decay of uranium-238 is when $^{210}_{83}Bi$ changes to $^{206}_{81}Tl$. The change occurs when one or more particles are emitted from the nucleus. The emitted particles travel away from the nucleus in straight lines (Figure 2.02).

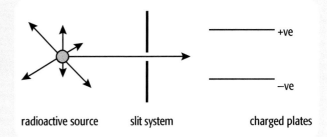

Figure 2.02

   i What particles are expelled from the $^{210}_{83}Bi$ nucleus when it forms $^{206}_{81}Tl$? [2]

   ii On the diagram, draw the path the particle takes after travelling through the slit system and between the charged plates. [2]

   b Another change which occurs during the decay is when $^{234}_{90}Th$ changes to $^{234}_{91}Pa$.

   i Complete the table.

Symbol	Number of protons	Number of neutrons	
$^{234}_{90}Th$			
$^{234}_{91}Pa$			[2]

   ii What is the relative charge and the relative mass of the emitted particle? [2]

   iii Suggest the identity of the emitted particle. [1]

   Total: 9

Unit 3

Electrons in atoms

Learning outcomes

You should be able to:

- describe the numbers and energies of s, p and d orbitals
- describe and sketch the shapes of s and p orbitals
- understand how the orbitals are filled by electrons and state the electronic configurations of atoms and ions
- explain and use ionisation energies and the factors which influence them
- explain how ionisation energy influences the chemical properties of an element and interpret how ionisation energy influences the position of elements in the Periodic Table
- deduce electron configurations of elements from ionisation energies

3.01 Principal quantum number and s, p and d orbitals

In Unit 2, we saw that the electrons in the atom spread out around the nucleus and, surprisingly, the nuclear protons are unable to attract them closer. The region of space in which an electron is likely to be found is an **orbital**. As well as being in an orbital, an electron can only have certain values of energy and cannot have 'in between' energy values.

The energy level is indicated by a **principal quantum number**, given the symbol n; $n = 1$ for the lowest energy level and then increases ... 2, 3, 4 ... for all integers up to infinity. The jumps between energy levels are not equal and get smaller and smaller so that by the time the energy level at infinity is reached, the jump is zero (Figure 3.01).

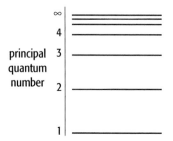

Figure 3.01 Energy levels in atoms

Each energy level has associated orbitals, which are given a letter, s, p, d or f, according to their shape. Every energy level has an associated spherical s orbital which gets larger as n gets larger (Figure 3.02). The electron is thought of as a mist of negative charge occupying the orbital.

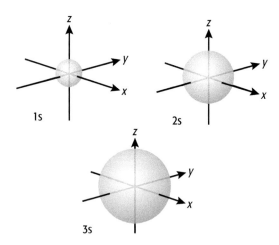

Figure 3.02 s orbitals

An electron in an atom is described by its principal quantum number and its orbital. For example, a 6s electron has energy corresponding to principal quantum number 6 and occupies a spherical s orbital.

Except for $n = 1$, every energy level has associated p orbitals; there are always three p orbitals together (Figure 3.03) pointing at right angles to each other.

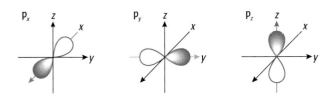

Figure 3.03 The p orbitals. The three p orbitals are aligned along perpendicular axes

When $n = 2$, there is a 2s orbital and three 2p orbitals available. Greater values of n have five associated d orbitals; these are more complex (see Unit 24).

You will be surprised that negatively charged electrons can occupy similar regions of space. The inter-electron repulsion is smaller than the energy jump into a higher energy level and so electrons occupy the lowest energy levels possible.

The detailed theory of the atom predicts that electrons have spin; they can be thought of as electric charge spinning around an axis. Some electrons spin one way and some the other way. Two electrons spinning in opposite directions are called **spin paired** and can occupy the same orbital, but then the orbital is full.

TIP
- An orbital can accommodate up to two electrons.
- Electrons occupy the lowest possible energy orbitals.
- Electrons spread between orbitals of the same energy.

The energy diagram for a hydrogen atom can show the orbitals in each energy level (Figure 3.04).

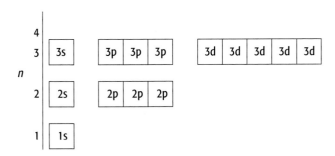

Figure 3.04 Energy levels and orbitals in a hydrogen atom

Each orbital can accommodate two electrons; the energy level $n = 2$ can accommodate up to eight electrons.

Progress check 3.01

1. a What shape is the 6s orbital?
   b How many electrons can a 6s orbital hold?
2. a How many orbitals are there in the level $n = 3$?
   b What is the maximum number of electrons that can be held in the level $n = 3$?

3.02 Ionisation energy

a **Information about the arrangement of electrons in orbitals comes from ionisation energy data**

The ionisation energy of an element is the amount of energy required to remove the outermost electron from each atom of a mole of gaseous atoms to form a mole of gaseous cations:

$$M(g) \longrightarrow M^+(g) + e^- \quad \Delta H = \text{ionisation energy}$$

TIP Make sure you know the definition of ionisation energy and the associated equation.

The periodicity of ionisation energy is marked. Group 1 metals have low ionisation energies and noble gases have high ionisation energies. Across a period, the ionisation energy rises; after each noble gas, there is a sharp drop to the low value of the alkali metal before the pattern is repeated.

For each successive element across a period, the nuclear charge increases as an electron is added to the outer energy level; electrons in this energy level are a similar distance from the nucleus. The attractive force between the electrons and the nucleus increases and so the ionisation energy increases.

When the first energy level is full, the next electron occupies the next energy level – which is further from the nucleus. The attractive force between the nucleus and the outer electron is now smaller; the outer electron is also **shielded** from the full charge of the nucleus by the negative charge of all the inner electrons. The ionisation energy shows a sharp decrease and then the pattern repeats.

The rise in ionisation energy across a period is not smooth. Electrons in p orbitals are slightly further from the nucleus than electrons in s orbitals and so their ionisation energy is lower. For similar reasons, d electrons take less energy to be removed than p electrons.

In the hydrogen atom, where there is only one electron, the 2s and 2p orbitals have the same energy. In atoms with more electrons, the p orbitals are higher in energy than the s orbital for any principal quantum number. Similarly, d orbitals are higher in energy than the p orbitals. Figure 3.05 shows the energy level diagram for Li.

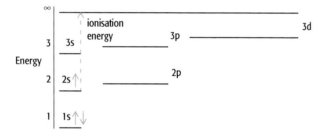

Figure 3.05 The energy level diagram for Li

The ionisation energy is the energy needed to remove an electron to the highest energy level where $n = \infty$ (see Figure 3.05). An electron occupying a higher energy orbital needs less energy to get to $n = \infty$ and the ionisation energy is smaller.

Compared with Li, Be has another proton and a fourth electron. This electron pairs up with the one already in the 2s orbital as more energy would be needed to jump into the 2p orbitals. The ionisation energy increases.

An ionisation energy decrease is shown by B as the new electron goes into a 2p orbital – which is further from the nucleus.

The ionisation energy increases for C and N as electrons are added to the 2p orbitals; the three 2p orbitals take an electron each (see Figure 3.06).

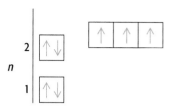

Figure 3.06 The electron diagram for N

The next added electron pairs up with one in a 2p orbital. Two electrons in the same space repel each other to some extent even though they are paired; this is enough to show a decrease in ionisation energy for O.

F and Ne show the expected rise in ionisation energy as the added electrons pair up with those already in the 2p orbitals.

Once the 2s and 2p orbitals are full, the next electron added occupies the 3s orbital – which is in a higher energy level, further from the nucleus. All the inner electrons shield the 3s electron from the full attractive force of the nucleus, making it easier to remove; there is a sharp decrease in ionisation energy. As the atoms get larger going down Group 1, the outer electron is further from the nucleus and the ionisation energy gets smaller.

Sample answer

Question:

Explain why the ionisation energy of Ga is lower than Ca and the ionisation energy of Se is lower than As. [7]

Answer:

Ga has more protons than Ca [1 mark] but the outer electron is in a 4p orbital [1 mark], which is further from the nucleus than the 4s orbital [1 mark] and shielded by the inner electrons [1 mark].

As has one electron in each of the 4p orbitals [1 mark] and the next electron added in Se has to pair [1 mark] so there is some repulsion [1 mark].

> **TIP** Make sure you know how and why the ionisation energy changes across a period and down a group.

Progress check 3.02

1. Describe and explain the changes in ionisation energy across Period 3, Na to Ar.
2. The first ionisation energy values of Li, Mg and Na are shown. Complete the table.

Element	First ionisation energy / kJ mol^{-1}
	496
Li	520
	738

The energy diagram is now extended to include $n = 4$ in Figure 3.07.

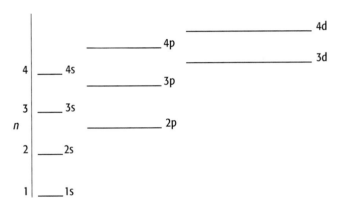

Figure 3.07 The energies of orbitals from $n = 1$ to $n = 4$

As you can see, 3d orbitals are above the 4s orbitals in energy. The five 3d orbitals are filled only after the 4s orbital has been filled with electrons at Ca. Up to ten electrons corresponding to the ten transition elements Sc to Zn can be accommodated by the five 3d orbitals. Only then will the 3p orbitals be filled.

b Consecutive ionisation energies

The arguments above concern the first ionisation only. It is possible to measure the energy needed to remove electrons one after the other from the same element:

first ionisation energy: $M(g) \longrightarrow M^+(g) + e^-$
second ionisation energy: $M^+(g) \longrightarrow M^{2+}(g) + e^-$
third ionisation energy: $M^{2+}(g) \longrightarrow M^{3+}(g) + e^-$

and so on.

Mg has 12 electrons and the 12 consecutive ionisation energies are shown in Table 3.01.

Number of electron removed	Ionisation energy / kJ mol^{-1}	x / kJ mol^{-1}
1	738	738
2	1451	713
3	7733	6282
4	10541	2808
5	13629	3088
6	17995	4366
7	21704	3709
8	25657	3953
9	31644	5987
10	35463	3819
11	169996	134533
12	189371	19375

Table 3.01 The consecutive ionisation energies of magnesium

The ionisation energy increases for each electron removed as the electron is taken from an increasingly positive species and the attractive forces are greater.

The third row in the table, x = (this ionisation energy − previous ionisation energy)

Thus, for the second electron, $713 = 1451 − 738$

For magnesium, x shows two large jumps in ionisation energy values (1451 to 7733 and 35463 to 169996), indicating jumps between the three occupied energy levels.

The first two electrons in the level $n = 3$ are relatively easy to remove, the next eight in the level $n = 2$ are more difficult to remove and the last two, in the level $n = 1$, are the most difficult to remove.

The pattern can be illustrated graphically. As two outer electrons are indicated, Mg is in Group 2 of the Periodic Table.

The wide difference in the ionisation energies makes them difficult to plot but the trends can be shown by plotting the $\log_{10}$[ionisation energy] against the number of the electron removed (see Figure 3.08).

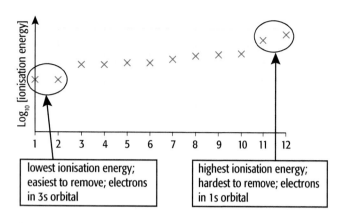

Figure 3.08

Progress check 3.03

1. Show on separate energy level diagrams the distribution of electrons in a fluorine atom and in a scandium atom.

2. Write an equation for the third ionisation energy of silicon.

3. The first six ionisation energies for an element are 590, 1145, 4912, 6474, 8144, 10496 kJ mol^{-1}. How many outer electrons does the element have?

3.03 Electronic configuration

The electron structure in an atom can be represented by its **electronic configuration**; the orbitals are written in order of increasing energy and the number of electrons in the orbital is written as a superscript.

- Hydrogen has one electron; its electronic configuration is $1s^1$.
- Helium has two electrons; its electronic configuration is $1s^2$.
- The 1s orbital is now full so the next electron has to occupy the 2s orbital; the electronic configuration for Li is $1s^2 2s^1$.

Worked example 3.01

What are the electronic configurations of Be, Ne and Ca?

How to get the answer:

Step 1: Use the Periodic Table to find the numbers of electrons. Be has 4 electrons, Ne has 10 and Ca has 20.

Step 2: Write down the orbitals in the correct order. 1s 2s 2p 3s 3p 3d

Step 3: Fill up the orbitals from left to right remembering that an s orbital can take up to 2 electrons and the p orbitals can take up to 6 electrons. Beryllium is $1s^2 2s^2$, neon is $1s^2 2s^2 2p^6$ and calcium is $1s^2 2s^2 2p^6 3s^2 3p^6 4s^2$.

The electronic configurations of ions can be shown in a similar way, remembering that positive ions have lost electrons and negative ions have gained electrons.

- Li$^+$ has two electrons; the electronic configuration is $1s^2$.
- Cl$^-$ has 18 electrons; the electronic configuration is $1s^2 2s^2 2p^6 3s^2 3p^6$.

Check that the number of electrons in an electronic configuration equals the number of electrons in the atom or ion.

Progress check 3.04

Write the electronic configurations of F, F$^-$, N, Ca^{2+}, S, S^{2-} and K.

Revision checklist

Check that you know the following:

- [ ] describe and sketch the shapes of s and p orbitals
- [ ] explain and use ionisation energy
- [ ] explain why the ionisation energy changes across a period and down a group
- [ ] how to draw a diagram of the energy levels in an atom and describe the number and type of orbitals associated with them
- [ ] Deduce which orbitals are filled and how to represent this by electron configurations

Exam-style questions

1 Figure 3.09 represents the energy levels in a nitrogen atom.

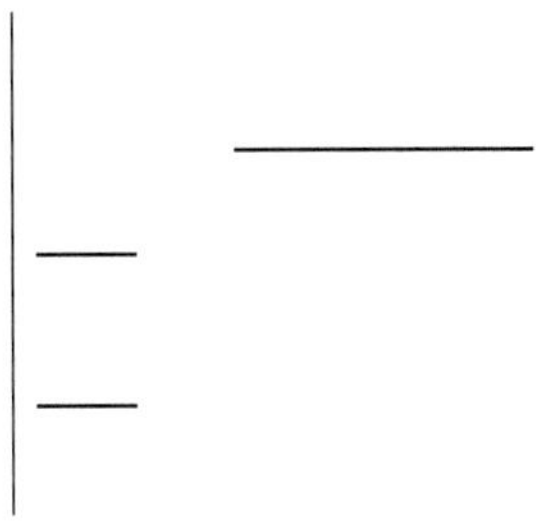

Figure 3.09

  a i Label the energy levels with the principal quantum number and the type of orbital it contains. [1]
     ii Show, by using half arrows to represent electrons, the electron distribution in a nitrogen atom. [2]
     iii Draw the shapes of two different types of orbital occupied by electrons in a nitrogen atom. [2]
  b i Give the electronic configurations of Cl and Cl^-. [2]
     ii Which atom has the same electronic configuration as the electronic configuration of Cl^-? [1]

Total: 8

2 a i Use a Data Book to state the values of the first ionisation energies of Be and B. [1]
     ii State how and explain why the first ionisation energies of Be and B differ. [2]
  b Write an equation for the second ionisation energy of an element, M. [1]
  c The first eight ionisation energies in kJ mol^{-1} for element X are:

  1140 2100 3500 4560 5760
  8549 9939 18600

  To which group in the Periodic Table does the element belong? [1]
  d i Write equations for the first ionisation energies of F^- and of F. [2]
     ii Suggest which of the two values is the larger and explain your answer. [2]

Total: 9

Unit 4: Chemical bonding

Learning outcomes

You should be able to:

- understand how ionic bonds and covalent bonds arise, including dative covalent bonds and describe them using dot and cross diagrams
- explain and predict the shapes and bond angles of molecules using electron pair repulsion theory
- describe orbital overlap and hybridisation and predict what effect σ and π bonds have on molecular shape
- describe hydrogen bonding and know its effects
- understand electronegativity and how to predict bond polarity
- explain the terms bond energy, bond length and bond polarity and use them to predict reactivity
- describe a suitable model of metallic bonding
- describe, interpret and predict what effect bonding and intermolecular bonding has on the properties of substances
- understand energy changes on making and breaking bonds

4.01 Ionic bonding

Atoms with low ionisation energies can lose outer electrons to become positive ions. The electrons are donated to atoms with unfilled spaces in their highest occupied energy level. This can be represented on a 'dot and cross' diagram (Figure 4.01).

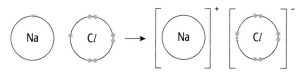

Figure 4.01 The dot-and-cross diagram for the formation of NaCl

Dots and crosses distinguish where the electron comes from, although in reality there is no difference between them. The charge on the ion should be shown as a superscript outside the square brackets.

The two ions that are formed usually have a noble gas electronic configuration. The ionic compound produced is neutral; the number of positive charges equals the number of negative charges (see Figure 4.02).

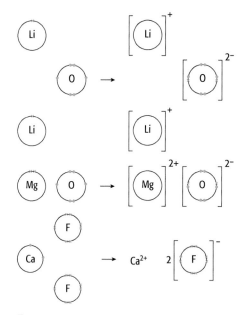

Figure 4.02 The formation of Li_2O, MgO and CaF_2

Ions pack together in a lattice with positive ions surrounded by negative ions and negative ions surrounded by positive ions. Strong electrostatic forces of attraction hold the structure together.

Progress check 4.01

1. Write down the electronic configurations of the ions in NaCl, Li_2O and MgO.

2. Draw dot-and-cross diagrams for the formation of KBr, AlF_3, $CaCl_2$, BaO and Cs_2O.

4.02 Covalent bonding

Atoms with high ionisation energies form bonds by sharing one or more pairs of electrons. One electron in a bond pair comes from each atom. In the simplest case, uncharged molecules are formed (Figure 4.03), held together by the attraction between the bond pair and the nuclei.

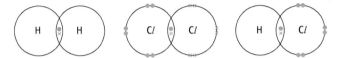

Figure 4.03 Dot-and-cross diagrams for covalent bond formation

In the H_2 molecule, the bond pair is equally shared and situated midway between the two atoms, which both now have the electronic configuration of helium.

A chlorine atom with seven electrons shares one electron with a second chlorine atom to form Cl_2. Each Cl now has eight outer electrons, one bond pair and three non-bonding (lone) pairs and ends up with the electronic configuration of argon.

One electron from H and one from Cl are shared in HCl and both atoms end up with a full outer shell. Sharing pairs of electrons and gaining full outer shells is seen in other molecules – see CH_4 in Figure 4.04. Usually the bond pair is represented by a line.

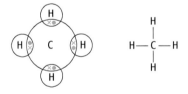

Figure 4.04 The bonding in methane, CH_4

Two or even three pairs of electrons can be shared. Oxygen atoms have six outer electrons and share two with another oxygen atom to form a double bond.

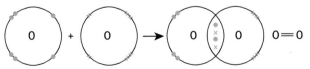

In CO_2, carbon atoms have four outer electrons and each oxygen has six outer electrons. For each atom to gain a full outer shell of four pairs of electrons, each O uses two electrons to form a double bond with two of the carbon electrons. The C ends up with two sets of shared pairs of electrons, each forming two double bonds; each O atom has two lone pairs and two bond pairs of electrons.

In the hydrocarbon, ethene, C_2H_4, each C uses two electrons to share with two electrons from the other C, forming a double bond. This leaves each C with two more electrons to share with two H atoms to form $H_2C=CH_2$.

Sometimes the shared pair is provided by only one atom, forming a **dative covalent bond**.

The five outer electrons of N form three bonds in NH_3, leaving a lone pair which can be used to bond with H^+.

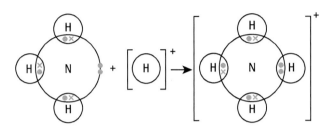

Dative covalent bonds are represented by arrows showing which way the lone pair is donated:

Worked example 4.01

Explain the bonding in Al_2Cl_6.

How to get the answer:

Step 1: In $AlCl_3$, aluminium shares its three outer electrons with Cl atoms to form $AlCl_3$. Each Cl now has eight outer electrons but Al has only six electrons so can accept an electron pair.

Step 2: Two $AlCl_3$ molecules bond together. Cl atoms in $AlCl_3$ have lone electron pairs and can donate one of them to Al to form a dative covalent bond; $Cl_2AlCl: \rightarrow AlCl_3$.

Step 3: This also occurs for the second Al:

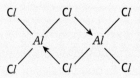

Step 4: Check every atom is surrounded by four electron pairs.

Progress check 4.02

1. Draw dot-and-cross diagrams for ethanol (C_2H_5OH), carbon dioxide (O=C=O) and ethene ($H_2C=CH_2$).
2. Show that all the atoms in carbon dioxide and ethene have full outer shells.
3. Draw the dot-and-cross diagrams of NH_3 and of BF_3.
4. The gases NH_3 and BF_3 form a white solid, NH_3BF_3, in which every atom has a full outer shell. Draw the dot-and-cross structure of NH_3BF_3.

Ionic bonding involves the complete transfer of electrons and covalent bonding involves a shared pair of electrons, represented in dot-and-cross diagrams.

4.03 Shapes of molecules and bond angles

A covalent bond has a definite direction. The outer electron pairs, both bond pairs and lone pairs, repel one another to be as far as possible from each other. This is the **electron pair repulsion theory** and leads to the shape and bond angles of covalent molecules.

The starting point for working out a molecular shape is the dot-and-cross diagram. Boron's three outer electrons are shared with fluorine atoms in BF_3 (Figure 4.05).

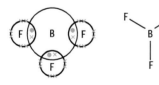

Figure 4.05 The bonding in BF_3

B is surrounded by three bond pairs which repel each other as far away as possible. As a result, BF_3 is a planar triangular shape with bond angles of 120° (Figure 4.06).

Figure 4.06 The molecule of BF_3

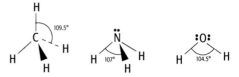

Figure 4.07

- The bond angle in CH_4 is the tetrahedral bond angle.
- In NH_3, the lone pair is closer to the N than the three bond pairs; repulsion between lone pair and a bond pair is larger than repulsion between bond pairs. This has the effect of pushing the bonds closer together and the bond angle is smaller.
- In H_2O, the two lone pairs on O have a greater effect; repulsion between lone pairs is greater than between a lone pair and a bond pair, pushing the bonds closer together.

Table 4.01 summarises this information.

Molecule	Number of electron pairs round the central atom			Structure	Bond angle
	Bond pairs	Lone pairs	Total pairs		
BF_3	3	0	3	Trigonal	120°
CH_4	4	0	4	Tetrahedral	109.5°
NH_3	3	1	4	Pyramidal	107°
H_2O	2	2	4	Non-linear	104.5°
SF_6	6	0	6	Octahedral	90°
PF_5	5	0	5	Trigonal bipyramidal	90° and 120°

Table 4.01 The electron pairs and shapes of molecules formed by Period Two and Period Three elements

Second-period atoms only have 2s and 2p orbitals to be filled but a central atom from the third period can use 3s, 3p and 3d orbitals. S in SF_6 has six bond pairs and P in PF_5 has five bond pairs, giving the structures shown in Figure 4.08.

Figure 4.08

A maximum of four pairs of electrons can surround Period 2 atoms but more pairs of electrons can surround Period 3 atoms.

A double bond has two electron pairs in the same region of space. We can consider that these two pairs take up the same amount of space as one set of electron pairs. The C at the centre of CO_2 forms double bonds with each O atom and is surrounded by two sets of bond pairs only. These two sets repel each other, making O=C=O linear.

The covalent bonds holding the atoms together in small molecules are strong but the attractive forces between the molecules are weak and easily overcome. A characteristic of small molecules is their low melting point.

4.04 σ and π bonds and molecular shape

Molecular shape can be explained by overlapping atomic orbitals from two different atoms forming a **molecular orbital**. If the two atomic orbitals each contained one electron, the resulting molecular orbital contains both electrons and is the bond keeping the atoms together.

CH_4 is puzzling. The electronic configuration for C is $1s^2 2s^2 2p^2$ but it forms four bonds. One 2s electron is promoted to a 2p orbital – the small amount of energy needed is supplied by the energy released on forming four bonds rather than two.

Both the energy and the regions of space of 2s and 2p orbitals are very close which allows the four orbitals to be combined and divided into four equivalent orbitals pointing to the corners of a tetrahedron. This is sp^3 **hybridisation** (sp^3 from one s and three p orbitals). (Figure 4.09).

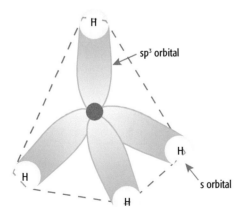

Figure 4.09 sp^3 hybridisation in CH_4

In CH_4, the four hybrid orbitals overlap with 1s orbitals of H atoms, forming single bonds called σ bonds. A slice taken through the electron density of the σ bond shows a circular electron density.

The double bond in ethene, $CH_2=CH_2$, confers some unexpected features (see Unit 15):

- The double bond easily breaks, forming a single bond.
- The molecule is planar and cannot easily twist.

Each C atom uses the 2s and only two of the 2p orbitals to form σ bonds directed to the corners of an equilateral triangle; this is sp^2 hybridisation. One electron is left in a 2p orbital perpendicular to the sp^2 plane; the neighbouring 2p orbitals can overlap sideways (Figure 4.10) to form a π **bond** containing the electron pair.

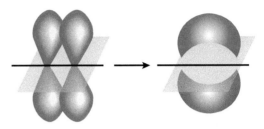

Figure 4.10 Sideways overlap of p orbitals

A π bond has regions above and below the molecular plane, preventing twisting of the molecule which would break the bond. Sideways overlap is less efficient than 'head on' overlap and so π bonds are weaker than σ bonds.

Sample answer

Question:

Explain why the carbonate ion, CO_3^{2-}, is trigonal planar and all the bonds are the same length. [6]

Answer:

The double bond is a σ bond plus a π bond [1 mark]. See Figure 4.11 [1 mark].

$$O=C\begin{matrix}O^-\\O^-\end{matrix}$$

Figure 4.11

A pair of electrons on a singly bonded O atom is put into a p orbital; that p orbital can overlap sideways with the π bond [1 mark]. A molecular π bond is formed covering all four atoms [1 mark]. Sideways overlap prevents the molecule twisting [1 mark]. Each bond is a '1⅓' bond [1 mark].

Progress check 4.03

1. Draw dot-and-cross diagrams for CH_4, NH_3 and H_2O; work out their shapes and then check them in Table 4.01.
2. Draw dot-and-cross diagrams for SF_6 and for PF_5.
3. Using hybridisation, describe the bonding in ethane (C_2H_6) and suggest the bond angles.
4. Using hybridisation, describe the bonding in propene ($CH_3CH=CH_2$) and suggest the bond angles.
5. Use dot-and-cross diagrams to determine the shape and bond angles of F_2O, CCl_4, NH_4^+, ClF_5 (Cl is at the centre) and XeO_3.

4.05 The hydrogen bond

The atoms of N, O and F in the second period are small, attract electrons strongly and also have lone pairs of electrons occupying a small region of space resulting in a high electron density in those regions. Bond pairs between H atoms and N, O or F are not equally shared. More of the electron density is closer to the N, O or F, which is the more electronegative atom. This leaves the H atom slightly positive (Figure 4.12).

$$H-F \quad H-N \quad H-O$$
$$\delta+ \ \delta- \quad \delta+ \ \delta- \quad \delta+ \ \delta-$$

Figure 4.12 The bond pair is not equally shared

The δ+ H atoms are strongly attracted to the lone pair electron density of a neighbouring atom, forming a **hydrogen bond** shown by the dotted lines (Figure 4.13). Hydrogen bonds are about a tenth as strong as covalent bonds.

$$H\!-\!\!-\!\!F: \cdots\cdots H\!-\!\!-\!\!F: \cdots\cdots H\!-\!\!-\!\!F:$$
$$\delta+ \ \ \delta- \qquad \delta+ \ \ \delta- \qquad \delta+ \ \ \delta-$$

Figure 4.13

The lone pair in a hybrid orbital has a definite direction, giving direction to the hydrogen bond. H_2O with two δ+ H atoms and two lone pairs of electrons forms two hydrogen bonds per molecule, accounting for many of the properties of water such as its relatively high melting and boiling point and the open structure of ice.

Hydrogen bonding is responsible for some structural aspects of proteins and DNA.

4.06 Electronegativity and bond polarity

Electronegativity is the ability of an atom to attract a bonding pair, which depends on the nuclear charge and the distance of the outer electrons from the nucleus. It increases across a period and decreases down a group as the outer electrons get further from the nucleus.

The attractive ability depends on the atoms bonded as the bond pair in Cl–Cl is shared equally, but in H–Cl the Cl has a higher electronegativity than hydrogen and so the bond pair is rather nearer to the Cl:

$$H-Cl$$
$$\delta+ \ \ \delta-$$

A difference in electronegativity leads to **bond polarisation**; one of the atoms is slightly positive and the other is slightly negative. A **dipole moment** occurs if one side of a molecule is more positive than the other; to decide whether there is a dipole moment, bond polarisation together with the three-dimensional structure need to be considered.

BF_3 is trigonal planar. The B–F bond is polarised so that B is δ+ and every F is δ− (Figure 4.14).

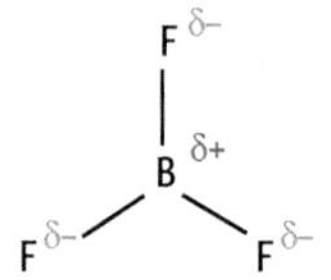

Figure 4.14

All the bonds are polarised to the same extent but the molecule is neutral and symmetrical and there is no dipole moment.

NF_3 is pyramidal with δ+N and δ−F (Figure 4.15).

Figure 4.15

One side of the molecule is δ+, the other is δ−. The crossed arrow shows the direction of electron movement and you will see NF_3 has a dipole moment.

4.07 Bond energy, bond length and bond polarity

Reactions of covalent compounds involve the breaking and making of bonds. Bonds have different strengths depending on which atoms are bonded together and, to some extent, which other atoms make up the molecule. Energy is required to break a covalent bond before any reaction can happen. If a large amount of energy is needed, then any reaction will be slow.

The bond length is the distance between the two nuclei of the bonding atoms. Multiple bonds are shorter than single bonds; as the bond length gets shorter, the bond gets stronger. The triple N≡N is very much stronger and harder to break than N–N.

A reaction also needs a method of attack; an electron pair can attack a $\delta+$ centre formed by bond polarisation. For example, bromoethane (CH_3CH_2Br) has a polarised C–Br bond, and attack can take place on the $^{\delta+}C$ by a lone pair on H_2O.

4.08 Intermolecular forces

Noble gases exist as single atoms. Their boiling points are low and their latent heats of evaporation are small but both increase down Group 18. The attractive forces between the molecules are weak but increase as the atoms get larger. They are called **van der Waals' forces** and exist between any two particles which attract each other because of instantaneous and induced dipoles. Instantaneous dipoles occur as atoms move around and the heavy positive nucleus and centre of negative charge of the electrons may not coincide, setting up an instantaneous dipole which attracts the electron clouds in surrounding particles (Figure 4.16).

Figure 4.16 An instantaneous dipole. + is the nucleus; − is the centre of negative charge of the electrons

The much lighter electrons are attracted to the positive end of the instantaneous dipole, forming an induced dipole. Larger atoms with more electrons experience larger van der Waals' forces and the boiling point and latent heat of evaporation both increase.

The boiling points of straight-chain alkanes get higher as the chain gets longer; the greater surface area of contact increases the van der Waals' forces as the chain gets longer and more energy is needed to boil the alkane.

Molecules with permanent dipole moments, such as $CHCl_3$ (Figure 4.17), experience permanent dipole–permanent dipole intermolecular attractions and have higher boiling points than expected (Table 4.02).

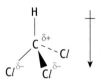

Figure 4.17

Molecule	M_r	Number of electrons	Boiling point / K
$CHCl_3$	119.5	58	335
Br_2	159.8	70	332

Table 4.02 The effect of a permanent dipole on boiling point

You can see from Table 4.02 that the number of electrons (and therefore induced dipole attractions) in $CHCl_3$ is lower than in Br_2 but the boiling points are similar as $CHCl_3$ molecules have extra permanent dipole attractions.

4.09 Metallic bonding

Metals are characterised by low ionisation energies and atoms lose their outer valence electrons with ease to form positive ions which are arranged in a regular lattice. The valence electrons are free to move throughout the structure – they are **delocalised** (see Unit 5). The electrostatic force of attraction between positive ions and delocalised electrons is strong, producing high melting point solids (apart from mercury). Metals with more outer electrons experience a larger electrostatic attraction and a stronger bonding, resulting in a higher melting point.

4.10 Effect of bonding on properties

> You need to be familiar with both the properties expected from different types of bonding and the associated reasons.

Type of bonding	Properties
ionic	• high melting and boiling points because of the strong electrostatic forces between the ions • no electrical conductivity when solid as the ions are unable to move • electrical conductivity when molten or in solution as the ions are able to move • often soluble in water as the ions can form ion–dipole forces with water molecules • brittle; when deformed, similarly charged ions come together and repel each other
covalent	• usually low melting and boiling points as only weak van der Waals' or permanent dipole forces attract the molecules together • no electrical conductivity when solid, molten or in solution as there are no charged particles • usually insoluble in water
metallic	• high melting and boiling points as there are strong electrostatic forces between the positive ions and delocalised electrons • conduct electricity when solid or molten because the delocalised electrons are free to move • malleable and ductile; even when deformed, the electrostatic force between positive ions and delocalised electrons holds the structure together

Progress check 4.04

1. Draw three-dimensional diagrams for the following molecules and show any bond polarisation: CO_2, CCl_4, NF_3, SF_6, SO_2, SO_3.
2. Which of the molecules in question 1 have a dipole moment?
3. Explain why $SnCl_4$ is a liquid at room temperature but $SnBr_4$ is a solid.
4. Explain why the boiling point of Na is 1156 K but the boiling point of Mg is 1380 K.
5. Three substances, A, B and C, have the following properties:

Substance	Boiling point / K	Electrical conductivity at room temperature
A	1757	✓
B	457	✗
C	1373	✗

The three substances are calcium, calcium iodide and iodine. Identify which is which. Explain your answer.

4.11 Breaking and making bonds

A bond holds particles together. In a chemical reaction, bonds are broken and energy is needed to:

- overcome the electrostatic attraction between bond pairs and nuclei of two atoms
- overcome the electrostatic attraction between oppositely charged ions
- overcome the electrostatic attraction between positive ions and delocalised electrons.

New bonds are then made with a release of energy.

Revision checklist

Check that you know the following:

- [ ] how to describe ionic and covalent bonding and draw dot and cross diagrams,
- [ ] how dative covalent bonding arises,
- [ ] use σ and π bonds to describe covalent bonding and understand hybridisation and its role in determining molecular shape
- [ ] the electron pair repulsion theory and its use in explaining and predicting the shapes of molecules and ions
- [ ] how to describe hydrogen bonding and its effect and identify situations where it (hydrogen bonding) arises
- [ ] the trends in electronegativity and understand how electronegativity explains bond polarity and molecular dipole moments,
- [ ] explain how bond energy, length and polarity affect reactivity
- [ ] describe how van der Waals forces arise and their effect on physical properties
- [ ] a description of metallic bonding
- [ ] describe, predict and interpret the effect of bonding on the properties of substances and how to deduce the type of bonding from the properties
- [ ] understand the energy transfers involved in making and breaking bonds

Exam-style questions

1. a For the following pairs of molecules, state or draw the shape of each molecule and explain any differences between them.
   - i CO_2 and SO_2 [4]
   - ii NH_3 and BF_3 [4]
   - iii CH_4 and SF_4 [4]

   b Consider the following molecules:
   SF_6 BeF_2 F_2O SO_3
   - i Which of the molecules has a dipole moment? [1]
   - ii Place the four molecules in order of decreasing bond angles. [2]

   Total: 15

2. a i Draw dot-and-cross diagrams of $NaCl$ and HCl. [2]
   ii Explain why $NaCl$ and HCl have different melting points. [3]

   b Explain why the bond between C atoms in ethane has a length of 0.154 nm but between C atoms in ethene is 0.143 nm. [2]

   c i Describe, using orbital theory, the bonding in ethene, $CH_2=CH_2$. [4]

   ii But-2-ene, $CH_3CH=CHCH_3$ exists as two isomers. Draw the two isomers and explain, using orbital theory, why two isomers can exist. [3]

   iii Does but-1-ene, $CH_3CH_2CH=CH_2$, exist as two isomers? Explain your answer. [1]

   Total: 15

3. The carbonate ion is CO_3^{2-}.
   a Describe the shape of the carbonate ion and explain your answer using orbital theory. [4]
   b How many outer electrons does each atom have in the carbonate ion? [1]

   total: 5

4. Both NaI and Cu exist as a structure of charged particles but only the Cu can conduct electricity at room temperature.
   a Describe the two structures and explain why NaI does not conduct electricity but Cu does. [6]
   b How could you get the NaI to conduct? [1]

   Total: 7

Unit 5: States of matter

Learning outcomes

You should be able to:

- ☐ use the kinetic theory to describe the characteristics of state and changes of state
- ☐ state the assumptions of the kinetic theory and explain how real gases approach ideality
- ☐ state and use the ideal gas equation
- ☐ describe the structures of ionic, metallic and giant covalent substances as well as small molecules
- ☐ outline the importance of hydrogen bonding and its effect on the properties of compounds
- ☐ discuss the uses of a range of materials and the economic importance of recycling
- ☐ suggest how structure and bonding affects the properties of materials

5.01 The kinetic theory and ideal gases

Particles move with a speed depending on their temperature.

- In a solid, particles can vibrate but not move from their positions.
- In a liquid, particles move around but remain close to neighbouring particles.
- In a gas, particles move randomly in straight lines at some distance from each other.

a **The kinetic theory of gases**
  - Gases are made of small particles in constant random motion; collisions with the container give the gas pressure.
  - Collisions are perfectly elastic (no energy is lost).
  - The particle volume is negligible compared to the container volume (particles are far apart).
  - There is no attractive force between the particles.

b **For real gases**
  - Collisions are not perfectly elastic.
  - The volume of the particles is not negligible.
  - There are attractive forces between the particles.

c **Real gases and ideal behaviour**
  Real gases approach ideal behaviour if:
  - the gas pressure is very low (the volume of the particles is negligible compared to the volume of the container)
  - the pressure of the gas is low so the particles are far apart (the attractive forces between particles are negligible)
  - the temperature is high, giving particles high kinetic energy and overcoming attractive forces

 Remember that at high temperature and low pressure, the behaviour of real gases approaches that of ideal gases.

Progress check 5.01

Consider the following gases: chlorine, hydrogen, hydrogen chloride and hydrogen fluoride.

1. At room temperature, which shows most ideal behaviour? Explain your answer.
2. At room temperature, which shows least ideal behaviour? Explain your answer.

5.02 $pV = nRT$

Experimentally, it is found that for a gas:

- at constant temperature, $V \propto \frac{1}{p}$ or $pV =$ constant, as shown in Figure 5.01.
- at constant pressure, $V =$ constant $\times T$, as shown in Figure 5.02

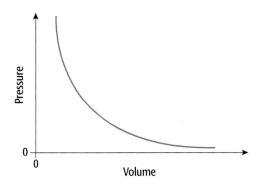

Figure 5.01 Pressure $\propto \frac{1}{\text{volume}}$

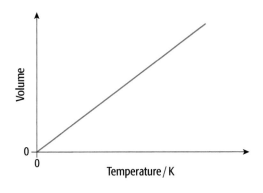

Figure 5.02 Volume $\propto$ the absolute temperature

Combining these laws gives $pV =$ constant $\times T$; the constant depends on the number of moles, n, giving the final ideal gas equation:

$$pV = nRT$$

for pressure in Pa, volume in m^3 and temperature in K, $R = 8.314 \, J \, mol^{-1} \, K^{-1}$.

For example, the volume occupied by 0.5 moles of ideal gas at a pressure of 0.95×10^5 Pa and a temperature of 400 K is found by using $pV = nRT$:

$$0.95 \times 10^5 \times V = 0.5 \times 8.314 \times 400$$
$$V = 0.0175 \, m^3 = 17.5 \, dm^3 \, (1 \, m^3 = 1000 \, dm^3)$$

Worked example 5.01

How many moles of an ideal gas occupy a volume of $100 \, cm^3$ if the temperature is 300 °C and the pressure is 130 kPa?

How to get the answer:

Step 1: Change the volume to m^3:
$100 \, cm^3 = 1.0 \times 10^{-4} \, m^3$

Step 2: Change the pressure to Pa:
$130 \, kPa = 130 \times 10^3 \, Pa$

Step 3: Change the temperature to K:
$300 \, °C = (273 + 300) \, K = 573 \, K$

Step 4: Insert these values into $pV = nRT$:
$$1.3 \times 10^5 \times 1.0 \times 10^{-4} = n \times 8.314 \times 573$$

Step 5: Find n: $n = 2.7 \times 10^{-3}$

> **TIP**
> Learn how to convert other units of volume, temperature and pressure to those required for the gas equation.

Progress check 5.02

1. How many moles of an ideal gas occupy a volume of $95 \, cm^3$ at a pressure of 100 kPa and a temperature of 300 K?

2. 0.142 g of an ideal gas in a gas syringe is kept in an oven at 95 °C. At atmospheric pressure (101 kPa), the volume of the gas is $76 \, cm^3$. What is the M_r of the gas?

5.03 The liquid state

Cooling decreases the energy of particles so they no longer overcome the attractive forces and move around. This is freezing and the energy removed is the latent heat of fusion.

Heating increases the temperature and the particles move faster until, at the boiling point, they have enough energy to overcome the attractive forces and escape.

The latent heat of vaporisation is energy supplied without a temperature change until all the particles have escaped from the liquid.

Further heating increases the temperature (Figure 5.03).

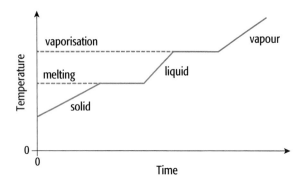

Figure 5.03 Change of water temperature with time (energy is supplied at a constant rate)

In the liquid state, the particles have a range of energies. As the temperature increases, more and more particles have high energies and may have sufficient energy to escape from the liquid surface to form vapour. An open vessel allows the particles to escape – the liquid evaporates, even at temperatures below the boiling point. At higher temperatures, more particles can vaporise and evaporation is quicker. Vapour particles move randomly; their collisions provide vapour pressure.

5.04 Crystalline solids

a Metallic solids

Elements with low ionisation energies show metallic bonding. The outer electrons are delocalised and the positive ions arranged in a lattice are held together by the strong electrostatic attraction between positive ions and delocalised electrons (Figure 5.04). Copper has a lattice of Cu^{2+} ions with $2e^-$ from each atom delocalised in the structure.

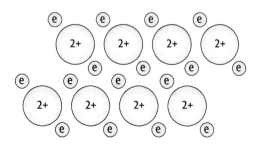

Figure 5.04 A metallic structure: a lattice of positive ions surrounded by delocalised electrons

The delocalised electrons can move; metals conduct electricity.

The strong electrostatic attraction between delocalised electrons and ions results in high melting and boiling points; it is also non-directional and holds the structure together even if the ions are pushed out of place; metals are malleable and ductile.

The sodium lattice of Na^+ has fewer delocalised electrons and so the electrostatic attraction is weaker, giving lower melting and boiling points than for copper.

b Ionic solids

Complete electron transfer between atoms forms ions. In $NaCl$, each Na^+ ion is surrounded by six of the larger Cl^- ions; each Cl^- ion is surrounded by six Na^+ ions (Figure 5.05).

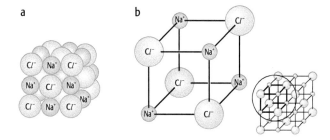

Figure 5.05 The structure of $NaCl$; Cl^- ions are larger than Na^+ ions

Ions are placed so that positive ions are next to negative ions and the structure is held together by strong attractive forces between oppositely charged ions.

Ionic structures have the following physical properties:

- high melting and boiling points; a lot of energy is needed to overcome the strong attractive forces
- brittle; deforming the structure pushes positive ions closer to positive ions and these repel one another
- a non-conductor when solid; the ions cannot move out of position
- a conductor when molten or in solution; the ions are able to move.

Higher ionic charges have greater electrostatic attraction. MgO exists as Mg^{2+} and O^{2-} ions and their electrostatic attraction is much greater than the electrostatic attraction between Na^+ and Cl^-, giving much higher melting and boiling points.

c **Giant molecular structures**

Diamond and silicon(IV) oxide have similar structures in which C and Si form four strong covalent bonds holding all the atoms together in a giant structure (Figure 5.06). The bond angle is the tetrahedral angle, 109.5°. Only carbon atoms are bonded in diamond but in silicon(IV) oxide, each Si atom bonds with four O atoms; each O atom is bonded to two Si atoms

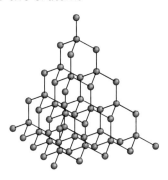

Figure 5.06 The structures of diamond and silica showing the tetrahedral arrangement around the C and the Si atoms. An O atom is between every two Si atoms in the silica structure

Giant covalent structures have the following properties:

- high melting and boiling points; a great deal of energy is needed to break the very large number of strong covalent bonds
- no electrical conductivity as there are no mobile charged particles

The giant structure of graphite has planes of hexagons; each C atom forms strong covalent bonds to three others (Figure 5.07).

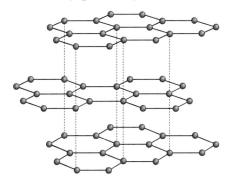

Figure 5.07 Hexagons in three separate planes of carbon atoms in graphite

Each C atom has a fourth electron in a p orbital perpendicular to the plane; neighbouring orbitals overlap sideways, allowing delocalisation of electrons over the whole plane, giving the following properties:

- high melting and boiling points; carbon atoms in the planes are held together by a very large number of strong covalent bonds
- slippery feel; a little pressure easily overcomes the weak van der Waals' forces between the planes
- conduction of electricity along the planes; the delocalised electrons are mobile

Graphite can be used as a high temperature lubricant, an electrode in electrochemical cells and as the 'lead' in pencils.

The carbon allotrope, graphene, has a graphite structure just one plane thick. With a high melting point and excellent conductivity of heat and electricity, it also has special properties of exceptional lightness, strength and flexibility.

d **Simple molecular structures**

Atoms in simple molecules are covalently bonded. The molecules are held in a lattice by intermolecular forces which can be van der Waals' forces, permanent dipole–dipole forces or hydrogen bonds. Melting and boiling requires the intermolecular forces to be overcome.

The attractive forces between the diatomic molecules of iodine are weak van der Waals' forces. These are strong enough to hold the molecules in a lattice at room temperature (Figure 5.08).

Figure 5.08 The structure of iodine

A small amount of heat is needed for the brown solid to become a purple vapour:

$$I_2(s) \longrightarrow I_2(g)$$

> **TIP**
> Simple molecular structures have low melting and boiling points; small amounts of energy convert solids into liquids or vapours.

The carbon allotropes, the fullerenes, can be hollow spheres (buckyballs), tubes (nanotubes) or cones (Figure 5.09). The molecules are held together by van der Waals' forces.

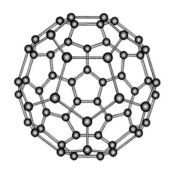

Figure 5.09 A buckyball

Water molecules contain fewer electrons than I_2; the van der Waals' forces are smaller. H_2O molecules also form the directional hydrogen bonds (see section 4.05).

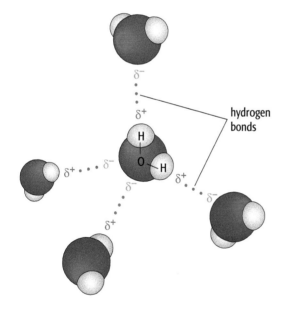

Figure 5.10 Hydrogen bonds are directional

Ice has two hydrogen bonds for each H_2O molecule. As ice melts, some hydrogen bonds are broken allowing the molecules to get closer. Ice has a lower density than water so icebergs float.

> **TIP**
> The hydrogen bonded structure of water gives it unusual properties. List as many as you can and give an explanation of each property.

5.05 The importance of hydrogen bonding

In the natural world, the most important hydrogen bonds are between the $^{\delta+}H$ and the lone pairs attached to O or N (Figure 5.11).

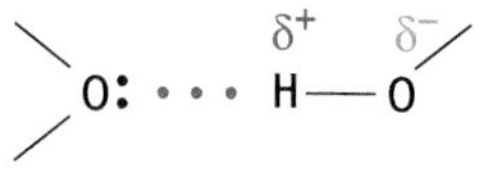

Figure 5.11

a **Hydrogen bonding in water**
- Ice floats on water, providing insulation and allowing fish to survive cold conditions.
- Water has much higher melting and boiling points compared to the other Group 16 hydrides. At normal temperatures, water is liquid – and we can drink it!

b **Hydrogen bonding in proteins and DNA**
Hydrogen bonding is important in forming the precise secondary and tertiary protein structures (and its strength in coupling the double helix of a nucleic acid is important for DNA replication).

c **Hydrogen bonds and solubility**
Using hydrogen bonds, many molecules can dissolve in water to be carried in plant sap or animal blood.

5.06 Materials as resources

a **Metals**
Extracting metals has both a financial and environmental cost. The extraction needs energy and an infrastructure and affects people in the surrounding area. Any effect of chemicals or industrial plant on water or land needs to be limited.

Metals are recycled for financial reasons, but also so they do not end up in landfill sites.

- Deep mining for copper is hazardous. Refining copper takes much electrical power. The electrolyte used in the refining cells is recycled and waste rock is used for building.

- Much electrical power is required to extract aluminium and the waste is caustic. Recycling aluminium takes only 5% of the extraction energy.

b **Ceramics**

'Ceramic' originally described the clay out of which pots were made. Ceramic materials are hard and unreactive so are widely used for:

- bricks, pipes and tiles
- kiln linings
- pottery
- crushed ceramics are used in road repair.

5.07 Structure, bonding and properties

Sample answer

Question:

State the structure and bonding of aluminium, diamond, graphene, magnesium oxide and buckminsterfullerene. Predict their relative melting points and electrical conductivities. [7]

Answer:

Substance	Structure and bonding	Melting point	Conduction of electricity
Al	giant metallic	high	good
diamond, graphene	giant covalent	high	poor for diamond, good for graphene
MgO	giant ionic	high	none in solid; good when molten
fullerenes	simple molecular	low	none
	[2 marks]	[2 marks]	[3 marks]

Progress check 5.03

1 Melting and boiling points of substances A, B, C, D and E are shown:

	Melting point / K	Boiling point / K
A	91	109
B	646	1189
C	195	240
D	1883	2503
E	922	1380

a Which one is a ceramic?

b Which two exist as simple molecules?

c The two substances in part b have the same number of electrons. Which one contains hydrogen bonding?

d One of the two substances not chosen in parts a, b and c conducts electricity at room temperature; both of them conduct at 800 K. Suggest the structure and bonding of each, explaining your answer.

2 Predict and explain which of K and Ca has the greatest electrical conductivity.

Revision checklist

Check that you know the following:

- [ ] state the kinetic theory assumptions and explain how real gases approach ideal behaviour,

- [ ] how to state and use the ideal gas equation with the appropriate units,

- [ ] how to describe liquids, melting and evaporation,

- [ ] the identification, description and properties of metallic, giant molecular, simple molecular and ionic lattices with suitable examples,

- [ ] describe the importance of hydrogen bonding and a range of examples,

- [ ] suggest the (identifying) structure and bonding from physical data,

- [ ] discuss the impact of using metals and ceramics.

Exam-style questions

1. a i What is an ideal gas? [3]

    ii Under what conditions can a real gas approach ideal behaviour? [2]

    iii Arrange the three gases, ammonia (NH_3), carbon monoxide (CO) and nitrogen (N_2) in order of increasing ideal behaviour and explain your answer. [4]

    b A $100\,cm^3$ gas syringe is held at 95 °C in an oven. Atmospheric pressure is 0.980×10^5 Pa. What is the maximum mass of ethanol (C_2H_5OH) which can be held in the gas syringe as a gas under these conditions? [3]

    Total: 12

2. a i Describe the lattices of I_2 and NaI. [4]

    ii Suggest what physical properties of I_2 and NaI are to be expected. [4]

    b Explain the following differences.

    i The boiling point of CH_3NH_2 is 267 K; the boiling point of CH_3CH_3 is 185 K. [3]

    ii $SnCl_4$ is a liquid but $SnBr_4$ is a solid at room temperature. [3]

    iii The boiling point of CO_2 is 195 K but the boiling point of SO_2 is 263 K. [3]

    c The melting point and electrical conductivity of three elements, X, Y and Z, are shown in the table.

Element	Melting point / K	Electrical conductivity
X	1683	very poor
Y	4900	very high
Z	386	none

X, Y and Z are graphene, silicon and sulfur. Identify X, Y and Z. [2]

Total: 19

Unit 6: Energy changes

Learning outcomes

You should be able to:

- explain *enthalpy of reaction* and *standard conditions*, identify exothermic and endothermic reactions and calculate enthalpy changes from experimental results
- explain and use enthalpy changes of combustion and formation in calculations
- explain Hess' Law and use it in enthalpy calculations with enthalpy changes or average bond energies
- explain and use enthalpy changes of atomisation, hydration and solution
- explain and use the enthalpy change of neutralisation
- explain and use *bond energy* in calculations
- construct and interpret a reaction pathway diagram

6.01 Exothermic and endothermic reactions

Exothermic reactions give out energy when reagents become products.

- On oxidising luciferin, light energy is emitted – enabling male fireflies to attract a mate.
- A reaction within a battery produces electrical energy.
- The combustion of fuel produces energy to thrust a rocket into the upper atmosphere.
- Sodium produces heat energy as it reacts with water and the solution gets warm.

Endothermic reactions take in energy when reagents become products.

- Energy is needed when ammonium chloride dissolves in water and the solution gets colder.
- Ammonium chloride and barium hydroxide octahydrate shaken together require so much energy to react that the temperature falls below the freezing point of water.
- Chlorophyll absorbs energy from sunlight to use in photosynthesis.
- Producing aluminium and oxygen from aluminium oxide needs energy for electrolysis.

Adding the energies of nucleus, electrons and bonding gives the energy content of a substance. We can compare energy contents by always using the same amount – the mole – of substance. The enthalpy, H, is the energy content per mole; **enthalpy changes**, ΔH, are caused by changes in bond energies and ionic attractions in a reaction and are made apparent by the exothermicity or endothermicity of reactions. This can be represented on energy diagrams (Figure 6.01).

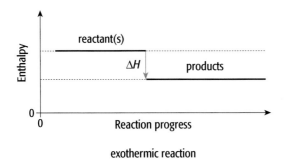
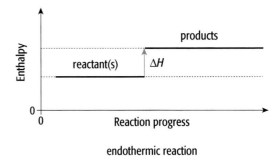

Figure 6.01 Energy diagrams for exothermic and endothermic reactions

As you can see, the products contain less enthalpy than the reactants for an exothermic reaction and for an endothermic reaction the products contain more enthalpy than the reactants.

ΔH is defined as [enthalpy of products] – [enthalpy of reactants]

ΔH is negative for exothermic reactions and positive for endothermic reactions.

> The –/+ sign with an enthalpy change needs to be included as it shows whether the reaction is exothermic or endothermic.

6.02 Enthalpy change of combustion

It is useful to classify some reactions and define the associated enthalpy change. Very many elements and compounds burn in air or oxygen; each of these reactions has its associated **enthalpy change of combustion**. Combustion reactions are exothermic but how exothermic depends on the temperature, pressure and what state the reactants and products are in.

For easy comparison, **standard conditions** are specified:

- 1 atmosphere pressure (1.01×10^5 Pa)
- a specified temperature (usually 298 K)
- reactants and products in their most stable physical state (solid / liquid / gas) under those conditions.

The **standard enthalpy change** of combustion, $\Delta H_c^\ominus$, is the energy released when 1 mole of element or compound burns completely in oxygen under standard conditions.

The symbol '$\ominus$' shows $\Delta H_c^\ominus$ is a standard value, calculated under standard conditions.

Every enthalpy of combustion applies to a balanced equation. Many have been accurately measured with values known to three or four significant figures.

Worked example 6.01

$\Delta H_c^\ominus[CH_4(g)] = -890 \text{ kJ mol}^{-1}$

How much energy is released when 0.75 dm^3 methane, measured at room temperature and 1.0 atmosphere pressure, is completely burnt in oxygen?

How to get the answer:

Step 1: The conditions are standard and so $\Delta H_c^\ominus[CH_4(g)]$ is relevant.

Step 2: Write the combustion equation:
$$CH_4(g) + 2O_2(g) \longrightarrow CO_2(g) + 2H_2O(l)$$

Step 3: The number of moles of methane in $0.75 \text{ dm}^3 = \dfrac{0.75}{24}$

Step 4: From the data, 1 mole of methane releases 890 kJ.

Step 5: The energy released from $\dfrac{0.75}{24}$ mol
$= \dfrac{0.75}{24} \times 890 \text{ kJ}$

Step 6: 0.75 dm^3 methane releases 27.8 kJ of energy.

Progress check 6.01

Write the equations for the combustion reactions illustrating the following enthalpy changes:

1. $\Delta H_c^\ominus[C_2H_6(g)]$
2. $\Delta H_c^\ominus[Mg(s)]$
3. $\Delta H_c^\ominus[CH_3CO_2H(l)]$
4. $\Delta H_c^\ominus[CH_3COOC_2H_5(l)]$

Often we carry out combustion reactions as described below (Figure 6.02), although the enthalpy change will not be standard.

A measured mass, x g, of fuel is burnt and the heat used to raise the temperature of a known mass, m g, of water; the temperature of the water is measured at the start and end of the reaction, giving the temperature change, ΔT.

The relative molecular mass of the fuel, M_r, and the specific heat capacity of water, c, are needed.

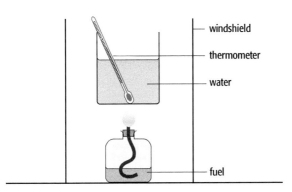

Figure 6.02

The **specific heat capacity** of water is the amount of heat required to raise the temperature of 1 g of water by 1 K. Its value is $4.18\,\text{J g}^{-1}\text{K}^{-1}$ (often taken as $4.2\,\text{J g}^{-1}\text{K}^{-1}$).

The calculations are as follows:

amount of heat given out by x g of fuel $= m \times c \times \Delta T$ J

amount of heat given out by 1 g of fuel
$$= \frac{m \times c \times \Delta T}{x} \text{ J}$$

amount of heat given out by 1 mol of fuel
$$= \frac{m \times c \times \Delta T}{x} \times M_r \text{ J}$$

$$\Delta H_c = -\frac{m \times c \times \Delta T}{1000 \times x} \times M_r \text{ kJ mol}^{-1}$$

This value of ΔH_c will be less exothermic than $\Delta H_c^\ominus$ because:

- energy is lost to the surrounding air
- the glass beaker absorbs some heat
- some incomplete combustion forms C (soot) and CO.

Progress check 6.02

Butanone, $CH_3CH_2COCH_3$, was burnt in a similar apparatus to that in Figure 6.02.

Calculate ΔH_c[butanone(l)] to three significant figures from the following information:

- mass of water = 100 g
- temperature of water at the start = 19.5 °C
- temperature of water at the end = 62.0 °C
- weight of fuel + burner at the start = 12.236 g
- weight of fuel + burner after flame is extinguished = 11.514 g

6.03 Enthalpy change of formation

These enthalpy changes indicate the stability of compounds compared to the elements from which they are made and can be endothermic or exothermic. Many are known accurately:

$\Delta H_f^\ominus[HI(g)] = +27\,\text{kJ mol}^{-1}$; HI is unstable with respect to $H_2(g)$ and $I_2(s)$.

$\Delta H_f^\ominus[PCl_5(s)] = -444\,\text{kJ mol}^{-1}$; PCl_5 is more stable than $P_4(s)$ and $Cl_2(g)$.

The standard **enthalpy change of formation**, $\Delta H_f^\ominus$, is the enthalpy change when 1 mole of a compound is made from its elements under standard conditions.

Note that the standard form of carbon is graphite.

Worked example 6.02

$\Delta H_f^\ominus[CH_3CH_2CH_2OH(l)] = -303\,\text{kJ mol}^{-1}$

Write the equation for the chemical change represented by this ΔH.

Work out the amount of energy released when 2.0 g $CH_3CH_2CH_2OH$ is formed from its elements.

How to get the answer:

Step 1: Analyse the formula and write the equation:

$$3C(\text{graphite}) + 4H_2(g) + \tfrac{1}{2}O_2(g) \longrightarrow CH_3CH_2CH_2OH(l)$$

Step 2: Work out $M_r(CH_3CH_2CH_2OH) = 60$

Step 3: Work out the number of moles:
$$2.0\,\text{g} = \frac{2.0}{60}\,\text{mol}$$

Step 4: Work out the energy released
$$= \frac{2.0}{60} \times 303\,\text{kJ}$$

Step 4. Include the units. Energy released = 10 kJ

Many formation reactions cannot be carried out directly. Graphite, hydrogen and oxygen do not react together to make propanol but the ΔH_f value can be calculated using Hess's law (see below).

6.04 Hess's law

Hess's law can be used in the calculation of enthalpy changes for reactions which cannot be carried out directly.

If energy could be created, then the world's energy crisis would be over ... Unfortunately, energy can neither be created nor destroyed. This is summed up in Hess's law.

The energy change from reactants to products is the same no matter which reaction path is followed, provided that the conditions at the start and finish are the same (Figure 6.03).

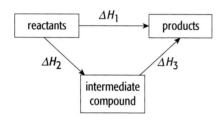

Figure 6.03

Figure 6.03 shows two routes to form the products. Using Hess's law:

$$\Delta H_1 = \Delta H_2 + \Delta H_3$$

$\Delta H_f^\ominus[CO(g)]$ cannot be found experimentally as burning carbon produces either a mixture of CO and CO_2 or just CO_2, depending on how much oxygen is present.

$\Delta H_c^\ominus[C(graphite)] = -394\,kJ\,mol^{-1}$

$\Delta H_c^\ominus[CO(g)] = -283\,kJ\,mol^{-1}$

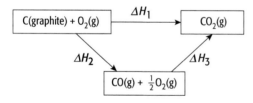

Figure 6.04

In the Hess law diagram (Figure 6.04), each side of the triangle contains a balanced equation. ΔH_1 and ΔH_3 are the two known enthalpies of combustion:

$\Delta H_1 = \Delta H_c^\ominus[C(graphite)]$

$\Delta H_3 = \Delta H_c^\ominus[CO(g)]$

ΔH_2 is $\Delta H_f^\ominus[CO(g)]$

Using $\Delta H_1 = \Delta H_2 + \Delta H_3$ gives

$-394 = \Delta H_f^\ominus[CO(g)] + (-283)$

$\Delta H_f^\ominus[CO(g)] = -111\,kJ\,mol^{-1}$

> ### Progress check 6.03
>
> 1. $\Delta H_c^\ominus[S(s)] = -297\,kJ\,mol^{-1}$
>
> $\Delta H_f^\ominus[SO_3(l)] = -441\,kJ\,mol^{-1}$
>
>   What is the enthalpy change for
> $2SO_2(g) + O_2(g) \longrightarrow 2SO_3(g)$?
>
> 2. Phosphorus forms two chlorides, PCl_5 and PCl_3.
>
> $\Delta H_f^\ominus[PCl_5(s)] = -444\,kJ\,mol^{-1}$
>
> $\Delta H_f^\ominus[PCl_3(l)] = -320\,kJ\,mol^{-1}$
>
>   What is the enthalpy change for
> $PCl_3(l) + Cl_2(g) \longrightarrow PCl_5(s)$?
>
> 3. Ethene can be hydrated to form ethanol:
>
> $CH_2{=}CH_2(g) + H_2O(l) \longrightarrow CH_3CH_2OH(l)$
>
>   Use the following enthalpy changes to draw a suitable Hess's law cycle and evaluate the enthalpy change for the hydration reaction.
>
> $\Delta H_f^\ominus[CH_2{=}CH_2(g)] = +52\,kJ\,mol^{-1}$
>
> $\Delta H_f^\ominus[H_2O(l)] = -286\,kJ\,mol^{-1}$
>
> $\Delta H_f^\ominus[CH_3CH_2OH(l)] = -277\,kJ\,mol^{-1}$

Enthalpies of formation can be found from a Hess's law cycle using enthalpies of combustion (Figure 6.05).

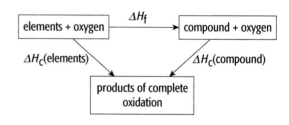

Figure 6.05

$\Delta H_f[\text{compound}] = \Sigma\Delta H_c[\text{elements}] - \Sigma\Delta H_c[\text{compound}]$

> **TIP:** The numbers of moles reacting and produced need to be taken into account when using Hess's law.

6.05 Enthalpy change of atomisation

The **enthalpy change of atomisation** is the amount of energy required for 1 mole of gaseous atoms to be formed from the element in its standard state.

½H_2(g) ⟶ H(g) $\Delta H = \Delta H_{at}[H] = +218$ kJ mol^{-1}

For hydrogen, this is half the amount of energy required to break the bonds in a mole of hydrogen molecules.

Solid or liquid elements may have other forces such as van der Waals' forces as well as covalent bonds which need to be overcome to form the gaseous atoms.

6.06 Enthalpy changes of hydration and solution

Enthalpy changes occur when substances dissolve. Molecules are pulled apart from each other, breaking intermolecular bonds; new intermolecular bonds between the substance and solvent are formed.

The **enthalpy change of solution**, ΔH_{sol}, is the enthalpy change when 1 mole of solute is dissolved in enough water so that adding any further water produces no further change.

As intermolecular forces vary widely in strength, substances vary widely in solubility. Strong intermolecular forces between solute and solvent molecules pull molecules into solution; strong intermolecular forces between the solute molecules themselves prevent the solute going into solution. The solubility depends on these two opposing forces.

If ΔH_{sol} is positive, the solute will have poor solubility; if ΔH_{sol} is negative, the solubility is likely to be high.

Ionic solutes usually have high solubility in water as positive ions become surrounded by the lone pairs on oxygen and negative ions by the $\delta+$ H atoms (Figure 6.06).

Figure 6.06

The ion–dipole forces more than compensate for the ionic attractions in the solid that are lost.

Small covalent molecules such as butane ($CH_3CH_2CH_2CH_3$) are insoluble in water as their weak van der Waals' forces cannot overcome the hydrogen bonds between water molecules.

The **enthalpy change of hydration**, ΔH_{hyd}, is the energy change when 1 mole of gaseous ions is hydrated.

The two enthalpy changes can be illustrated by the following:

NH_4Cl(s) + aq ⟶ NH_4Cl(aq) $\Delta H_{sol}[NH_4Cl]$
Na^+(g) + aq ⟶ Na^+(aq) $\Delta H_{hyd}[Na^+(g)]$

ΔH_{hyd} depends on the ion–dipole attractions in the solution. It becomes more exothermic for ions with higher charges or for smaller ions.

6.07 The enthalpy change of neutralisation

The **enthalpy change of neutralisation** is the energy released when solutions of an acid and an alkali react together to produce 1 mole of water under standard conditions.

Strong acids reacting with strong bases release more energy than weak acids reacting with weak bases.

Strong acids and bases are completely ionised and the reaction is the formation of water from the ions:

H^+(aq) + OH^-(aq) ⟶ H_2O(l)

The other ions, for example Na^+(aq) and Cl^-(aq) when hydrochloric acid and sodium hydroxide react, are 'spectator ions'. The O–H bond formation leads to a release of energy.

Weak acids and bases are only partially ionised; energy is needed to ionise them before the reaction to form H_2O molecules and so the overall release of energy is less.

ΔH_{neut} can be found by reacting known volumes of acid and base of known concentrations in a calorimeter and measuring the temperature change.

Sample answer

Question:

50 cm³ of 2.0 mol dm⁻³ HCl is reacted with 50 cm³ of 2.0 mol dm⁻³ NaOH in a polystyrene cup (Figure 6.07).

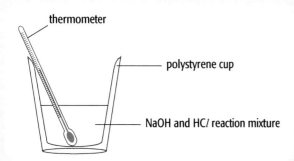

Figure 6.07

initial temperature of HCl = 19.0 °C

initial temperature of NaOH = 21.0 °C

final temperature of mixture = 31.5 °C

a What is the value of ΔH_{neut} from this experiment? [7]

b What errors might arise during the experiment? [2]

Answer:

a Assume that the density of the final solution = density of water = 1.0 g cm⁻³ and that the specific heat capacity of the solution = specific heat capacity of water = 4.18 J g⁻¹ K⁻¹ [1 mark].

As equal volumes of solutions are taken, the starting temperature is taken as the average of 19.0 °C and 21.0 °C, i.e. 20.0 °C.

$\Delta T = 31.5 - 20.0 = 11.5$ °C [1 mark]

energy released = mass of solution × specific heat capacity × ΔT [1 mark]

= 100 × 4.18 × 11.5 J

= $\dfrac{100 \times 4.18 \times 11.5}{1000}$ kJ [1 mark]

50 cm³ of the HCl solution contains $\dfrac{2 \times 50}{1000}$ moles and this reacts completely with the same number of moles of NaOH. [1 mark]

energy released per mole of HCl

= $\dfrac{\text{energy released}}{\text{number of moles of HCl}}$

= $\dfrac{100 \times 4.18 \times 11.5}{2 \times 50}$ kJ [1 mark]

$\Delta H_{neut} = -48.1$ kJ mol⁻¹ [1 mark]

b Possible errors:

- there is a heat loss to the environment – no lid or lagging to prevent this nor any correction factor to allow for it
- the specific heat capacity of the solution may not be the same as that of water
- the density of the solution may not be the same as that of water.

[any two for **2 marks**]

Progress check 6.04

1. Explain why sodium hydroxide produces −32 kJ mol⁻¹ when it neutralises H₂S but −58 kJ mol⁻¹ when it neutralises HNO₃.

2. 100 cm³ of 1.0 mol dm⁻³ HBr(aq) is mixed with 100 cm³ of 1.0 mol dm⁻³ KOH(aq).

   initial temperature of HBr(aq) = 19.5 °C

   initial temperature of KOH(aq) = 21.0 °C

   final temperature of mixture = 26.8 °C

   Work out the enthalpy of neutralisation for this reaction.

3. What final temperature would be achieved if only 50 cm³ of the acid and 50 cm³ of the base in question 2 were mixed? Explain your answer.

4. What final temperature would be achieved if 100 cm³ of the acid and 50 cm³ of the base in question 2 were mixed?

TIP Learn the definitions of the six different enthalpy changes dealt with so far in the unit.

6.08 Bond energies

Energy is needed to break a covalent bond.

The **bond energy** is the energy required to break 1 mole of similar bonds in a mole of gaseous element or compound to form gaseous atoms or radicals.

For hydrogen gas, the bond energy, $E[\text{H–H}]$ is twice $\Delta H_{at}[\text{H(g)}]$.

$$H_2(g) \longrightarrow 2H(g) \qquad \Delta H = E[\text{H–H}] = 436 \text{ kJ mol}^{-1}$$

Bond energies are not known as accurately as enthalpies of combustion because the values are affected by the other bonded atoms. Even in similar molecules, the same bonds may have different bond enthalpies; the value for the C–H bond in CH_4 is different to the value in CH_3CH_3. For this reason, average bond energy values are usually used.

Although bond energies refer to gases, they are used with a Hess's law cycle (see Figure 6.03) to estimate enthalpies of reaction. The intermediate is the individual gaseous atoms.

- ΔH_1 is the enthalpy of reaction.
- ΔH_2 is the sum of all the bond energies in the reactants.
- ΔH_3 is the sum of all the bond energies in the products.

Bond breaking, ΔH_2, is endothermic and bond making, ΔH_3, is exothermic.

$$\Delta H_1 = \Delta H_2 + \Delta H_3$$

Worked example 6.03

Use bond energies to calculate the $\Delta H_{reaction}$ for this reaction:

$$CH_3CH_2OH(g) \longrightarrow CH_2=CH_2(g) + H_2O(g)$$

$E(\text{C–H}) = 413 \text{ kJ mol}^{-1}$
$E(\text{C–C}) = 347 \text{ kJ mol}^{-1}$
$E(\text{C–O}) = 358 \text{ kJ mol}^{-1}$
$E(\text{O–H}) = 464 \text{ kJ mol}^{-1}$
$E(\text{C=C}) = 612 \text{ kJ mol}^{-1}$

How to get the answer:

Step 1: List the bonds that are broken:

$$5E(\text{C–H}) + E(\text{C–O}) + E(\text{C–C}) + E(\text{O–H})$$

Step 2: Evaluate the bond energies for the bonds that are broken:

$$2065 + 358 + 347 + 464 = 3234 \text{ kJ mol}^{-1}$$

Step 3: List the bond energies for the bonds that are made:

$$4E(\text{C–H}) + E(\text{C=C}) + 2E(\text{O–H})$$

Step 4: Evaluate the bond energies for the bonds that are made:

$$1652 + 612 + 928 = 3192 \text{ kJ mol}^{-1}$$

Step 5: $\Delta H_{reaction} = 3234 - 3192$

Step 6: Evaluate $\Delta H_{reaction} = +42 \text{ kJ mol}^{-1}$

Progress check 6.05

Work out enthalpies of reaction for the following reactions, using the bond energies given below and in the Worked examples elsewhere in this unit:

$E(\text{I–I}) = 151 \text{ kJ mol}^{-1}$
$E(\text{H–I}) = 298 \text{ kJ mol}^{-1}$
$E(\text{O=O}) = 498 \text{ kJ mol}^{-1}$
$E(\text{C=O}) = 805 \text{ kJ mol}^{-1}$
$E(\text{N≡N}) = 945 \text{ kJ mol}^{-1}$
$E(\text{N=O}) = 587 \text{ kJ mol}^{-1}$
$E(\text{H–Br}) = 366 \text{ kJ mol}^{-1}$
$E(\text{C–Br}) = 290 \text{ kJ mol}^{-1}$

1. $H_2(g) + I_2(g) \longrightarrow 2HI(g)$
2. $CH_4(g) + 2O_2(g) \longrightarrow CO_2(g) + 2H_2O(g)$
3. $N_2(g) + O_2(g) \longrightarrow 2NO(g)$
4. $CH_3CH_2OH(g) + HBr(g) \longrightarrow CH_3CH_2Br(g) + H_2O(g)$

6.09 Reaction pathway diagrams

Some very exothermic reactions do not seem to occur at room temperature.

For example, $\Delta H_c^\ominus[CH_4] = -890\,kJ\,mol^{-1}$ but a mixture of methane and air undergoes no reaction at room temperature. An energy barrier, the **activation energy**, E_a, is in the way of the reaction. Colliding molecules need energy above E_a for reaction to happen.

The energy profile diagrams from Figure 6.01 are amended in Figure 6.08.

Activation energies for endothermic reactions tend to be larger than for exothermic reactions. Reactions with high values of E_a either do not proceed or proceed very slowly even for exothermic reactions; reactions with very low values seem to be instantaneous. A spark may provide enough energy to overcome E_a, as in the combustion of methane; once started, the heat the reaction generates is enough to provide E_a for further reaction.

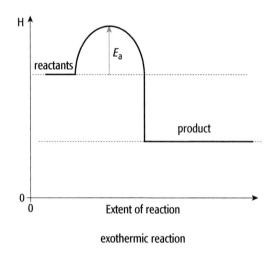
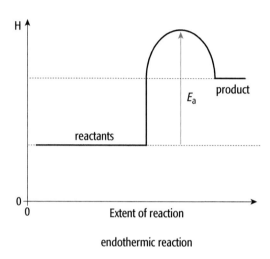

Figure 6.08 Energy profile diagrams

Revision checklist

Check that you know the following:

- [ ] an explanation that energy changes arise through breaking and making bonds,
- [ ] explain and use the standard conditions for reactions,
- [ ] explain and use the terms enthalpy changes of formation, combustion, atomisation, solution, hydration and neutralisation,
- [ ] understand and use bond energy,
- [ ] a description of an experiment to find an enthalpy of combustion and understand where errors may arise,
- [ ] how to calculate enthalpy changes using $m \times c \times \Delta T$,
- [ ] apply (the use of) Hess' Law to construct energy cycles to evaluate unknown enthalpy changes,
- [ ] how to draw and interpret reaction pathway diagrams.

Exam-style questions

1. a i Using the enthalpies of formation given below, evaluate the enthalpy change for the reaction: [2]

   $CH_4(g) + CH_3CH_2CH_3(g) \longrightarrow CH_3CH_2CH_2CH_3(g) + H_2(g)$

   $\Delta H_f^\theta[CH_4(g)] = -75\,kJ\,mol^{-1}$
   $\Delta H_f^\theta[C_3H_8(g)] = -105\,kJ\,mol^{-1}$
   $\Delta H_f^\theta[C_4H_{10}(g)] = -127\,kJ\,mol^{-1}$

   ii Sketch a clearly labelled energy profile for the reaction in part i. [4]

   b i Evaluate the enthalpy change for the reaction in part a using the following bond energies: [2]

   $E[C–H] = 413\,kJ\,mol^{-1}$
   $E[C–C] = 347\,kJ\,mol^{-1}$
   $E[H–H] = 436\,kJ\,mol^{-1}$

   ii Explain any difference in the values from part a i and from part b i. [1]

   c i Using the given enthalpies of combustion, evaluate the enthalpy change for the following isomerisation reaction: [2]

   $CH_3CH_2CH_2CH_2CH_3(l) \longrightarrow C(CH_3)_4(g)$

   $\Delta H_c^\theta[CH_3CH_2CH_2CH_2CH_3(l)] = -3509\,kJ\,mol^{-1}$
   $\Delta H_c^\theta[(CH_3)_4C(g)] = -3493\,kJ\,mol^{-1}$

   ii Explain why this reaction does not occur at room temperature. [1]

   Total: 12

2. a What is the minimum mass of propane, $CH_3CH_2CH_3$, needed to bring $100\,cm^3$ of water at $18.5\,°C$ to its boiling point at $100\,°C$? [2]

   $\Delta H_c^\theta[CH_3CH_2CH_3(g)] = -2219\,kJ\,mol^{-1}$

   b Why would more propane be required than your answer in part a? [1]

   Total: 3

Unit 7

Redox reactions

Learning outcomes

You should be able to:

☐ recognise redox processes and calculate oxidation numbers of elements in compounds and ions

☐ describe redox processes and balance redox equations by using electron transfer

☐ use oxidation numbers to balance redox equations

7.01 Redox processes and oxidation numbers

Oxidation and reduction occur together in a reaction. Originally, definitions of **oxidation** included gain of oxygen, loss of hydrogen or an increase in the non-metallic part of a formula and definitions of **reduction** included loss of oxygen, gain of hydrogen or a decrease in the non-metallic part of a formula. Some reactions seem to be **redox** processes but fall outside these definitions; the **oxidation number** system was introduced as a way of including all redox processes.

We assign an oxidation number (ox. no.), to each element in a compound. The ox. no. has both a sign and a number based on which atoms are combined and their electronegativity.

- All atoms in pure elements are given the oxidation number zero (ox. no. = 0).

- Simple ions such as M^{2+} or X^- are given an oxidation number equal to the sign and charge on the ion.

- For ionic or covalent compounds, the oxidation numbers of all the atoms added together equal zero.

- For a complex ion, the oxidation numbers of the atoms added together equal the sign and numerical charge on the ion.

There are further rules to follow:

- Fluorine, F, is the most electronegative element and always given the ox. no. = -1.

- Oxygen, O, is given the ox. no. = -2 unless combined with F (or in peroxides where it becomes -1).

- Hydrogen, H, is always assigned the ox. no. = $+1$ (unless in metal hydrides where it is assigned -1).

- A more electronegative atom takes the negative oxidation number.

Worked example 7.01

What are the oxidation numbers of each atom making up $KMnO_4$?

How to get the answer:

Step 1: Identify any ionic bonding. Here we have K^+ and MnO_4^-. The overall charge = 0.

Step 2: Pick out known oxidation numbers. ox. no. $K^+ = +1$.

Step 3: Identify the charge on the complex ion, $MnO_4^- = -1$.

Step 4: ox. no. O $= -2$.

Step 5: Remembering to include every atom appearing in the ion, use the overall charge and known ox. no.'s:

$-1 = $ ox. no. Mn $+ 4 \times (-2)$

Step 6: ox. no. Mn $= +7$.

Progress check 7.01

Calculate the oxidation numbers of each atom in the following:

CO_2 CH_4 NH_3 HCl
P_2O_5 ClO_2 $KClO_3$ HNO_3
CrO_4^{2-} $Ni(OH)_2$ CO_3^{2-} NH_4VO_3

a **Oxidation**

When an atom or ion is oxidised, it loses one or more electrons:

$$Fe^{2+} \rightarrow Fe^{3+} + e^-$$

$$Mg \rightarrow Mg^{2+} + 2e^-$$

Losing an electron also means an increase in oxidation number. When you cannot see electrons are lost, oxidation numbers can be used:

$$\underline{S}O_2 + \tfrac{1}{2}O_2 \rightarrow SO_3$$
$$\underline{C}H_4 + 2O_2 \rightarrow CO_2 + 2H_2O$$

The oxidation numbers of the underlined atoms increase and so they have been oxidised:

ox. no. S changes from +4 to +6

ox. no. C changes from −4 to +4

b Reduction

When an atom or ion is reduced, it gains one or more electrons:

$$Cl_2 + 2e^- \rightarrow 2Cl^-$$
$$O_2 + 4e^- \rightarrow 2O^{2-}$$
$$VO_2^+ + e^- \rightarrow VO^{2+}$$

Here there is a decrease in oxidation number:

ox. no. Cl changes from 0 to −1

ox. no. O changes from 0 to −2

ox. no. V changes from +5 to +4

In the following reactions

$$\underline{Cl}_2 + H_2 \rightarrow 2HCl$$
$$\underline{O}_2 + 2H_2 \rightarrow 2H_2O$$

the oxidation number of the underlined atoms decreases and they have been reduced:

ox. no. Cl changes from 0 to −1

ox. no. O changes from 0 to −2

> **TIP** Every atom in a molecule or compound ion has its own oxidation number.

7.02 Balancing redox equations using electron transfer

Oxidation implies a loss of electrons and reduction implies a gain of electrons; redox reactions require a transfer of one or more electrons from the oxidised atom to the reduced atom. The decrease in the number of electrons in the oxidised species must exactly balance the increase in the number of electrons in the reduced species.

For example, when Na and Cl_2 react:

$$Na \rightarrow Na^+ + e^- \qquad [\text{eqn 1}]$$
$$Cl_2 + 2e^- \rightarrow 2Cl^- \qquad [\text{eqn 2}]$$

The two equations shown are **half-equations** – one is showing an oxidation and the other is showing a reduction. To produce a complete equation, the two half-equations need to be combined so that the loss of electrons balances the gain of electrons. Equation 1 needs to be multiplied by two before the two half-equations can be added together:

$$2Na \rightarrow 2Na^+ + 2e^- \qquad Cl_2 + 2e^- \rightarrow 2Cl^-$$

Complete redox equation:

$$2Na + Cl_2 + 2e^- \rightarrow 2Na^+ + 2Cl^- + 2e^-$$

Cancelling the electrons:

$$2Na + Cl_2 \rightarrow 2Na^+ + 2Cl^-$$

Worked example 7.02

Write the balanced equation for MnO_4^- oxidising Fe^{2+}.

How to get the answer:

Step 1: Write down the two half-reactions, one the oxidation and the other the reduction:

$$Fe^{2+} \rightarrow Fe^{3+} + e^- \qquad [\text{eqn 3}]$$
$$MnO_4^- + 8H^+ + 5e^- \rightarrow Mn^{2+} + 4H_2O \qquad [\text{eqn 4}]$$

Step 2: Multiply the equations so that the number of electrons released equals the number of electrons used. Equation 3 needs to be multiplied by five, to give:

$$5Fe^{2+} \rightarrow 5Fe^{3+} + 5e^-$$

Step 3: Add the equations:

$$MnO_4^- + 8H^+ + 5Fe^{2+} + 5e^- \rightarrow Mn^{2+} + 4H_2O + 5Fe^{3+} + 5e^-$$

Step 4: Cancel the electrons:

$$MnO_4^- + 8H^+ + 5Fe^{2+} \rightarrow Mn^{2+} + 4H_2O + 5Fe^{3+}$$

TIP: In a redox reaction, the number of electrons released on oxidation equals the number of electrons used up in the reduction.

Progress check 7.02

1. a. Write the two half-equations for the oxidation and reduction which occur when $FeCl_3$ reacts with KI to produce $FeCl_2$ and I_2 (tip – ignore the spectator ions).
   b. Write the complete balanced equation for the redox reaction which occurs.

2. a. In acidic solution, blue $VO^{2+}(aq)$ ions react with tin metal to produce green $V^{3+}(aq)$ and $Sn^{2+}(aq)$ ions. Write the half-equations for the oxidation and reduction that occur.
   b. Write the complete balanced equation for the redox reaction which occurs.

7.03 Balancing redox reactions using oxidation numbers

An increase in oxidation number implies oxidation and a decrease implies reduction. An increase in oxidation number of one atom must be balanced by a decrease in oxidation number of another atom in a reaction.

TIP: In a balanced redox equation, the total increase in oxidation number equals the total decrease in oxidation number.

Progress check 7.03

Balance the following equations using oxidation numbers:

1. $H^+ + VO_2^+ + Zn \rightarrow V^{2+} + Zn^{2+} + H_2O$
2. $H_2S + SO_2 \rightarrow H_2O + S$
3. $MnO_4^- + SO_2 + H_2O \rightarrow Mn^{2+} + SO_4^{2-} + H^+$
4. $S_2O_8^{2-} + Fe^{2+} \rightarrow SO_4^{2-} + Fe^{3+}$

Worked example 7.03

Use oxidation numbers to write a balanced equation for the reaction between Cl_2 and cold aqueous KOH to form KCl, $KOCl$ and H_2O.

How to get the answer:

Step 1: Identify which atoms have an ox. no. change. O and H do not change ox. no. but Cl does.

Step 2: Work out relevant ox. no.'s. The ox. no. of Cl in $Cl_2 = 0$; $KCl = -1$; $KOCl = +1$.

Step 3: Work out changes in ox. no. Cl increases from 0 to +1 and decreases from 0 to −1.

Step 4: Decide how many atoms are involved in the oxidation and reduction to balance the ox. no. changes. For every Cl in Cl_2 that produces KCl, one Cl must also produce $KOCl$.

Step 5: Write down the partial equation showing the redox changes:

$$Cl_2 \rightarrow KCl + KOCl$$

Step 6: Balance the other atoms in the equation. To balance the two potassium ions on the right, two KOH need to be on the left:

$$Cl_2 + 2KOH \rightarrow KCl + KOCl + H_2O$$

Step 7: Check the numbers of other atoms balance. The number of H and O on each side balance and the overall equation is now balanced.

Sample answer

Question:

The solution formed when Cl_2 dissolves in cold KOH(aq) contains KCl and $KOCl$. When this solution is boiled, $KClO_3$ is formed. Write a balanced equation for the reaction which occurs on boiling. [4]

Answer:

ox. no.'s of Cl in $KCl = -1$, $KOCl = +1$, $KClO_3 = +5$ (no other ox. no. changes) [1 mark].

Some Cl is oxidised so some must be reduced to Cl^-:

$$KOCl \rightarrow KClO_3 + KCl$$

Increase in ox. no. $= +1$ to $+5$; decrease $= +1$ to -1 [1 mark].

One $KClO_3$ produced for $2KCl$ [1 mark].

$$3KOCl \rightarrow KClO_3 + 2KCl \text{ [1 mark]}$$

Revision checklist

Check that you know the following:

- [ ] how to calculate the oxidation number of atoms in molecules and ions
- [ ] how to describe and explain oxidation in terms of electron loss or increase in oxidation number
- [ ] how to describe and explain reduction in terms of electron gain or reduction in oxidation number
- [ ] write balanced equations from half reactions by balancing electrons gained and lost
- [ ] write balanced equations by balancing oxidation numbers

Exam-style questions

1. a Using the following two half-equations, write the full equation for the reaction which occurs when MnO_4^- acts as an oxidising agent: [2]

    $$XeO_6^{4-} + 5H^+ + 2e^- \rightarrow HXeO_4^- + 2H_2O$$
    $$MnO_4^- + 5e^- + 8H^+ \rightarrow Mn^{2+} + 4H_2O$$

   b The following equation shows the reaction occurring when ammonium dichromate is heated:

    $$(NH_4)_2Cr_2O_7 \rightarrow Cr_2O_3 + N_2 + 4H_2O$$

   i Which species has been reduced? Write down the oxidation number change. [2]

   ii Which species has been oxidised? Write down the oxidation number change. [2]

   Total: 6

2. a Write down the two half-equations from the following equation: [2]

    $$BrO_3^- + F_2 + 2OH^- \rightarrow BrO_4^- + 2F^- + H_2O$$

   b The equation for the reaction between tin and sulfuric acid is given by:

    $$Sn + H_2SO_4 \rightarrow Sn(SO_4)_2 + H_2O + SO_2$$

   i Which species has been oxidised? State the oxidation number change. [1]

   ii Which species has been reduced? State the oxidation number change. [1]

   iii Write the balanced equation. [2]

   Total: 6

Unit 8: Equilibria

Learning outcomes

You should be able to:

- explain the dynamic nature of reversible reactions
- state Le Chatelier's Principle and use it to decide what effect changes in temperature, concentration or pressure have on reactions at equilibrium
- deduce expressions for equilibrium constants and state what changes affect them
- calculate values of equilibrium constants
- describe the conditions used in some important industrial processes where chemical equilibrium is a feature
- understand and use the Brønsted Lowry theory
- explain the difference between strong and weak acids and bases

8.01 Reversible reactions and dynamic equilibrium

Although many reactions go to completion, many stop before all the reactants are changed into products – no matter how long you wait. Here are two examples:

- Liquid bromine in a closed flask evaporates to produce both liquid and vapour together.
- Heating calcium carbonate in a closed container produces a mixture of calcium carbonate, calcium oxide and carbon dioxide.

If the final conditions are the same, the resultant mixture will be the same whether you start from the reactants or the products; heating calcium oxide in carbon dioxide produces the same mixture of calcium carbonate, calcium oxide and carbon dioxide.

These reactions that stop before completion are **reversible reactions**. To show the reversible nature, the equation is written with a double arrow:

- $Br_2(l) \rightleftharpoons Br_2(g)$
- $CaCO_3(s) \rightleftharpoons CaO(s) + CO_2(g)$
- $Co(H_2O)_6^{2+} + 4Cl^- \rightleftharpoons CoCl_4^{2-} + 6H_2O$

When no further reaction is seen to occur, the reaction is at **equilibrium**. However, at the molecular level, reactions still occur:

- Br_2 molecules leave the surface of the liquid and become vapour at exactly the same rate that gaseous Br_2 molecules condense.
- $CaCO_3$ breaks up to form CaO and CO_2 at exactly the same rate that CaO and CO_2 recombine to form $CaCO_3$.

At equilibrium, the rate of the forward reaction equals the rate of the backward reaction and it is a **dynamic equilibrium**.

8.02 Le Chatelier's principle

This section looks at Le Chatelier's principle and the effect of changes in temperature, concentration or pressure on reactions at equilibrium.

Reversible reactions at equilibrium respond to changes in the conditions such as temperature or pressure. If the following reaction at equilibrium changes, it can be followed visually as the reactants are colourless but the product is a dark brown gas:

$$2NO(g) + O_2(g) \rightleftharpoons 2NO_2(g) \qquad \Delta H_r = -114 \text{kJ mol}^{-1}$$

- If the temperature is increased, the mixture becomes paler – indicating more reactants are present; the equilibrium has moved to the left.
- If the temperature is decreased, the mixture becomes darker brown – indicating there is more product present; the equilibrium has moved to the right.
- If the pressure is increased, the mixture becomes darker; if the pressure is decreased, the mixture becomes paler.

These changes in equilibrium position in response to changes in conditions are summed up in **Le Chatelier's principle**:

> If a chemical system at equilibrium experiences a change in conditions, the equilibrium will move to minimise the change and establish a new equilibrium.

- For a reaction at equilibrium, increasing the pressure makes the reaction move in a direction to decrease the pressure.
- Decreasing the temperature makes the reaction move in a direction to increase the temperature.
- Increasing the concentration of a reactant or product makes the reaction move in a direction to decrease the concentration of the reactant or product.

For the exothermic reaction considered above, increasing the temperature causes the equilibrium to move to the left. As there are fewer molecules on the right-hand side, increasing the pressure moves the equilibrium to the right.

Use this information to follow through the colour changes.

> **TIP**
> Remember that if you try to alter a reaction at equilibrium, it will move in a direction to oppose you.

Progress check 8.01

1. Write balanced equilibrium equations for the following gaseous reactions:
   a. phosphorus pentachloride producing phosphorus trichloride and chlorine
   b. hydrogen iodide producing hydrogen and iodine
   c. nitrogen and hydrogen producing ammonia
   d. hydrogen and carbon dioxide producing steam and carbon monoxide
   e. sulfur dioxide and oxygen producing sulfur trioxide.

2. For the following reactions at equilibrium, which way does the equilibrium move when the temperature is increased?

   a. $PCl_5(g) \rightleftharpoons PCl_3(g) + Cl_2(g)$
      $\Delta H = +93 \, kJ\,mol^{-1}$
   b. $2HI(g) \rightleftharpoons H_2(g) + I_2(g)$
      $\Delta H = +10 \, kJ\,mol^{-1}$
   c. $N_2(g) + 3H_2(g) \rightleftharpoons \underline{2NH_3(g)}$
      $\Delta H = -92 \, kJ\,mol^{-1}$
   d. $H_2(g) + \underline{CO_2(g)} \rightleftharpoons H_2O(g) + CO(g)$
      $\Delta H = +41 \, kJ\,mol^{-1}$
   e. $2SO_2(g) + O_2(g) \rightleftharpoons \underline{2SO_3(g)}$
      $\Delta H = -197 \, kJ\,mol^{-1}$

3. For the above reactions at equilibrium, which way does the equilibrium move, if at all, when the pressure is increased?

4. If the underlined reactants in the above reactions are removed from the reactions at equilibrium, what effect does this have?

8.03 The equilibrium constant and factors affecting it

For the general reaction between A and B forming C and D, at equilibrium:

$$aA + bB \rightleftharpoons cC + dD$$

a, b, c and d are the number of moles in the balanced equation.

It is found experimentally that $\dfrac{[C]^c[D]^d}{[A]^a[B]^b} =$ a constant value, K.

K is called the **equilibrium constant**. The concentrations used are those at equilibrium in $mol\,dm^{-3}$.

> **TIP**
> Remember, when writing K expressions, that product concentrations are in the numerator (top line of the fraction) and reactant concentrations are in the denominator (bottom line).

For the reaction $PCl_5(g) \rightleftharpoons PCl_3(g) + Cl_2(g)$

$\Delta H = +93 \, kJ\,mol^{-1}$

$$K = \dfrac{[PCl_3][Cl_2]}{[PCl_5]}$$

Increasing the pressure moves the equilibrium to the left where there are fewer gaseous molecules. Although the number of moles of PCl_5 increases and the number of moles of both PCl_3 and Cl_2 decrease, the volume also decreases so the *concentration relationship* remains the same – K is unchanged.

At equilibrium, adding more Cl_2 moves the reaction to the left but again K does not change. Increasing the temperature moves the reaction to the right as the reaction is endothermic; this time the concentration of PCl_5 decreases and those of both PCl_3 and Cl_2 increase but the volume does not change so K increases.

The value of K is different for different equilibria and also if the temperature changes.

8.04 K_c and K_p and calculations of values

When the equilibrium constant expression contains concentrations in $mol\,dm^{-3}$, as for the examples above, the equilibrium constant is given the symbol K_c.

$$K_c = \frac{[PCl_3][Cl_2]}{[PCl_5]}$$

The units of K_c are found by substituting the concentration units in the expression:

$$\text{units of } K_c = \frac{(mol\,dm^{-3})(mol\,dm^{-3})}{mol\,dm^{-3}} = mol\,dm^{-3}$$

K_c expressions relate to a correctly stated balanced equation. Thus for the equation

$$PCl_3(g) + Cl_2(g) \rightleftharpoons PCl_5(g)$$

$$K_c = \frac{[PCl_5]}{[PCl_3][Cl_2]}\, mol^{-1}\,dm^3$$

TIP Equilibrium constant expressions are written from the balanced equation.

K_c can be evaluated from known equilibrium concentrations.

Worked example 8.01

For $H_2(g) + I_2(g) \rightleftharpoons 2HI(g)$, at 500 K

$[HI(g)]_{eq} = 0.0086\,mol\,dm^{-3}$

$[H_2(g)]_{eq} = 7.0 \times 10^{-4}\,mol\,dm^{-3}$

$[I_2(g)]_{eq} = 7.0 \times 10^{-4}\,mol\,dm^{-3}$

Evaluate K_c and include the units.

How to get the answer:

Step 1: Write the expression for K_c:

$$K_c = \frac{[HI]^2}{[H_2][I_2]}$$

Step 2: Insert the given concentrations:

$$K_c = \frac{(0.0086)^2}{(7.0 \times 10^{-4})(7.0 \times 10^{-4})}$$

Step 3: Work out the answer: 151

Step 4: Insert the concentration units:

$$\frac{(mol\,dm^{-3})^2}{(mol\,dm^{-3})(mol\,dm^{-3})}$$

Step 5: Tidy the expression. No units.

Sample answer

Question:

0.05 moles N_2O_4 are put into a 1 dm^3 vessel and allowed to reach equilibrium at 600 K:

$$N_2O_4(g) \rightleftharpoons 2NO_2(g)$$

0.095 mol NO_2 are found to be present at equilibrium.

Evaluate the equilibrium concentrations of $N_2O_4(g)$ and $NO_2(g)$.

What is the value of K_c? Include the units in your answer. [6]

Answer:

$$N_2O_4 \rightleftharpoons 2NO_2$$

Initial concentrations 0.05

Equilibrium concentrations 0.05 − x   2x   [1 mark]

where x = no of moles reacted

$2x = 0.095$, $x = 0.0475$ [1 mark]

$[NO_2] = 0.095\,mol\,dm^{-3}$

$[N_2O_4] = 0.0025\,mol\,dm^{-3}$

$K_c = \dfrac{[NO_2]^2}{[N_2O_4]}$ [1 mark]

$= \dfrac{(2x)^2}{0.05 - x} = \dfrac{(0.095)^2}{0.0025}$ [1 mark]

$= 3.61$ [1 mark] $mol\,dm^{-3}$ [1 mark]

Progress check 8.02

1. Write K_c expressions for the following equilibria:

   a $H_2(g) + I_2(g) \rightleftharpoons 2HI(g)$

   b $HI(g) \rightleftharpoons \frac{1}{2}H_2(g) + \frac{1}{2}I_2(g)$

   c $N_2(g) + 3H_2(g) \rightleftharpoons 2NH_3(g)$

   d $2NO_2(g) \rightleftharpoons N_2O_4(g)$

2. Include the units in each example above.

3. N_2, H_2 and NH_3 are allowed to come to equilibrium in a 0.50 dm³ vessel. At equilibrium, the number of moles were:

   $N_2 = 0.0025$; $H_2 = 0.0075$; $NH_3 = 2.0 \times 10^{-4}$

   Evaluate K_c, including the units.

4. For the reaction $H_2(g) + CO_2(g) \rightleftharpoons H_2O(g) + CO(g)$, $K_c = 0.288$ at 800 K.

   0.05 mol H_2 is mixed with 0.05 mol CO_2 and allowed to reach equilibrium.

   How many moles of each gas are present at equilibrium?

For a gas at constant temperature, the pressure is proportional to its concentration so a different equilibrium constant, K_p, for gases uses the **partial pressure** of each gas. The partial pressure of a gas is the pressure that the gas exerts in a mixture and is the same pressure that the gas would exert if it was alone in the same volume.

For $H_2(g) + I_2(g) \rightleftharpoons 2HI(g)$

$$K_p = \frac{p_{HI}^2}{p_{H_2} \times p_{I_2}}$$

For a gaseous reaction with the same number of molecules on the left and right of the equation, K_p has no units and is numerically equal to K_c.

For each gas:

partial pressure = **mole fraction** × total pressure

$$\text{mole fraction} = \frac{\text{number of moles of the gas}}{\text{total number of gas moles in the mixture}}$$

To find partial pressures of gases from the equilibrium numbers of moles, start with a correctly balanced equation.

Worked examples 8.02

1. $SO_2(g)$ and $O_2(g)$ react to form $SO_3(g)$ at a total pressure of 21 atm. Equilibrium pressures of SO_2 and O_2 are 0.5 atm.

   What is the equilibrium SO_3 pressure?

   Write the equilibrium equation, the expression for K_p and find its value. Include the units in your answer.

2. For the equilibrium reaction,
   $H_2(g) + I_2(g) \rightleftharpoons 2HI(g)$, $K_p = 25$ at 1100 K

   If $p_{I_2} = 1$ atm, what partial pressure of hydrogen is needed to give a partial pressure of $HI = 5$ atm?

How to get the answers:

1. $2SO_2(g) + O_2(g) \rightleftharpoons 2SO_3(g)$

   As $p_{SO_2} + p_{O_2} + p_{SO_3} = 21$ atm, $p_{SO_3} = 20$ atm

   $$K_p = \frac{p_{SO_3}^2}{p_{SO_2}^2 \times p_{O_2}} = \frac{20^2}{0.5^2 \times 0.5} = 3200 \text{ atm}^{-1}$$

2. $$K_p = \frac{p_{HI}^2}{p_{I_2} \times p_{H_2}} = \frac{5^2}{1 \times p_{H_2}} = 25$$

   $p_{H_2} = 1$ atm

Progress check 8.03

For the gaseous reaction $A + B \rightleftharpoons C$, 1.0 mol A reacts with 1.0 mol B at a total pressure of 5.0 atm. At equilibrium, it is found that 0.7 mol of A is present.

1. How many moles of B and C are present at equilibrium?

2. What is the total number of moles present at equilibrium?

3. What are the partial pressures of A, B and C present at equilibrium?

4. Write the expression for K_p.

5. Evaluate K_p and include the units.

8.05 Some important industrial processes

a The Haber process

Many tonnes of ammonia for use in fertilisers, explosives and polymers are produced each year. In the Haber process, nitrogen reacts with hydrogen:

$$N_2(g) + 3H_2(g) \rightleftharpoons 2NH_3(g) \quad \Delta H^\ominus = -92 \text{ kJ mol}^{-1}$$

To get as much ammonia as possible, there are two considerations.

- The equilibrium: the reaction is exothermic and so a low temperature is needed; the number of moles decreases and so a high pressure is needed.
- The rate: a high temperature, a high pressure and the use of a catalyst are needed.

From this you will see a high pressure, a catalyst and an optimum temperature (between the low temperature needed for the equilibrium and the high temperature needed for the rate) are the best conditions. A very high pressure is not used as this would cause issues of expense for high pressure equipment and safety.

The actual conditions used are a pressure of 15 000 kPa, a temperature of 700 K and an iron catalyst.

b The Contact process

The Contact process converts sulfur to sulfuric acid; about 200 million tonnes of sulfuric acid are produced each year, much of it used in copper mining. Sulfur and sulfide ores are mined but much sulfur comes from removing it from fossil fuels before combustion.

Sulfur dioxide is produced by burning sulfur or by roasting ores. It is then oxidised:

$$2SO_2(g) + O_2(g) \rightleftharpoons 2SO_3(g) \quad \Delta H^\ominus = -197 \text{ kJ mol}^{-1}$$

The exothermic reaction produces fewer molecules so a low temperature and a high pressure are needed for the maximum amount of SO_3 in the equilibrium mixture. The rate requires a high temperature, high pressure and a catalyst. Combining these, a high pressure and a catalyst are needed but the temperature needs to be at an optimum. As high pressures require expensive equipment, they are not used as there is a good yield at atmospheric pressure.

The temperature used is 700 K and the catalyst is vanadium(V) oxide (V_2O_5).

$SO_3(g)$ reacting with water produces so much heat that a sulfuric acid mist is produced. Instead $SO_3(g)$ is dissolved in 98% sulfuric acid to produce oleum, and the oleum is then diluted:

$$SO_3(g) + H_2SO_4(l) \rightarrow H_2S_2O_7(l)$$

$$H_2S_2O_7(l) + H_2O(l) \rightarrow 2H_2SO_4(l)$$

8.06 The Brønsted–Lowry theory

Acids contain at least one $\delta+$ H, often connected to an electronegative O:

$$\overset{\delta+}{H}-Cl \qquad CH_3C\underset{\underset{O}{\|}}{-}O-H^{\delta+} \qquad \overset{\delta+}{H}-O-\underset{\underset{O}{\|}}{\overset{\overset{O}{\|}}{S}}-O-H^{\delta+}$$

The **Brønsted–Lowry** theory states that acids are proton donors and bases are proton acceptors. Brønsted–Lowry acids and bases act in pairs.

When HCl dissolves in water:

$$HCl(aq) + H_2O(l) \rightleftharpoons Cl^-(aq) + H_3O^+(aq)$$

HCl donates H^+ to H_2O; H_2O acts as a base.

H_2O can act as both a base (as above) or as an acid, for example with the base NH_3:

$$NH_3(aq) + H_2O(l) \rightleftharpoons NH_4^+(aq) + OH^-(aq)$$

As this reaction is reversible, H_2O and NH_4^+ are acids and NH_3 and OH^- are bases.

By donating its proton, an acid produces its conjugate base; by accepting a proton, a base produces its conjugate acid. Cl^- is the conjugate base of the acid HCl. NH_4^+ is the conjugate acid of the base NH_3. These related acids and bases are known as **conjugate acid–base pairs**.

Progress check 8.04

Identify the six Brønsted–Lowry acids in the following equations:

$HNO_3 + OH^- \rightleftharpoons NO_3^- + H_2O$

$CH_3COOH + H_2O \rightleftharpoons CH_3COO^- + H_3O^+$

$HNO_3 + H_2SO_4 \rightleftharpoons H_2NO_3^+ + HSO_4^-$

8.07 pH and strong and weak acids and bases

Pure water ionises slightly:

$H_2O(l) + H_2O(l) \rightleftharpoons H_3O^+(aq) + OH^-(aq)$

In pure water at 298 K, $[H_3O^+] = 1.0 \times 10^{-7}\,\text{mol dm}^{-3}$

As $[H_3O^+] = [OH^-]$, in pure water at 298 K, $[OH^-] = 1.0 \times 10^{-7}\,\text{mol dm}^{-3}$

In $1.0\,\text{mol dm}^{-3}$ HCl, $[H_3O^+] = 1.0\,\text{mol dm}^{-3}$

In $1.0\,\text{mol dm}^{-3}$ NaOH, $[H_3O^+] = 1.0 \times 10^{-14}\,\text{mol dm}^{-3}$

The acidity of a solution is defined by $[H_3O^+]$, which varies between about $1.0\,\text{mol dm}^{-3}$ and $1.0 \times 10^{-14}\,\text{mol dm}^{-3}$. The **pH** scale makes these numbers easier to manage and runs between 0 (high acidity) and 14 (low acidity). The actual relationship, $pH = -\log_{10}[H_3O^+]$, is discussed in Unit 21.

When $[H_3O^+] = 1.0\,\text{mol dm}^{-3}$, pH = 0.

When $[OH^-] = 1.0\,\text{mol dm}^{-3}$, pH = 14.

Strong acids, mainly the mineral acids, are so good at donating protons that they are completely ionised in aqueous solution:

$HCl(aq) + H_2O(l) \rightarrow Cl^-(aq) + H_3O^+(aq)$

Note the one way arrow. pH values are between about 0 to 3, depending only on the concentration.

Weak acids, mainly the organic acids, are poor proton donors and only ionise partially in aqueous solution:

$CH_3COOH(aq) + H_2O(l) \rightleftharpoons CH_3COO^-(aq) + H_3O^+(aq)$

This dissociation is reversible as the base formed is reasonably good at accepting protons. pH values are between about 3 and 7, depending on both the concentration and the weakness of the acid.

Strong bases, alkalis with pH values of 13 or 14, are completely ionised in solution. Weak bases, for example organic amines and ammonia, are little ionised and give pH values of about 7 to 10.

Revision checklist

Check that you know the following:

- [ ] explain what is meant by a reversible reaction and dynamic equilibrium,
- [ ] be able to state Le Chatelier's Principle and use it to predict what happens to an equilibrium when the pressure, concentration or temperature changes,
- [ ] be able to deduce and evaluate expressions for K_c and K_p from suitable data,
- [ ] state the changes which affect the value of the equilibrium constant,
- [ ] how to calculate quantities present at equilibrium from suitable data,
- [ ] describe and explain the conditions used in the Haber and Contact processes,
- [ ] show understanding of the Brønsted-Lowry theory of acids and bases,
- [ ] state the pH scale and explain the values of pH expected for strong and weak acids and bases.

Exam-style questions

1. $50\,cm^3$ of N_2O_4 was put into a gas syringe held at 350 K and 101 kPa and the equilibrium allowed to establish:

   $N_2O_4 \rightleftharpoons 2NO_2$

   The volume of the mixture at equilibrium = $85\,cm^3$

   a How many moles of N_2O_4 are present initially? [1]

   b How many moles of mixture are present at equilibrium? [1]

   c What is the change in number of moles (number of moles at equilibrium − number of moles initially)? [1]

   d How many moles of each gas are present at equilibrium?

      i N_2O_4

      ii NO_2 [2]

   e What is the partial pressure of each gas at equilibrium?

      i N_2O_4

      ii NO_2 [2]

   f i Write an expression for K_p. [1]

      ii Find a value for K_p, including the units in your answer. [2]

   Total: 10

2. The Haber process involves the reaction of nitrogen and hydrogen gases:

   $N_2(g) + 3H_2(g) \rightleftharpoons 2NH_3(g)$

   $\Delta H = -92\,kJ\,mol^{-1}$

   a i Choose from the list the best operating conditions to get the most ammonia in the equilibrium mixture:

   low temperature and low pressure

   high temperature and high pressure

   low temperature and high pressure

   high temperature and low pressure [1]

      ii Explain your choice. [2]

      iii The actual operating conditions are $1.5 \times 10^4\,kPa$ and 700 K. Explain any difference in these and your choice in part a i. [2]

   b Write an equation to show the ionisation of the Brønsted–Lowry base, NH_3, in water. [1]

   c When ammonium chloride is boiled with aqueous sodium hydroxide, ammonia is expelled and sodium chloride solution is left behind.

      i Write an ionic equation for the reaction. [2]

      ii Identify the Brønsted–Lowry acid and base in the reactants. [2]

   Total: 10

Rates of Reaction

Learning outcomes

You should be able to:

- explain and use the terms *rate of reaction* and *activation energy*

- explain why the concentration of a reactant and the temperature affect the rate of reaction

- use a Boltzmann distribution to explain the effect of temperature and adding a catalyst on the rate of reaction

- explain the different ways that catalysts, including enzymes, work

9.01 Rate of reaction and activation energy

Ionic reactions seem to be instantaneous but other reactions are so slow that years go by before any product is seen. Reactions falling between these two extremes have measurable rates; for these, the **rate of reaction** is defined as the number of moles of reagent used up or the number of moles of product produced in a given time.

We will consider the following reaction:

$$CaCO_3(s) + 2HCl(aq) \rightarrow CaCl_2(aq) + H_2O(l) + CO_2(g)$$

If the acid is in excess, the rate could be followed by measuring the volume of $CO_2(g)$ produced or the mass of solid $CaCO_3$ left at known times.

For a reaction to occur, the reacting molecules collide, bonds break and new bonds form. Bond breaking needs energy and so the reacting molecules not only need to collide but also need sufficient energy to weaken or break bonds. The minimum energy that the colliding molecules must have for reaction to occur is called the **activation energy**, E_a. The activation energy is an energy barrier to the reaction taking place; a high activation energy means that few molecules will have the minimum energy required and so the reaction will be slow. A low activation energy, however, means that many molecules will have sufficient energy to react on collision and so the reaction is fast. The activation energy can be represented on an energy profile diagram (Figure 9.01).

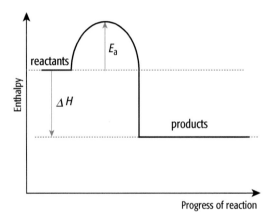

Figure 9.01 An energy profile diagram for an exothermic reaction showing E_a

Catalysts speed up chemical reactions without being changed chemically at the end. They provide a new pathway with a lower activation energy so many more molecules now have energy greater than the activation energy and will be able to react on collision; the reaction is faster (Figure 9.02). The activation energy is lowered for both the forward and reverse reactions.

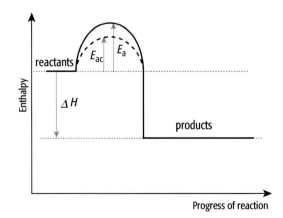

Figure 9.02 The activation energy is lower in the presence of a catalyst

E_a is the activation energy without a catalyst; E_{ac} is the activation energy with a catalyst.

Worked example 9.01

For the reaction $X + Y \rightarrow P$

$\Delta H = +32\,kJ\,mol^{-1}$, $E_a = +53\,kJ\,mol^{-1}$

What is the value of E_a for $P \rightarrow X + Y$?

How to get the answer:

Step 1: A positive ΔH means an endothermic reaction; draw an energy profile as in Figure 9.01 with P above $X + Y$.

Step 2: Insert arrows in the correct direction for the two energy changes.

Step 3: Label the arrows with the correct values.

Step 4: Insert and label the E_a arrow for the back reaction, starting from the 'products' line.

Step 5: Calculate the value from the numbers on your energy profile:

$53 - 32 = 21\,kJ\,mol^{-1}$

9.02 Effect of concentration on rate

As solutions become more concentrated or the pressure of a gas is increased, the molecules or ions are pushed together and collide more often. There will be a greater rate of collisions in which the particles have more energy than the activation energy and so will be able to react. More particles will react per second and so the rate of reaction increases.

For a solid reactant, the reaction can only occur on the surface where collisions between the reactant particles can take place. The greater the surface area, the more collisions there can be per second and so again the rate of reaction increases.

For our example reaction:

$CaCO_3(s) + 2HCl(aq) \rightarrow CaCl_2(aq) + H_2O(l) + CO_2(g)$

the reaction rate increases if the concentration of the $HCl(aq)$ is increased and also if the $CaCO_3(s)$ is powdered rather than in one lump.

The product concentration has no effect on the reaction rate.

> **TIP**
>
> Reaction rate is increased by increasing the concentration of a soluble reactant, increasing the pressure of a gaseous reactant or increasing the surface area of a solid reactant.

9.03 The Boltzmann distribution

The kinetic theory of gases assumes that the molecules move randomly with a kinetic energy which depends on their speed. In any sample of gas, different molecules move with different speeds and so have different kinetic energies. At any particular temperature, we can work out how many molecules there are for each value of kinetic energy and show this on a diagram called the Boltzmann distribution (Figure 9.03).

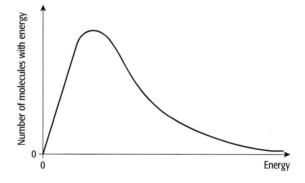

Figure 9.03 The Boltzmann distribution at one temperature

The Boltzmann distribution shows that, although most molecules have average energies, some molecules have very low energies and a few have very high energies. We could include E_a, as in Figure 9.04.

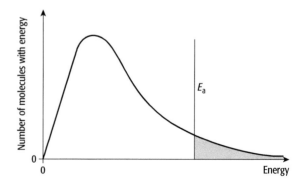

Figure 9.04 The Boltzmann distribution, showing E_a

The shaded area represents the number of molecules with energy greater than E_a and so able to react on collision.

If E_a is very high, then only a few molecules will have sufficient energy to react.

9.04 The effect of temperature on rate

Increasing the temperature of a gas increases the kinetic energy of the molecules. The collision rate increases so the reaction rate increases but this effect is small as a temperature increase of 10 K only increases the collision rate by 2%.

Experimentally, we find that many reaction rates double for a 10 K rise in temperature. The main explanation for this great increase in rate is how the Boltzmann distribution changes with temperature (Figure 9.05).

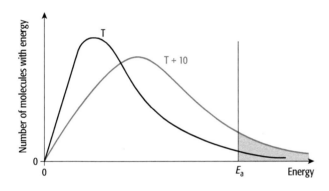

Figure 9.05 Boltzmann distributions for a gas at T K and the same gas at $(T + 10)$ K

In Figure 9.05 the area under the two distribution curves is the same as it represents the total number of molecules. With a 10 K rise in temperature, the maximum of the curve is lower and moves to a higher energy value. Fewer molecules have very low energies and more molecules have higher energies. The activation energy has not changed but you can see there are more molecules with energy higher than E_a and so more molecules can react on collision. The rate increases substantially.

Sample answer

Question:

The reaction to form carbonyl chloride, $COCl_2$, is slow. Explain what happens to the rate as the temperature is raised. [6]

Answer:

For a reaction to happen, the molecules must collide [1 mark] with energy greater or equal to E_a [1 mark]. As the temperature is raised, the molecules gain kinetic energy and move faster [1 mark] and so collide more often [1 mark]. More collisions have energy greater or equal to E_a [1 mark] so the reaction rate increases [1 mark].

Progress check 9.01

1. For the reaction R → P, $\Delta H_r = -48 \text{ kJ mol}^{-1}$; $E_a = 96 \text{ kJ mol}^{-1}$; $E_{ac} = 40 \text{ kJ mol}^{-1}$.

   a Draw an energy profile diagram for R → P, showing the three energy changes by labelled arrows.

   b What is the value of ΔH_r for the reverse reaction P → R?

   c What is the value of E_a for P → R?

2. 1.0 g zinc powder reacts with 50 cm³ of 1.0 mol dm⁻³ HCl at room temperature:

   $Zn(s) + 2HCl(aq) \rightarrow ZnCl_2(aq) + H_2(g)$

   a How could the reaction rate be followed experimentally? Suggest suitable apparatus.

   b Using the rate from part a) as a basis for your comparisons.

   i how would the rate change if the acid is changed to 2.0 mol dm⁻³ HCl? Explain your answer.

   ii how would the rate change if a single piece of zinc weighing 1.0 g reacts with 1.0 mol dm⁻³ HCl? Explain your answer.

iii how would the rate change if 2.0 g zinc powder reacts with 1.0 mol dm^{-3} HCl? Explain your answer.

3 A process, no longer used, for producing nitrogen monoxide (NO) passes air through an electric arc where the temperature can reach 3000 °C. About 4% NO is produced at this higher temperature. $\Delta H_f^\ominus$ [NO(g)] = +90 kJ mol^{-1}.

a Sketch an energy profile diagram for the production of NO from air.

b Sketch Boltzmann distributions on the same axes for air at room temperature and at 3000 °C and explain why some NO is produced at the higher temperature but none at room temperature.

TIP

When drawing a Boltzmann distribution, do not forget to label the axes and also to indicate that the origin is at 0,0.

9.05 Catalysis

A catalyst lowers the activation energy. It may provide a surface on which the reaction takes place or it may provide an intermediate compound; either way, it provides another route by which the reaction can proceed (Figure 9.06).

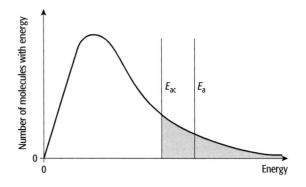

Figure 9.06 The Boltzmann distribution showing the activation energy with and without a catalyst

As you can see in Figure 9.06, by lowering the activation energy to E_{ac} in the presence of a catalyst, many more molecules have energy greater than the new lowered activation energy and so can react on collision.

a Heterogeneous catalysts

A **heterogeneous catalyst** is in a different phase to the reactants. Often the catalyst is a solid and the reactants are gases or liquids.

- Solid iron is used with the gaseous mixture in the Haber process:

$$N_2(g) + 3H_2(g) \rightleftharpoons 2NH_3(g)$$

- Manganese(IV) oxide, MnO$_2$(s), is used to decompose hydrogen peroxide:

$$2H_2O_2(aq) \rightarrow 2H_2O(l) + O_2(g)$$

b Homogeneous catalysts

A **homogeneous catalyst** is in the same phase as the reactants. Sometimes the catalyst and the reactants are all in aqueous solution; sometimes the catalyst and the reactants are all gases.

- NO$_2$(g) catalyses the oxidation of SO$_2$(g), a process which can lead to the formation of acid rain:

$$2SO_2(g) + O_2(g) \rightarrow 2SO_3(g)$$

- FeSO$_4$(aq) catalyses the reaction between aqueous peroxodisulfate and iodide ions:

$$S_2O_8^{2-}(aq) + 2I^-(aq) \rightarrow 2SO_4^{2-}(aq) + I_2(aq)$$

In this reaction, repulsion between the two negative ions slows down the reaction but the positively charged Fe^{2+}(aq) ions are able to interact with the negative ions and speed the reaction up.

9.06 Enzymes

Enzymes are protein molecules held in a definite shape by the many intramolecular bonds along the length of the protein chain. Enzymes are biological catalysts which control the rate of the many reactions, such as digestion, respiration and synthesis of complex molecules necessary to living organisms.

They act, like inorganic catalysts, by lowering the activation energy of a reaction.

Unlike inorganic catalysts:

- they act at relatively low temperatures and are destroyed at high temperatures; many reactions they catalyse are near body temperature
- they are sensitive to pH

- they are specific to one reaction; the particular shape of an enzyme allows only one type of molecule to be catalysed
- they are highly efficient; the rate is often controlled by how quickly the molecules can move to the enzyme

> **TIP**
> Make sure you can compare and contrast enzymes and inorganic catalysts.

Progress check 9.02

1. Ammonia and air do not react together at room temperature in the absence of a catalyst. They do react if a hot platinum wire is inserted into the mixture:

   $4NH_3(g) + 5O_2(g) \rightarrow 4NO(g) + 6H_2O(g)$

   By using Boltzmann distribution diagrams, explain the effect on the mixture of:
   a. heating the mixture
   b. the platinum wire.

2. What type of catalysts (homogeneous or heterogeneous) are the following?
   a. chlorine atoms in the upper atmosphere, decomposing ozone
   b. nickel in the hydrogenation of unsaturated hydrocarbons to produce margarine
   c. vanadium(V) oxide to speed up the oxidation of sulfur dioxide to sulfur trioxide.

3. The enzyme catalase is produced in animal and plant cells; it catalyses the breakdown of hydrogen peroxide, H_2O_2:

   $2H_2O_2(aq) \rightarrow 2H_2O(l) + O_2(g)$

   a. The reaction is exothermic; what effect does this have on the rate of the reaction as the reaction proceeds? Explain your answer.
   b. What is the effect on the rate of the reaction if the mixture is boiled? Explain your answer.
   c. Why does catalase not break down the H_2O molecules produced into H_2 and O_2?

Revision checklist

Check that you know the following:

- [ ] explain and use the term 'rate of reaction' and explain the effect of concentration changes
- [ ] explain the effect of temperature and a catalyst on the rate of reaction using a Boltzmann distribution
- [ ] explain the different actions of homogeneous and heterogeneous catalysts
- [ ] explain the catalytic activity of enzymes

Exam-style questions

1. a. The decomposition of the nitrogen(V) oxide, N_2O_5, is slow:

    $$2N_2O_5(g) \rightarrow 4NO_2(g) + O_2(g)$$

    Suggest how the rate of reaction could be followed. [1]

    b. The reaction to form a precipitate of silver chloride seems to be instantaneous:

    $$NaCl(aq) + AgNO_3(aq) \rightarrow AgCl(s) + NaNO_3(aq)$$

    Explain why this reaction is very quick but the decomposition of N_2O_5 is very slow. [5]

    Total: 6

2. The Haber process is carried out at 1.5×10^4 kPa, 700 K and with an Fe catalyst:

    $$N_2(g) + 3H_2(g) \rightleftharpoons 2NH_3(g)$$

    a. Draw a labelled Boltzmann distribution for the reaction and indicate the activation energy, labelling it E_a. [4]

    b. On your Boltzmann distribution, draw a line to indicate the activation energy if the iron is not used and label it C. [1]

    c. Describe how the distribution would change if the temperature is changed to 600 K. [2]

    Total: 7

Periodicity

Learning outcomes

You should be able to:

- [ ] describe the periodicity in physical and chemical properties, including the structure and bonding, of Period 3 elements
- [ ] describe and write equations for some of the reactions of the elements
- [ ] interpret the periodicity of the physical and chemical properties of the oxides and chlorides of Period 3 using electronegativity and bonding
- [ ] describe and explain the reactions of Period 3 oxides and chlorides with water

10.01 Periodicity of Period 3 elements

Moving across Period 3, the atoms have one more proton in their nucleus and another electron in their outer shell. Firstly the 3s orbital is completely filled before the three 3p orbitals are occupied. As the electrons are added to the same shell, they are attracted more strongly to the increasingly positive nucleus and therefore the atoms get smaller and the ionisation energy increases.

The increase in ionisation energy is not smooth and is explained by the 3s orbitals having lower energy than the 3p orbitals and also that there is some electron repulsion when electrons are required to pair in an orbital (see Unit 3).

The **atomic radius** is half the distance between two neighbouring nuclei in the solid element (except for argon, which exists as separate atoms so the radius is half the closest distance that two atoms can approach).

Across the period, the electrical conductivity increases from Na to Al but then is negligible for the rest of the elements – indicating the bonding change from metallic to covalent. As more electrons are able to be delocalised from Na to Mg to Al, the conductivity increases.

The melting point increases from Na to become very high at Si and then plunges, corresponding to the change in structure from giant lattices to small molecules.

The structure and bonding information is gathered together in Table 10.01.

Element	Structure and bonding
Na	giant metallic
Mg	giant metallic
Al	giant metallic
Si	giant covalent
P_4	small molecules
S_8	small molecules
Cl_2	small molecules
Ar	unreactive atoms

Table 10.01 The structure and bonding of Period 3 elements

Na, Mg and Al have metallic bonding and relatively low ionisation energies so they lose electrons in reactions to become positive ions. The outer shell of electrons is lost and the ions are therefore much smaller than the corresponding atoms.

As the electronegativity increases across the period, P, S and Cl gain electrons to produce negative ions in ionic compounds. The extra electrons occupy spaces in the outer energy level and you might expect that the ionic radius would therefore not change; however, the nuclear charge has less positive charge per electron and so does not hold them so close, so the negative ions are larger than the parent atoms.

> **TIP**
> Make sure you know the outer electron configurations of the atoms as they indicate the chemical and physical properties of the element.

> **Worked example 10.01**
>
> Explain why the ionisation energy of Mg is lower than that of S.
>
> How to get the answer:
>
> Step 1: State the electron configurations: Mg is [Ne]$3s^2$ and S is [Ne]$3s^2 3p^4$.
>
> Step 2: Decide on the relative numbers of protons: S has four more protons.
>
> Step 3: Decide which electron is removed: they come from the same energy level.
>
> Step 4: Decide on how strongly the electrons are held: the attraction by the nucleus for the outer electron is greater for S than for Mg.

10.02 Reactions of Period 3 elements

a **Reaction with oxygen**

Most Period 3 elements can be burnt in oxygen to form oxides (Table 10.02). Chlorine does not react directly with oxygen; silicon will only react at a very high temperature.

Element	Notes	Product	Nature of product
Na	when lit, burns with a yellow flame	Na_2O	white solid
Mg	when lit, burns with a brilliant white flame	MgO	white solid
Al	finely divided Al is needed and it then burns with a white flame	Al_2O_3	white solid
P_4	when lit, burns vigorously with a white flame	P_4O_{10}	white solid
S_8	when lit, burns with a blue flame	SO_2	colourless gas

Table 10.02 The reactions of Period 3 elements with oxygen

With extra oxygen and a catalyst, SO_2 can be oxidised further to SO_3.

b **Reaction with chlorine**

When lit, Period 3 elements burn in chlorine (Table 10.03).

Element	Notes	Product	Nature of product
Na	burns with an orange flame	NaCl	white solid
Mg	burns with a brilliant white flame	$MgCl_2$	white solid
Al	pass dry chlorine gas over heated aluminium foil and the Al glows red	Al_2Cl_6	pale yellow solid
Si	pass dry chlorine gas over heated silicon powder	$SiCl_4$	colourless liquid
P_4	phosphorus burns in excess chlorine gas with a yellow flame	PCl_5	pale yellow solid

Table 10.03 The reactions of the Period 3 elements with chlorine

c **Reaction with water**

Sodium reacts vigorously with cold water:

$$2Na(s) + 2H_2O(l) \rightarrow 2NaOH(aq) + H_2(g)$$

The sodium floats as it has a low density. A solution of sodium hydroxide is formed and hydrogen gas bubbles out, often catching fire from the heat of reaction.

Magnesium reacts very slowly with cold water:

$$Mg(s) + 2H_2O(l) \rightarrow Mg(OH)_2(s) + H_2(g)$$

It takes a long time to collect a few tiny bubbles of hydrogen; the magnesium hydroxide formed is insoluble.

Hot magnesium reacts with steam; the Mg glows with a bright white light:

$$Mg(s) + H_2O(g) \rightarrow MgO(s) + H_2(g)$$

The white solid MgO is formed and one colourless gas is replaced by another colourless gas.

> **Progress check 10.01**
>
> 1. Explain the variation in the ionisation energy across the elements of Period 3.
> 2. Write balanced equations for the reactions of the Period 3 elements with oxygen.
> 3. Using the formulae in Table 10.02, work out the oxidation number of the Period 3 elements in their oxides.
> 4. Write balanced equations for the reactions of the Period 3 elements with chlorine and work out the oxidation numbers of the Period 3 elements in the products.

10.03 Oxides and chlorides of Period 3

a The oxides

The electrical conductivity is good for the molten oxides, Na_2O, MgO and Al_2O_3, indicating ionic bonding, but the other oxides are non-conductors and so covalent.

The melting points are high and increase from Na_2O to SiO_2, indicating giant structures. P_4O_{10} has a much lower melting point and SO_2 and SO_3 are gases at room temperature, indicating they exist as small molecules.

The metals produce ionic oxides. The semiconductor Si, which has a giant covalent structure, produces a giant covalent oxide. The small molecules of phosphorus and sulfur produce small molecular oxides.

The ionic structures of MgO and Al_2O_3 contain small, highly charged ions which attract each other strongly. SiO_2 has a lattice in which the atoms are held together by strong covalent bonds. A great deal of energy is needed to separate these structures, giving them particularly high melting points. These are the ceramics: materials which combine strength with chemical stability and good electrical insulation.

b The chlorides

The melting points of $NaCl$ and $MgCl_2$ are high, indicating giant structures, but the others have low melting points, indicating they exist as small molecules.

The good electrical conductivity of molten $NaCl$ and $MgCl_2$ indicate their ionic nature; the other chlorides have no conductivity as their bonding is covalent.

The chlorides show a trend from giant ionic lattices to small molecules as we go across the period. There are no giant covalent lattices as Cl can only form one covalent bond (to a limited extent it can form a dative covalent bond in Al_2Cl_6, see Unit 4).

The point in the period where the change from ionic to covalent occurs is at a different place for the oxides and the chlorides because the electronegativity of O is greater than that of Cl; Al_2O_3 is ionic but Al_2Cl_6 is covalent.

> **TIP** Note the gradual change in the normal oxidation number of the element in its oxide and chloride as you move across the Periodic Table.

10.04 Reactions of Period 3 oxides with water

Not all the oxides react with water, but of those that do, the metal oxides form alkalis and the non-metal oxides form acids.

Sodium oxide, Na_2O, reacts exothermically to form a colourless solution of sodium hydroxide with a pH = 1:

$$Na_2O(s) + H_2O(l) \rightarrow 2NaOH(aq)$$

Magnesium oxide, MgO, has a very limited solubility but the small amount of reaction to form magnesium hydroxide gives a pH of approximately 9.

The insoluble impervious layer of Al_2O_3 covering aluminium protects the metal from corrosion.

Water cannot attack the giant covalent structure of silica, so we continue to enjoy sandy beaches around the world.

Phosphorus oxide, P_4O_{10}, reacts exothermically with water producing phosphoric acid with pH of 1:

$$P_4O_{10}(s) + 6H_2O(l) \rightarrow 4H_3PO_4(aq)$$

Sulfur dioxide gas dissolves and reacts with cold water to form sulfurous acid, pH = 3, but heating the solution reverses the reaction:

$$SO_2(g) + H_2O(l) \rightleftharpoons H_2SO_3(aq)$$

Sulfur trioxide dissolves in water in an extremely exothermic reaction to form sulfuric acid with a pH = 1:

$$SO_3(g) + H_2O(l) \rightarrow H_2SO_4(aq)$$

> **TIP** The ionic oxides of elements at the left of Period 3 are alkalis and bases; the covalent oxides of elements at the right are acidic.

a **Reactions of the oxides with aqueous acid**

The three ionic oxides, or their aqueous solutions, react with aqueous acids to form salt solutions:

$$NaOH(aq) + HCl(aq) \rightarrow NaCl(aq) + H_2O(l)$$
$$Mg(OH)_2(aq) + 2HCl(aq) \rightarrow MgCl_2(aq) + 2H_2O(l)$$
$$Al_2O_3(s) + 6HCl(aq) \rightarrow 2AlCl_3(aq) + 3H_2O(l)$$

b **Reactions with aqueous sodium hydroxide**

The acids formed by the covalent oxides react with aqueous sodium hydroxide to form salt solutions:

$$H_3PO_4(aq) + 3NaOH(aq) \rightarrow Na_3PO_4(aq) + 3H_2O(l)$$
$$H_2SO_3(aq) + 2NaOH(aq) \rightarrow Na_2SO_3(aq) + 2H_2O(l)$$
$$H_2SO_4(aq) + 2NaOH(aq) \rightarrow Na_2SO_4(aq) + 2H_2O(l)$$

c **Amphoteric oxides**

Aluminium oxide, Al_2O_3, reacts with acids but also reacts with hot concentrated sodium hydroxide solution to form a salt. It is behaving like one of the covalent oxides:

$$Al_2O_3(s) + 2NaOH(aq) \rightarrow 2NaAlO_2(aq) + H_2O(l)$$

Al_2O_3 is an **amphoteric** oxide as it reacts with both aqueous acids and alkalis.

10.05 Reactions of Period 3 chlorides with water

Chloride	With water	Approximate pH of solution
NaCl	dissolves	7
$MgCl_2$	dissolves	6
Al_2Cl_6	dissolves and hydrolyses	3
$SiCl_4$	hydrolyses	1
PCl_5	hydrolyses	1

Table 10.04 Reactions of Period 3 chlorides with water

There is a clear trend across Period 3 in the pH of the solutions produced when the chlorides react with water, as shown in Table 10.04. In general, ionic chlorides dissolve to produce hydrated ions. The positive and negative ions are surrounded by water molecules and are taken into solution by the formation of ion–dipole attractions between the ions and water molecules. The covalent chlorides react with water to produce acidic solutions and clouds of steamy HCl fumes:

$$SiCl_4(l) + 2H_2O(l) \rightarrow SiO_2(s) + 4HCl(aq)$$
$$PCl_5(s) + 4H_2O(l) \rightarrow H_3PO_4(aq) + 5HCl(aq)$$

Al_2Cl_6 is a covalent molecule. The reaction with water can be thought of in two stages.

- Solution:

$$Al_2Cl_6(s) + aq(l) \rightarrow 2Al^{3+}(aq) + 6Cl^-(aq)$$

- Hydrolysis: the Al^{3+} is surrounded by water molecules and strongly attracts the electron pairs on the O of H_2O to form the octahedral complex, $[Al(H_2O)_6]^{3+}$. The lone pairs are so strongly attracted that one of the H_2O molecules loses H^+ and the solution becomes acidic:

$$[Al(H_2O)_6]^{3+} \rightarrow [Al(H_2O)_5(OH)]^{2+} + H^+$$

Sample answer

Question:

Explain how the chlorides of sodium, aluminium and silicon behave when shaken with water. Write equations for any reactions. [10]

Answer:

NaCl dissolves in water [1 mark], pH = 7 [1 mark].

$SiCl_4$ hydrolyses in water [1 mark], HCl fumes are produced [1 mark], pH = 2 [1 mark].

$SiCl_4(l) + 2H_2O(l) \rightarrow SiO_2(s) + 4HCl(g)$ [1 mark], a white precipitate [1 mark].

$AlCl_3$ hydrolyses [1 mark], pH = 3 [1 mark].

$Al(H_2O)_6^{3+} \rightarrow [Al(H_2O)_5OH]^{2+} + H^+$ [1 mark]

Progress check 10.02

1. Work out the oxidation number of Al in Al_2O_3 and in $NaAlO_2$.

2. Draw out a three-dimensional sketch of the octahedral complex $[Al(H_2O)_6]^{3+}$.

3. Explain why the metal chloride, aluminium chloride, produces an acidic solution when dissolved in water although sodium chloride produces a neutral solution.

4. Suggest why the solution of $MgCl_2$ in water is weakly acidic.

Revision checklist

Check that you know the following:

- [ ] understand and describe how the periodicity in physical and chemical properties of Period 3 elements arises
- [ ] describe and write equations for some of the reactions of the elements
- [ ] understand and describe how the periodicity of the physical properties of the oxides and chlorides of Period 3 elements arises
- [ ] describe and write equations for the reactions of Period 3 oxides and chlorides with water

Exam-style questions

1. a i Name an element from Period 3 which has both an ionic chloride and an ionic oxide. [1]

   ii Describe the structure and bonding of your stated element in part a i. [1]

   iii Describe and explain what is observed when the chloride is shaken with water. [3]

   iv Describe and explain what is observed when the oxide is shaken with water. [3]

   b i Name an element from Period 3 which has both a covalent chloride and a covalent oxide. [1]

   ii Describe the structure and bonding of your stated element in part b i. [1]

   iii Describe and explain what is observed when the chloride is shaken with water. [3]

   iv Describe and explain what is observed when the oxide is shaken with water. [3]

   Total: 16

2. a i Describe how the electronegativity of the atoms changes across Period 3. [1]

   ii Explain what effect the electronegativity has on the bonding of the Period 3 oxides. [2]

   b i Explain why the boiling point of Al is higher than the boiling point of Mg. [2]

   ii Explain why the melting point of S is higher than the melting point of P. [2]

   Total: 7

Group 2

Learning outcomes

You should be able to:

- [ ] describe the main reactions of the elements
- [ ] describe some of the physical and chemical properties of the oxides, hydroxides, carbonates and sulfates
- [ ] describe the thermal decomposition of the carbonates and nitrates
- [ ] describe and explain some of the uses of the compounds

11.01 About Group 2

a Outer electronic configuration

Group 2 elements have an outer electronic configuration of ns^2. The first two ionisation energies are low but the third ionisation energy is much higher and they form ions with a 2+ charge. Going down the group, the ionisation energy gets smaller as the electron is removed from an orbital in a higher energy level, further from the nucleus and shielded by the inner electrons.

For similar reasons, the atomic radius increases as you go down the group but the M^{2+} ion is much smaller than the atom as the outer shell has been lost.

The melting points are high, indicating giant structures; the good electrical conductivity of the solids indicates metallic bonding. The elements have giant metallic lattices.

b Reaction with oxygen

All the elements burn, when lit, to produce oxides; there is no clear trend in reactivity:

$$2M(s) + O_2(g) \rightarrow 2MO(s)$$

Magnesium burns with an intense white light which is used in marine flares and fireworks; calcium and strontium burn with red flames. Barium oxidises easily and needs to be kept under oil; it burns with a green flame.

The oxides produced are white solids with high melting points.

c Reaction with water

There is a clear trend down the group; the reaction becomes more vigorous as the outer electrons are further from the nucleus and less tightly held. The solutions produced are alkaline.

- Be does not react with water or steam even when red hot.
- Mg reacts in steam (see Unit 10):

$$Mg(s) + H_2O(g) \rightarrow MgO(s) + H_2(g)$$

and very slowly in cold water:

$$Mg(s) + 2H_2O(l) \rightarrow Mg(OH)_2(s) + H_2(g)$$

- Ca reacts vigorously with cold water:

$$Ca(s) + 2H_2O(l) \rightarrow Ca(OH)_2(s) + H_2(g)$$

Down the group, the hydroxides produced become more soluble.

d Reaction with aqueous acid

All the metals react with aqueous hydrochloric acid, producing colourless solutions of the chlorides and bubbles of hydrogen gas:

$$M(s) + 2HCl(aq) \rightarrow MCl_2(aq) + H_2(g)$$

The reaction gets more vigorous down the group.

If sulfuric acid is used, the sulfates formed get more insoluble down the group – which interferes with the reaction.

$$M(s) + H_2SO_4(aq) \rightarrow MSO_4(aq) + H_2(g)$$

Beryllium and magnesium react steadily, with hydrogen bubbling out of the colourless sulfate solutions. Calcium reacts but the sulfate is produced as a white precipitate. The insoluble sulfates of strontium and barium coat the metal and prevent further reaction.

> ### Sample answer
>
> **Question:**
>
> Explain how the electronic configuration is related to the ionisation energies and ionic radii of Group 2 elements. [8]
>
> **Answer:**
>
> The outer electronic configuration is ns^2 [1 mark].
>
> The outer two electrons are in a new shell [1 mark]. Going down the group, these electrons are further from the nucleus [1 mark], and so the first ionisation energy decreases [1 mark].
>
> The first two ionisation energies are low but the third is high [1 mark] so two electrons are lost to form M^{2+} [1 mark].
>
> Forming the ion means a whole outer shell is lost so the ion is smaller than the atom [1 mark] but as there are more (inner) shells going down the group, the ions get larger down the group [1 mark].

11.02 Group 2 oxides, hydroxides and carbonates

a **The oxides and hydroxides**

The oxides are white solids with high melting points. There is a clear trend in the reaction with water (Table 11.01). The equation for any reaction is:

$$MO(s) + H_2O(l) \rightarrow M(OH)_2(aq)$$

Oxide	Comments
BeO	No reaction.
MgO	The solid oxide is apparently unchanged; the pH of the mixture is approximately 9, indicating some reaction.
CaO	There is an exothermic reaction to produce the slightly soluble slaked lime; the solution has a pH of approximately 12.
SrO	The oxide reacts to produce a colourless solution of pH approximately 14.
BaO	The oxide reacts to produce a colourless solution of pH approximately 14.

Table 11.01 The reactions of the Group 2 oxides with water

All the oxides and hydroxides are soluble in aqueous hydrochloric acid to produce solutions of the chlorides:

$$MO(s) + 2HCl(aq) \rightarrow MCl_2(aq) + H_2O(l)$$

As the sulfates become more insoluble and the hydroxides become more soluble going down the group, the situation is complicated if aqueous sulfuric acid is used. The insoluble oxides of beryllium, magnesium and calcium dissolve to produce solutions of the sulfates; the soluble strontium and barium hydroxides produce white precipitates of the sulfates.

b **The carbonates**

The carbonates are insoluble in water but react with dilute aqueous acids to produce carbon dioxide gas:

$$MCO_3(s) + 2H^+(aq) \rightarrow M^{2+}(aq) + CO_2(g) + H_2O(l)$$

HCl(aq) produces colourless chloride solutions. H_2SO_4(aq) produces colourless sulfate solutions for $BeCO_3$, $MgCO_3$ and $CaCO_3$ but the insoluble sulfates will prevent much reaction for $SrCO_3$ and $BaCO_3$.

11.03 The decomposition of Group 2 carbonates and nitrates

a **The carbonates**

As the ionic radius increases, the carbonates become more stable to heat:

$$MCO_3(s) \rightarrow MO(s) + CO_2(g)$$

$BeCO_3$ is unstable at room temperature. You can readily decompose $MgCO_3$ using the heat from a Bunsen burner but $CaCO_3$ requires a higher temperature. The decomposition temperature increases as the group is descended. We can test for any CO_2(g) being produced using limewater.

b **The nitrates**

All the nitrates are decomposed by heat from a Bunsen burner to produce the oxide (see Unit 19 for an explanation):

$$2M(NO_3)_2 \rightarrow 2MO(s) + 4NO_2(g) + O_2(g)$$

Brown fumes which relight a glowing splint are also produced; nitrogen dioxide gives the brown colour and the oxygen relights a glowing splint.

> **TIP** Make sure you know the trends in chemical properties of the elements and their compounds.

Worked example 11.01

Three white solids are known to be BaO, BaCO$_3$ and Ba(NO$_3$)$_2$ but have lost their labels. Explain how you would distinguish them.

How to get the answer:

Step 1: Revise the reactions of Group 2 oxides, carbonates and nitrates.

Step 2: Select 'heat' and 'dissolving in dilute HC*l*' as two possible reactions.

Step 3: Heat. Only barium nitrate will produce brown fumes which relight a glowing splint.

Step 4: Dilute hydrochloric acid. Only barium carbonate will produce bubbles of carbon dioxide.

Step 5: Arrange the tests logically. Heat the samples and distinguish the nitrate. Then add the two remaining to dilute hydrochloric acid to distinguish the carbonate.

11.04 Uses of Group 2 compounds

- The extremely high melting point and thermal stability of magnesium oxide makes it suitable for crucibles and furnace linings.
- Calcium carbonate reacts with acid and is used to raise the pH of soil as well as providing calcium for uptake by plants.

Progress check 11.01

1. Samples of BaO, BaCO$_3$ and Ba(NO$_3$)$_2$ are in separate unlabelled containers. Suggest **one** test that could be carried out on each of the three in turn to decide which is which, explain what you would expect to see and write equations for any reactions.

2. Explain how you could prepare a pure sample of MgSO$_4$ from MgO.

3. Explain how you could prepare a pure sample of BaSO$_4$ from BaO.

Revision checklist

Check that you know the following:

- [ ] describe how the group 2 elements react with oxygen, water and dilute acids,
- [ ] describe and write equations for the reactions of the oxides, hydroxides and carbonates of Group 2 elements with water and with dilute acids,
- [ ] describe and write equations for the thermal decomposition of the nitrates and carbonates,
- [ ] describe the trends in properties of the elements and their compounds,
- [ ] state the variation in the solubilities of the hydroxides and sulfates,
- [ ] describe and explain some of the uses of the compounds.

Exam-style questions

1. Some reactions of barium oxide are shown:

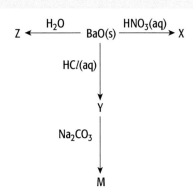

   a i Give the formula of each of the compounds **X**, **Y**, **Z** and **M**. Include state symbols. [4]

   ii Write equations for the four reactions. [4]

   b **X** and **M** are heated with a Bunsen burner until there is no further reaction. Write an equation for any reaction which occurs. If there is no reaction, state 'no reaction'. [3]

   Total: 11

2. a i State and explain any difference in the radii of Mg^{2+} and Mg. [2]

   ii State and explain how the radius of M^{2+} changes going down Group 2. [2]

   b A mixture of equal amounts of $MgCO_3$ and $CaCO_3$ powders is heated with a Bunsen burner, until no further reaction occurs, to form a new mixture, N. Aqueous HCl is then added to N.

   i What is observed when the mixture is heated to form N? [1]

   ii What is observed when the HCl(aq) is added? [1]

   iii Explain the observations. [4]

   Total: 10

Unit 12: Group 17

Learning outcomes

You should be able to:

- describe the physical properties of the halogens and interpret the volatility
- describe the trend in oxidising power of the halogens
- describe, explain and interpret the stability of the hydrides
- describe and explain the reactions of the halide ions with silver nitrate solution and with concentrated sulfuric acid
- describe and interpret the oxidising properties of chlorine with aqueous sodium hydroxide
- explain the use of chlorine in water purification and state the industrial importance of halogen compounds

12.01 Physical properties of the halogens

The outer electron configuration of a halogen atom is $ns^2 np^5$; in the elemental state, two atoms form a single bond to produce the small molecules, X_2 (Table 12.01).

Halogen	Colour and state at room temperature
fluorine, F_2	pale yellow gas
chlorine, Cl_2	green/yellow gas
bromine, Br_2	dark red liquid → brown vapour on slight warming
iodine, I_2	dark grey solid with a metallic lustre → purple vapour on heating

Table 12.01 The state and colours of the halogens

You can see that, on descending the group, the colour of the halogens gets deeper and the intermolecular forces get larger so that there is a change from gas to liquid to solid.

The covalent bond joining the two atoms together becomes weaker going down the group, but the van der Waals' forces between the molecules grow stronger as there are more electrons in the molecule.

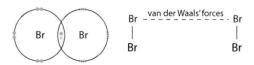

All the halogens are volatile as the only forces to be broken are weak van der Waals' forces. Going down the group, more energy is needed to form the vapour.

The halogens are not very soluble in water as they cannot form bonds strong enough to overcome the strongly hydrogen bonded structure of water. The dilute solutions produced have distinctive colours. The halogens are, however, much more soluble in organic solvents (Table 12.02).

Colour of halogen in water	Colour of halogen in organic solvent
Cl_2(aq), very pale green	Cl_2(org), very pale green
Br_2(aq), orange	Br_2(org), orange
I_2(aq) [I_2 dissolves well in KI(aq)], brown	I_2(org), purple

Table 12.02 The colours of halogen solutions

It can be difficult to tell the difference between dilute Br_2(aq) and I_2(aq) but shaking a little with an organic solvent allows them to be distinguished.

12.02 Halogens as oxidising agents

When halogens act as oxidising agents, they are reduced to halide ions. Most chlorides, bromides and iodides form colourless solutions.

a Halogens and halide ions in solution

Chlorine can oxidise bromides to bromine and iodides to iodine:

$$Cl_2(aq) + 2NaBr(aq) \rightarrow Br_2(aq) + 2NaCl(aq)$$

Bromine can oxidise iodides:

$$Br_2(aq) + 2KI(aq) \rightarrow 2KBr(aq) + I_2(aq)$$

The oxidising power of halogens lessens going down Group 17.

b The oxidation of Fe^{2+}

Both chlorine and bromine can oxidise Fe^{2+} to Fe^{3+} ions:

$$Cl_2(aq) + 2Fe^{2+}(aq) \rightarrow 2Fe^{3+}(aq) + 2Cl^-(aq)$$

However, Fe^{3+} oxidises iodide ions to iodine:

$$2Fe^{3+}(aq) + 2I^-(aq) \rightarrow 2Fe^{2+}(aq) + I_2(aq)$$

c The production of halogens

Chlorine is produced by the oxidation of chloride ions in brine by electrolysis.

Both bromine and iodine are produced using chlorine gas to oxidise the bromide and iodide ions present in seawater.

> **TIP**
> Remember that the halogens are oxidising agents but their oxidising power decreases as you go down the group. The halide ions are reducing agents and they become stronger reducing agents as you go down the group.

d The reaction of chlorine with sodium hydroxide solution

i With cold dilute sodium hydroxide

$$Cl_2(aq) + 2NaOH(aq) \rightarrow NaCl(aq) + NaOCl(aq) + H_2O(l)$$

As an ionic equation:

$$Cl_2(aq) + 2OH^-(aq) \rightarrow Cl^-(aq) + OCl^-(aq) + H_2O(l)$$

The oxidation number of Cl changes from 0 (in Cl_2) to -1 (in $NaCl$) and $+1$ (in $NaOCl$). None of the other species change their oxidation number. This is a **disproportionation reaction** as the Cl has been both oxidised and reduced. The two chlorine-containing products are called sodium chloride and sodium chlorate(I).

ii With hot concentrated sodium hydroxide

$$3Cl_2(aq) + 6NaOH(aq) \rightarrow 5NaCl(aq) + NaClO_3(aq) + 3H_2O(l)$$

The oxidation number of Cl changes from 0 to -1 and $+5$ (in $NaClO_3$, sodium chlorate(V)). Again Cl has been both oxidised and reduced in the same reaction.

12.03 The hydrides

The halogen hydrides (or hydrogen halides) are colourless gases at room temperature and pressure, except for the liquid HF.

a Formation

Hydrogen and chlorine gases explode when mixed in direct sunlight but react quietly in the dark.

Hydrogen and bromine gases require heat and a platinum catalyst for reaction to occur.

Hydrogen and iodine gases also need heat and a platinum catalyst but the reaction is slow and reversible:

$$X_2(g) + H_2(g) \rightleftharpoons 2HX(g)$$

b Physical properties

The small molecules of hydrogen halides have permanent dipole attractions between them. Hydrogen fluoride is also extensively hydrogen bonded. Down the group, the bond lengths increase as the radius of the halogen atom increases; the bond strengths decrease because the bond pair of electrons is further from the nucleus down the group.

c Thermal stability

The thermal stability decreases down the group as the bond strength decreases. This can be shown by plunging a hot platinum wire into separate test tubes of the colourless hydrogen halide gases.

- HCl is unchanged
- HBr may show a slight brown coloration due to some bromine being formed
- HI shows a cloud of purple vapour showing the decomposition of the gas into hydrogen and iodine.

A test tube of HI does not decompose at room temperature as the reaction is too slow but as the temperature is increased, the reaction moves to the equilibrium position.

d Solution in water

The hydrogen halides form acidic solutions in water. The acids are strong except for the weak acid HF(aq):

$$HCl(aq) + H_2O(l) \rightarrow H_3O^+(aq) + Cl^-(aq)$$

The halide ions, apart from F^-, are very weak bases. The extra electron enters the space in the p orbitals of the highest occupied energy level and all eight s and p electrons in this energy level are strongly attracted by the nuclear charge; the H^+ forms a stronger bond with the lone pair on O of H_2O rather than a lone pair on Cl^-.

Progress check 12.01

1. Write ionic equations for the reactions of chlorine with sodium iodide solution and for bromine with calcium iodide solution.
2. Explain why hydrogen fluoride has a higher boiling point than expected.
3. Suggest why the boiling points of HCl, HBr and HI increase as the group is descended.
4. Explain why NaCl is ionic but Al_2Cl_6 is covalent.
5. Write the ionic equation for the reaction of chlorine with hot concentrated sodium hydroxide.

12.04 Reactions of halides

The s-block metals form ionic halides which are white, soluble solids. Most of the p-block halides are covalent as there is less of an electronegativity difference between the two atoms. Thus the metal aluminium forms the covalent chloride Al_2Cl_6.

a Ease of oxidation of halide ions

Sulfuric acid has three important chemical properties; it is a strong acid, a dehydrating agent and an oxidising agent. To illustrate the ease of oxidation of halide ions, we are interested in sulfuric acid acting as an acid and as an oxidising agent.

Sulfuric acid is such a strong acid that it donates a proton to the chloride ion to form hydrogen chloride, which bubbles out of the mixture as steamy fumes:

$$H_2SO_4(l) + NaCl(s) \rightarrow NaHSO_4(s) + HCl(g)$$

A similar reaction occurs for the bromides and iodides, but sulfuric acid is a sufficiently strong oxidising agent to oxidise the HBr and HI to the respective halogens:

$$2HBr(g) + H_2SO_4(l) \rightarrow Br_2(g) + 2H_2O(l) + SO_2(g)$$

Iodide is such a strong reducing agent that it can reduce sulfuric acid to $H_2S(g)$; the purple vapour of iodine is seen and $H_2S(g)$ will give the smell of rotten eggs.

Figure 12.01 The effect of conc. sulfuric acid on halide ions: chloride, bromide and iodide

b Reaction of the ionic halides with a non-oxidising acid

Phosphoric acid is a non-oxidising acid and will protonate halide ions to produce hydrogen halides without further redox reactions taking place. All three hydrogen halides can be seen as steamy fumes during the reaction of phosphoric acid with a suitable halide:

$$H_3PO_4(l) + KBr(s) \rightarrow KH_2PO_4(s) + HBr(g)$$

Worked example 12.01

What are the oxidation number changes when hydrogen iodide reduces sulfuric acid to hydrogen sulfide, H_2S? By balancing the oxidation number changes, write the equation for the reaction of concentrated sulfuric acid with hydrogen iodide.

How to get the answer:

Step 1: Find ox. no. [S] in H_2SO_4 and H_2S: +6 and −2.

Step 2: Reduction of S means I is oxidised; ox. no. [S] has decreased, ox. no. [I] must increase.

Step 3: Work out ox. no. [I]. In HI it is −1 and in I_2, ox. no. [I] = 0.

Step 4: Work out ox. no. changes. ox. no. [S] +6 to −2 = −8; ox. no. [I] −1 to 0 = +1.

Step 5: ox. no. changes must balance. For each S that changes eight I must change.

Step 6: Write the balanced redox part of the equation: $8HI + H_2SO_4 \rightarrow 4I_2 + H_2S$

Step 7: Balance the H and O atoms: $8HI + H_2SO_4 \rightarrow 4I_2 + H_2S + 4H_2O$

c **Tests for halide ions**

Most ionic halides are soluble in water but those of silver and lead are insoluble. The silver halides, $AgCl$, $AgBr$ and AgI, have different colours and are used to identify which halide ion is present. An extra aid to identification is given by the different solubilities of the silver halides in aqueous ammonia (Table 12.03).

Aqueous halide ion	Observation in $AgNO_3$(aq)	Observation with NH_3(aq)
chloride	white precipitate, $AgCl$	dissolves in dilute ammonia solution to form a colourless solution
bromide	cream precipitate, $AgBr$	insoluble in dilute ammonia, but dissolves in concentrated ammonia solution to form a colourless solution
iodide	yellow precipitate, AgI	does not dissolve in dilute or concentrated ammonia solution

Table 12.03 Tests for aqueous halide ions

The solutions need to be acidic, so the silver nitrate is acidified with HNO_3(aq).

> **TIP**
> The silver nitrate test for halide ions does not work for covalently bonded halogens but can be adapted for use in organic chemistry. The halogen-containing organic compound is warmed with water or sodium hydroxide solution. The halogen is then released as a halide ion and this can then be tested (see also Unit 16).

If the silver halide precipitates are filtered off, washed and dried, they can be weighed to find the amount of halide ion present.

The colourless solution formed when either $AgCl$ or $AgBr$ dissolves in ammonia solution contains the same complex cation. The difference in the way the three precipitates behave is due to their differing solubilities. Each one has a very small solubility; AgI is the least soluble and $AgCl$ is the most soluble. Dilute ammonia is able to react with the most soluble salt; concentrated ammonia is needed to dissolve the $AgBr$ but AgI is too insoluble.

For $AgCl$(s), a very small amount of the solid dissolves to produce small concentrations of the ions in solution:

$$AgCl(s) \rightleftharpoons Ag^+(aq) + Cl^-(aq)$$

On addition of the ammonia solution, the Ag^+ ions present in solution can react with NH_3 molecules in solution to produce a complex ion:

$$Ag^+(aq) + 2NH_3(aq) \rightleftharpoons [Ag(NH_3)_2]^+(aq)$$

Sample answer

Question:

$100 cm^3$ of an $NaCl$ solution was shaken with excess acidified silver nitrate solution. The precipitate was filtered off, washed and dried before weighing. 0.140 g of silver chloride was formed. What was the concentration of the $NaCl$ solution in $g dm^{-3}$? [5]

Answer:

no. of moles $AgCl$ = no. of moles Cl^-

$$= \frac{0.140}{107.9 + 35.5} = 9.76 \times 10^{-4}$$

[1 mark]

no. of moles Cl^- = no. of moles $NaCl$ in $100 cm^3$ solution

$$= 9.76 \times 10^{-4}$$ [1 mark]

concentration of $NaCl$ solution = 9.76×10^{-3} mol dm^{-3}

[1 mark]

$$= 9.76 \times 10^{-3} \times (23.0 + 35.5) g dm^{-3}$$ [1 mark]

concentration of $NaCl$ solution = $0.571 g dm^{-3}$ [1 mark]

Progress check 12.02

1. Work out all the oxidation number changes for the reaction when hydrogen bromide reduces sulfuric acid to sulfur dioxide.

2. What would be observed when concentrated sulfuric acid reacts with calcium iodide?

3. A solution is equimolar (has the same molar concentration) in two sodium halides. When 50.0 cm³ of the solution is shaken with an excess of acidified silver nitrate solution, the precipitate, after washing and drying, weighed 3.79 g. The precipitate is then shaken with an excess of concentrated ammonia solution and the remaining precipitate, after washing and drying, weighed 2.35 g. Which two halides were in the solution and what were their concentrations in mol dm^{-3}?

12.05 Uses of halogen compounds

a Water purification

Chlorine dissolves in water and reacts:

$$Cl_2(aq) + H_2O(l) \rightleftharpoons HCl(aq) + HOCl(aq) \quad [Eqn\ 1]$$

Two acids are produced – HCl is a strong acid and fully ionised; HOCl is a weak acid and only partially ionised:

$$HOCl(aq) + H_2O(l) \rightleftharpoons H_3O^+(aq) + OCl^-(aq) \quad [Eqn\ 2]$$

In an acidic solution, the equilibrium [Eqn 2] lies to the left-hand side, which pushes Eqn 1 to the left-hand side. The species present are mainly Cl_2(aq) and HOCl(aq) as well as Cl^-(aq).

In alkaline solution, the H_3O^+(aq) are removed and the equilibria move to the right-hand side. The main species present are Cl^-(aq) and OCl^-(aq).

Both Cl_2 and HOCl are oxidising agents and uncharged molecules; they can enter the negatively charged surface of a pathogen and, by oxidation, kill the bacteria and viruses which might be found in water reservoirs.

Some disadvantages of using chlorine is that it gives a taste and odour to the water and also it is volatile and so is eventually lost from the water.

b Bleach

Pale yellow bleaching powder is formed when chlorine is absorbed by slaked lime (calcium hydroxide). This behaves as though it is a mixture of $Ca(OCl)_2$ and basic calcium chloride ($CaCl_2 \cdot Ca(OH)_2$). The OCl^- produces Cl_2 when in neutral or acidic solution by pushing the following equilibrium to the left:

$$Cl_2(g) + 2OH^-(aq) \rightleftharpoons Cl^-(aq) + OCl^-(aq) + H_2O(l)$$

The released chlorine gas can be used to bleach cotton, linen and wood pulp.

c PVC

Chlorine is used to make chloroethene ($CH_2=CHCl$) the monomer of the polymer polychloroethene or PVC.

d Halogenated hydrocarbons

When different numbers and types of halogen atoms replace hydrogen atoms in hydrocarbons, compounds with a wide range of properties can be produced. The C–F bond is particularly strong and gives stability; Cl atoms can give a molecule anaesthetic properties; the heavier Br atom can alter the volatility and also give flame-retardant effects. The liquid halogenated hydrocarbons are used as solvents, refrigerants and aerosols as well as anaesthetics (see Unit 16).

Revision checklist

Check that you know the following:

- [ ] describe the colours and volatility trend of the halogens
- [ ] how to interpret the volatility of the halogens
- [ ] describe the elements as oxidising agents and the halide ions as reducing agents
- [ ] describe and explain the reactions of the elements with hydrogen and the thermal stability of the hydrides
- [ ] describe and explain the reactions of halide ions with aqueous silver nitrate
- [ ] describe and explain the reactions of halide ions with concentrated sulfuric acid
- [ ] describe and interpret the redox reactions of chlorine with aqueous sodium hydroxide
- [ ] an explanation of why chlorine can be used in water purification
- [ ] state some uses of halogens and their compounds and their environmental significance

Exam-style questions

1. a Explain the differences in the boiling points of the halogens and halogen hydrides shown in the table. [6]

Element or compound	Boiling point / K
chlorine	238
bromine	332
hydrogen chloride	188
hydrogen fluoride	293

    b State and explain the difference in pH between aqueous solutions of HI and KBr. [3]

    c Mixture X contains $AgNO_3(aq)$ acidified with $HNO_3(aq)$. Equal volumes of hydrogen halides are bubbled through X successively to form mixture Y and the mixture Z, as shown. The acidified $AgNO_3(aq)$ is always in excess.

    $$X \xrightarrow{HCl(g)} Y \xrightarrow{HI(g)} Z$$

    An excess of concentrated aqueous ammonia is then added to Z to form mixture W. What is the colour of the precipitate in the following mixtures. If there is no precipitate, write 'none'.

    i Y ii Z iii W [4]

    d Describe what is seen when a few drops of concentrated sulfuric acid is dropped onto the pale yellow crystals of anhydrous calcium iodide. Write an equation for the reaction. [3]

    Total: 16

2. a Complete the following electronic configurations: [1]

    Br $1s^2 2s^2 2p^6 3s^2 3p^6 3d^{10}$ _____

    Br^+ $1s^2 2s^2 2p^6 3s^2 3p^6 3d^{10}$ _____

    b Draw dot-and-cross diagrams for NaBr and PBr_3 showing the outer electrons only. [2]

    c Write equations, including state symbols, for the reaction of water with:

    i Cl_2 [1]

    ii the interhalogen compound BrCl. [2]

    d When the oxide Cl_2O_7 dissolves in water, a strong monoprotic acid is the only product. A solution made by dissolving 0.458 g Cl_2O_7 in 25.0 cm³ water requires 50.0 cm³ of 0.100 mol dm⁻³ NaOH for complete reaction. Construct an equation for the formation of the monoprotic acid and state the oxidation number of the Cl in the acid. [7]

    Total: 13

Unit 13: Nitrogen and sulfur

Learning outcomes

You should be able to:

- explain why nitrogen is so unreactive
- describe and explain ammonia's basic character and how the ammonium ion is formed
- state and explain how atmospheric nitrogen oxides arise, their role as pollutants and how to limit or remove them
- describe the formation of sulfur dioxide and its role in acid rain formation

13.01 The unreactivity of nitrogen

The N–N bond strength is much lower than the C–C bond strength. However, the electron configuration of N is $1s^2 2s^2 2p^3$; the three single electrons in the 2p orbitals can be shared with the three single electrons from another N atom and form a triple bond – a σ bond and two π bonds.

The N≡N bond is nearly six times as strong as the N–N bond and so the π bonds do not easily break to form a single bond as it is not energetically favourable. Any reaction involving nitrogen has a very large activation energy and so the reaction will be slow.

Sample answer

Question:

Explain why the oxidation of nitrogen only occurs at high temperatures but sulfur easily burns when lit. [6]

Answer:

$N_2(g) + O_2(g) \rightarrow 2NO(g)$ [1 mark]

$S_8(s) + 8O_2(g) \rightarrow 8SO_2(g)$ [1 mark]

N_2 has a strong triple bond [1 mark], E_a is very high / strong triple bond has to be broken and much energy is needed [1 mark].

S_8 has much weaker single bonds [1 mark] so less energy is needed to break them [1 mark].

13.02 Ammonia

a **The manufacture of ammonia**

The reactants for the Haber process are hydrogen and nitrogen:

$$N_2(g) + 3H_2(g) \rightleftharpoons 2NH_3(g) \quad \Delta H \ -ve$$

To get the most NH_3 in the equilibrium mixture, we need:

- a high pressure, as four moles produce two moles
- a low temperature, as the reaction is exothermic

To speed up the reaction, we need:

- a high temperature, to increase the collision rate
- a high pressure, to increase the concentration
- a catalyst, to lower the activation energy

Putting the two sets of conditions together, we need:

- a high pressure, although the equipment is costly for too high a pressure
- a compromise temperature
- an iron catalyst

b **Ammonia and the ammonium ion**

NH_3 is a Brønsted–Lowry base as it has a lone pair of electrons on the N which can be donated to a proton (Figure 13.01):

$$NH_3 + H^+ \rightarrow NH_4^+$$

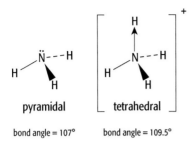

Figure 13.01 The shapes and bond angles of ammonia and ammonium ion

c **Ammonia as a weak base**

The pH of aqueous ammonia is about 11:

$$NH_3(aq) + H_2O(l) \rightleftharpoons NH_4^+(aq) + OH^-(aq)$$

The [OH^-] in 0.1 mol dm^{-3} NH_3 solution is sufficient to precipitate the transition metal hydroxides as well as magnesium and calcium hydroxides from aqueous solutions of their salts:

$$Cu^{2+}(aq) + 2OH^-(aq) \rightarrow Cu(OH)_2(s)$$

Ammonia can also act as a **ligand** to form complexes with many transition metal ions. The pale blue precipitate of copper hydroxide dissolves on further addition of aqueous ammonia to form a dark blue solution containing the complex ion [$Cu(NH_3)_4(H_2O)_2$]$^{2+}$.

d **Producing ammonia from salts**

Ammonium salts contain NH_4^+, which is the complementary acid to the weak base NH_3. Stronger bases such as NaOH or Ca(OH)$_2$ provide a great enough concentration of OH^- ions to steal the H^+ from ammonium ions and leave ammonia:

$$NH_4^+(aq) + OH^-(aq) \rightarrow NH_3(g) + H_2O(l)$$

e **Nitrate fertilisers**

About 80% of the atmosphere is nitrogen but the triple bond is too strong for it to be easily utilised by plants as a nutrient. Ammonium nitrate is soluble and contains readily accessible nitrogen, and so can be used as a fertiliser. It occurs naturally in desert regions but most is synthesised by the exothermic reaction of ammonia with nitric acid:

$$NH_3(g) + HNO_3(aq) \rightarrow NH_4NO_3(aq)$$

The solution is concentrated and crystallised. Ammonium nitrate absorbs moisture from the air and needs to be kept in an airtight container.

Ammonium phosphate or ammonium sulfate, made by reacting ammonia with the appropriate acid, are also used as fertilisers.

f **Uncontrolled use of nitrate fertilisers**

Soluble nitrate fertilisers have given a much needed boost to food plant production. However, using nitrates in too great a quantity, at an unsuitable time of year for plants to be able to take up nutrients or before heavy rain costs farmers a great deal of money as the excess fertiliser is washed off the land and into water sources such as lakes and rivers.

As well as the economic cost, the fertiliser in water sources may then stimulate the growth of organisms such as phytoplankton, which are small bodies which drift in the water and have an important role to play in food chains; an increase in their numbers may produce an algal bloom, making the water bright green or red. This is **eutrophication** and the effect is:

- to block sunlight and so prevent water plants from respiring
- to use up oxygen in the water so some types of aquatic animals cannot survive
- sometimes to produce toxins

Plants and animals in the water die as a result of eutrophication (Figure 13.02).

Figure 13.02 Eutrophication

Reversing the process is difficult, so the problem needs to be tackled at source by using less fertiliser and taking care with its application. Sewage farms have stringent controls as to what they can release into water sources; perhaps nitrate fertilisers should also be strictly controlled.

> **TIP**
> As nitrogen is unreactive, converting it to ammonia makes it more accessible to plants as ammonium salts are soluble.

Progress check 13.01

1. Explain why an iron catalyst is needed in the Haber process.

2. Explain why a pressure of only 150 atm is used in the Haber process even though a higher pressure would produce a quicker reaction.

3. When aqueous nickel chloride, $NiCl_2$, reacts with aqueous sodium hydroxide, the green solution produces a green precipitate. When a separate portion of $NiCl_2(aq)$ reacts with $NH_3(aq)$, a green precipitate is first observed but, on the addition of more ammonia solution, a blue solution is formed. Explain these observations.

4. Write a balanced equation for the reaction between hot aqueous solutions of ammonium nitrate and barium hydroxide.

5. Write a balanced equation for the formation of ammonium phosphate, $(NH_4)_3PO_4$, from ammonia and an appropriate acid.

Worked answer 13.01

Explain why ammonia is a Brønsted–Lowry base but nitrogen is not.

How to get the answer:

Step 1: Define Brønsted–Lowry base: a species with a lone pair which can accept a proton.

Step 2: Is ammonia a Brønsted–Lowry base? N is surrounded by three bond pairs and a lone pair of electrons.

Step 3: Why can the lone pair be donated? N is more electronegative than H and there is bond polarisation.

Step 4: How are the electrons arranged in N_2? Each N has a lone pair and three bond pairs with the other N atom.

Step 5: Why are the lone pairs not donated to protons? There is no electronegativity difference and so neither N is negative enough.

13.03 Nitrogen oxides

a **The nitrogen cycle**

As the triple bond in nitrogen has such a high bond energy, reactions involving nitrogen usually require high temperatures:

$$N_2(g) + O_2(g) \rightleftharpoons 2NO(g) \quad \Delta H = +269 \, kJ\,mol^{-1}$$

You can see that the reaction with oxygen to produce nitrogen monoxide is strongly endothermic. The high temperatures required for reaction are produced by lightning. The NO is then oxidised to NO_2; this dissolves in rainwater to produce nitric acid, which is then deposited on the ground. Nitrates are a source of nitrogen for plants and so nitrogen is recycled:

$$2NO_2(g) + H_2O(l) \rightarrow HNO_3(aq) + HNO_2(aq)$$

b **The internal combustion engine**

The amount of atmospheric nitrogen oxides has increased because of burning fuels in cars and other vehicles and also because of burning oil or natural gas in power stations. The air drawn through a car engine or through the burners in a power station reaches the high temperature needed for nitrogen and oxygen to combine and produce the colourless gas NO; when this cools on its release into the atmosphere, it reacts with further oxygen to produce the dark brown gas, NO_2.

Nitrogen oxides are pollutants.

- In sunny conditions and with volatile organic compounds such as those contained in fuels, they produce smog and low-level ozone. This pollutant soup is damaging to lung tissue; ozone is toxic and causes irritation of the respiratory tract as well as being a greenhouse gas.

- In rainy conditions, an acid is produced which damages buildings by eroding limestone and metal; it also damages rivers, which may become too acidic for fish and plants to survive.

- If SO_2 is present in the atmosphere, NO_2 oxidises it to SO_3, which also forms acid rain; SO_3 forms the strong acid H_2SO_4 whereas SO_2 produces the weaker acid H_2SO_3.

A catalytic converter fitted in the exhaust system of vehicles, generators and aeroplanes limits the amount of nitrogen oxides released into the atmosphere. The most usual type is a 'three-way' converter; so called because it oxidises carbon monoxide, oxidises unburnt hydrocarbons and reduces nitrogen oxides.

The catalyst is a ceramic honeycomb support covered by platinum, palladium and rhodium. Palladium catalyses the oxidation of CO to CO_2 and unburnt hydrocarbons to CO_2 and H_2O; rhodium speeds up the reduction of NO_x to N_2; platinum catalyses both the oxidation and reduction reactions.

13.04 Sulfur dioxide

a Naturally occurring sulfur

Sulfur is required by plants and animals to build some of the proteins necessary for them to function. Volcanic eruptions emit sulfur dioxide, SO_2, into the atmosphere but much environmental sulfur exists as minerals in rocks or deep in the oceans. In the atmosphere, sulfur dioxide can be oxidised to sulfur trioxide, SO_3, which dissolves in rainwater and is then readily washed to the ground to be absorbed by plants. In the oceans, the action of plankton produces volatile dimethyl sulfide, $(CH_3)_2S$.

b Sulfur from human activity

As we saw above, SO_2 is emitted into the atmosphere when sulfur-containing compounds are burnt. The oxidation of SO_2 to SO_3 is normally a slow reaction but nitrogen oxides can act as a catalyst:

$$NO_2(g) + SO_2(g) \rightarrow NO(g) + SO_3(g)$$
$$2NO(g) + O_2(g) \rightarrow 2NO_2(g)$$

H_2SO_4 produced by SO_3 dissolving in rainwater is a much stronger acid than the H_2SO_3 produced from SO_2 dissolving. Rainwater thus becomes much more acidic when nitrogen oxides are present. Lakes and streams become acidic and the growth of plants is affected. Limestone buildings and statues are eroded (Figure 13.03).

$$H_2SO_4(aq) + CaCO_3(s) \rightarrow$$
$$H_2O(l) + CO_2(g) + CaSO_4(aq)$$

The calcium sulfate produced is soluble and so is washed away, leaving further surfaces to be eroded.

Figure 13.03 Acid rain erodes limestone statues.

c Uses of sulfur dioxide

Sulfur dioxide has antimicrobial properties and is used as a preservative for dried fruits such as apricots. In small amounts, it does not alter the taste of the fruit but helps to maintain the appearance.

It is also used in wine making to clean equipment and also as an additive to the wine itself where it acts as an antibiotic. The sulfur dioxide is reasonably soluble, much of it existing in the molecular form:

$$SO_2(aq) + H_2O(l) \rightleftharpoons HSO_3^-(aq) + H^+(aq) \rightleftharpoons SO_3^{2-}(aq) + 2H^+(aq)$$

The active species is the molecular SO_2; as the wine becomes more acidic, the equilibria move to the left and there is more active ingredient present.

> **TIP**
> Sulfur dioxide is a reducing agent and itself is oxidised to sulfur trioxide. It turns acidified potassium manganate(VII) from purple to colourless.

Revision checklist

Check that you know the following:

- [ ] explain nitrogen's lack of reactivity
- [ ] describe and explain the structure of ammonia and ammonium ion
- [ ] describe and explain why ammonia acts as a base and how to release it from an ammonium salt
- [ ] understand and state the importance of ammonia and ammonium compounds
- [ ] state and explain what effect the uncontrolled use of nitrate fertilisers have on the environment
- [ ] state and explain how atmospheric nitrogen oxides are produced and how they can be removed
- [ ] describe how atmospheric sulfur dioxide is produced and state its role in acid rain formation and the environmental consequences

Exam-style questions

1.
   a
   i Suggest values for the bond angles in N_2H_4 and N_2O (the **N** atoms are bonded together in each case). [2]

   ii Explain why N_2H_4 acts as a Brønsted–Lowry base but N_2O does not. [4]

   b 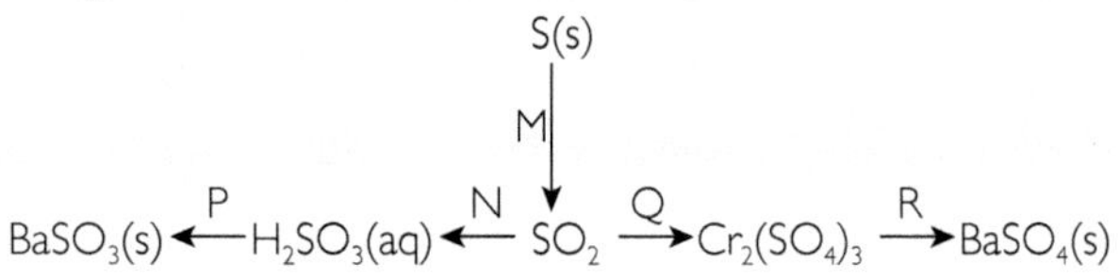

   $$BaSO_3(s) \xleftarrow{P} H_2SO_3(aq) \xleftarrow{N} SO_2 \xrightarrow{Q} Cr_2(SO_4)_3 \xrightarrow{R} BaSO_4(s)$$

   with $S(s) \xrightarrow{M}$ into SO_2

   i What kind of reagent is SO_2 in stage **Q**? [1]

   ii What reactants and conditions are needed for stages **M, N, P, Q** and **R**? [5]

   Total: 12

2.
   a A farmer limed his field with $Ca(OH)_2$ to raise the pH of the soil. What effect does this have on the ammonium sulfate fertiliser he then used on the field? [2]

   b
   i Describe a way in which nitrogen oxides can get into the atmosphere. [2]

   ii What particular problem occurs when NO_2 and SO_2 occur together in the atmosphere? Explain your answer with the aid of an equation. [3]

   iii What effects does the problem described in part b ii have? [2]

   c
   i How does eutrophication occur? [2]

   ii Suggest one way to prevent eutrophication. [1]

   Total: 12

Unit 14: Introduction to organic chemistry

Learning outcomes

You should be able to:

- interpret and use the general, structural, displayed and skeletal formulae of alkanes, alkenes, halogenoalkanes, alcohols, aldehydes, ketones, carboxylic acids and esters

- understand and use systematic nomenclature of simple aliphatic organic molecules with the above functional groups

- deduce the possible isomers for an organic molecule of known molecular formula

- deduce the molecular formula of a compound, given its structural, displayed or skeletal formula

- interpret and use the following terminology: functional group, homolytic and heterolytic fission, free radical, nucleophile, electrophile, addition, substitution, elimination, hydrolysis, oxidation and reduction

- describe structural isomerism, and its division into chain, positional and functional group isomerism

- describe stereoisomerism, and its division into geometrical (*cis–trans*) and optical isomerism

- explain what is meant by a chiral centre, and be able to identify chiral centres and/or *cis–trans* isomerism in a molecule knowing its structural formula

14.01 The different classes of organic compounds

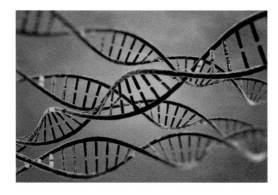

Figure 14.01 The substances that form the basis of all living things are organic compounds

Organic compounds are covalent compounds containing carbon atoms. There are over 10 million different organic compounds known today: they outnumber inorganic ones by 80:1. Why are there so many?

One reason is that carbon forms strong bonds to other carbon atoms, so they can join together in chains and rings. Another reason is that organic compounds tend to be kinetically stable: there are large activation barriers associated with their reactions, so these reactions tend to be slow, or require strong heat, or need strong reagents to get them going.

Many organic compounds can be considered as being made up of two parts: a hydrocarbon chain (R) and a functional group (Fg):

$$R—Fg$$

Most of the reactions of an organic compound are reactions of the functional group it contains – the hydrocarbon chain stays the same during the reactions.

For example, in the hydrolysis of bromoethane:

$$CH_3CH_2Br + OH^- \rightarrow CH_3CH_2OH + Br^-$$

the CH_3CH_2- chain (called the *ethyl* group) stays the same. Furthermore, the conditions required for the reaction (heating under reflux with aqueous sodium hydroxide) are the same no matter how long the chain is, for example:

$$CH_3CH_2CH_2CH_2Br + OH^- \rightarrow CH_3CH_2CH_2CH_2OH + Br^-$$

If two compounds contain the same functional group, and differ only in the length of their carbon chain, they are said to belong to the same **homologous series**.

Table 14.01 lists the functional groups you need to know about, and the names of the classes of compounds that contain them.

Name of functional group	Formula	Class of compound	Formula and name of an example
bromo	–Br	bromoalkanes	CH_3Br, bromomethane
chloro	–Cl	chloroalkanes	CH_3CH_2Cl, chloroethane
hydroxy	–OH	alcohols	CH_3CH_2OH, ethanol
carbonyl: aldehyde	–CH=O	aldehydes	CH_3CH_2CHO, propanal
carbonyl: ketone	>C=O	ketones	CH_3COCH_3, propanone
carboxylic acid	–CO_2H	carboxylic acids	CH_3CO_2H, ethanoic acid
ester	–CO_2R	esters	$CH_3CO_2CH_2CH_3$, ethyl ethanoate
acyl chloride	–COCl	acyl chlorides	CH_3COCl, ethanoyl chloride
amide	–$CONH_2$	amides	$CH_3CH_2CONH_2$, propanamide
amino	–NH_2	amines	CH_3NH_2, methylamine
nitrile	–C≡N	nitriles	CH_3CH_2CN, propanonitrile

Table 14.01 The functional groups you need to know about

14.02 The different formulae used to describe organic compounds

Organic compounds can be represented by six different types of formulae. Which one you use depends on the level of detail about the structure of the molecule you want to convey. The simplest formula is the **empirical formula**. This tells us the simplest ratio of elements the compound contains.

The **molecular formula** tells us the number of atoms of each element that are in one molecule of the compound. The molecular formula of a compound is always a multiple of its empirical formula (e.g. ×1, ×2, ×3).

The **structural formula** indicates which atoms are joined to which in the molecule of the compound.

The **displayed formula** shows all the bonds and atoms within a molecule of the compound.

The **stereochemical formula** is a displayed formula that shows the three-dimensional shape of the molecule.

The **skeletal formula** is a simplified version of the displayed formula: it shows all the bonds in a molecule, except those to hydrogen atoms, but omits all carbon atoms and their attached hydrogen atoms.

These six formulae are illustrated in Figure 14.02 for the compound methylpropanoic acid.

C_2H_4O — empirical
$C_4H_8O_2$ — molecular
$CH_3CH(CH_3)CO_2H$ or $(CH_3)_2CHCO_2H$ — structural

displayed, stereochemical, skeletal

Figure 14.02 The different types of formulae for methylpropanoic acid

> **TIP**
> Whenever you write the structural or displayed formula of an organic compound, remember to include *all* the hydrogen atoms. Also check that the valency of each element is correct.
> - Hydrogen, bromine and chlorine always have a valency of 1.
> - Oxygen always has a valency of 2.
> - Nitrogen always has a valency of 3, except in the nitro (–NO_2) group.
> - Carbon always has a valency of 4.

> **TIP** Remember that the *only* hydrogen atoms shown in a skeletal formula are those joined to oxygen or nitrogen: C–H atoms are *not* shown.

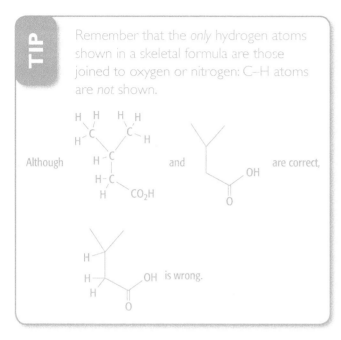

Progress check 14.01

1. Draw the displayed formulae of *three* compounds that have the molecular formula C_3H_6O.

2. Write down **i** the molecular, and **ii** the empirical formulae of the following compounds:

   a, b, c

14.03 How to name organic compounds

The names of organic compounds tell us what carbon chain and which functional group(s) they contain. All names contain a stem and a stem-end. Most also contain either a suffix or a prefix, or both. Tables 14.02–14.05 list some of these.

The *stem* specifies how many carbon atoms are in the longest chain.

> **TIP** The longest chain may not be the most obvious one from the way the formula is drawn out. You should count each of the possible chains contained in the compound before you make up you mind which is the longest.

Stem name	Number of carbon atoms
meth-	1
eth-	2
prop-	3
but-	4
pent-	5
hex-	6

Table 14.02 The first six stem names

The *stem-end* shows the type of carbon–carbon bonds that occur in the compound (Table 14.03).

Stem-end	Type of bond present
-an	all C–C single bonds
-en	one C=C double bond
-dien	two C=C double bonds
-trien	three C=C double bonds
-yn	one C≡C triple bond

Table 14.03 Some of the stem-ends used

The *suffix* describes any oxygen-containing group that might be present (Table 14.04).

Suffix	Group
-al	an aldehyde group: (C=O with H)
-one	a ketone group: (C=O)
-ol	an alcohol group: —C—OH
-oic acid	a carboxylic acid group: —C(=O)OH

Table 14.04 Four suffixes, all involving oxygen

The *prefix* indicates non-oxygen-containing groups, which may be other carbon (alkyl) groups, or halogen atoms which are substituted for H along the carbon chain (Table 14.05).

Prefix	Group
methyl-	CH_3-
ethyl-	CH_3CH_2-
propyl-	$CH_3CH_2CH_2-$
butyl-	$CH_3CH_2CH_2CH_2-$
bromo-	$Br-$
chloro-	$Cl-$

Table 14.05 Some of the prefixes used

Prefixes can be combined with each other, and they might in turn be prefixed by 'di' or 'tri' to show multiple substitution.

Numbers can be added before the stem-end or the suffix or the prefix to show which carbon atom the functional group is attached to. The carbon atoms are numbered from the end nearest the functional group.

Table 14.06 contains some examples to show how naming works. Check that you can see how the names are constructed for each compound.

Structural formula	Explanation	Name
$CH_3CH_2CH_3$	3 carbon atoms, no functional group	propane
$CH_3CH=CHCH_3$	4 carbon atoms, C=C starts at second atom	but-2-ene
$CH_3CH(OH)CH_3$	3 carbon atoms, with OH on second atom	propan-2-ol
$CH_3CH_2COCH_2CH_3$	5 carbon atoms, with C=O on third atom	pentan-3-one
$CH_3CH_2CH(CH_3)CO_2H$	4 carbon atoms in chain, CH_3 on second atom, carboxylic acid group (assumed to be carbon 1)	2-methylbutanoic acid
$ClCH_2CH_2CHO$	3 carbon atoms in chain, Cl on third carbon atom from the aldehyde group (assumed to be carbon 1)	3-chloropropanal

Table 14.06 Illustrating how the systematic names are built up

Progress check 14.02

1. Draw the structural formulae of the following compounds (their molecular formulae have been given to help you):

   a 2-methylbutanal ($C_5H_{10}O$)

   b 3-ethylpent-2-en-1-ol ($C_7H_{14}O$)

   c 3,3-dimethylpentan-2,4-dione ($C_7H_{12}O_2$)

2. Name the following compounds:

   a, b, c (structural formulae shown)

14.04 Isomerism

Isomers are compounds that have the same molecular formulae, but have different arrangements of atoms.

There are two main types of isomerism: structural and stereoisomerism. Structural isomers have different structural formulae from each other, but stereoisomers have the same structural formulae.

a **Structural isomers**

There are three types of **structural isomerism**.

Chain isomers have the same number of carbon atoms as each other, but their carbon backbones are different.

For example, pentane and 2-methylbutane are isomers with the molecular formula C_5H_{12}:

$CH_3-CH_2-CH_2-CH_2-CH_3$ $CH_3-CH_2-CH(CH_3)-CH_3$

Note that $CH_3-CH_2-CH_2-CH_2$ with CH_3 branch is not another isomer: this is just another way of writing the formula of pentane.

Note also that $CH_3-CH_2-CH(CH_3)-CH_3$ is not another isomer. This is another way of writing the formula of 2-methylbutane.

Positional isomers have the same carbon backbone, but the positions of their functional group(s) differ.

Their systematic names reflect this. For example, 1-bromobutane and 2-bromobutane are positional isomers:

$CH_3CH_2CH_2CH_2Br$ and $CH_3CH_2CHBrCH_3$

Functional group isomers contain different functional groups, and so they often have very different reactions.

Examples would be the carboxylic acid $CH_3CH_2CO_2H$ (propanoic acid), the ester $CH_3CO_2CH_3$ (methyl ethanoate) and the ketone–alcohol CH_3COCH_2OH (hydroxypropanone). These are all isomers with the molecular formula $C_3H_6O_2$.

b **Stereoisomers**

There are two types of **stereoisomerism**.

Geometrical isomerism, or *cis–trans* isomerism, occurs with alkenes that do not have two identical groups on either end of the C=C double bond.

Unlike the case of a C–C single bond, it is not possible to rotate one end of a C=C double bond with respect to the other. For example, there are two different structures of but-2-ene:

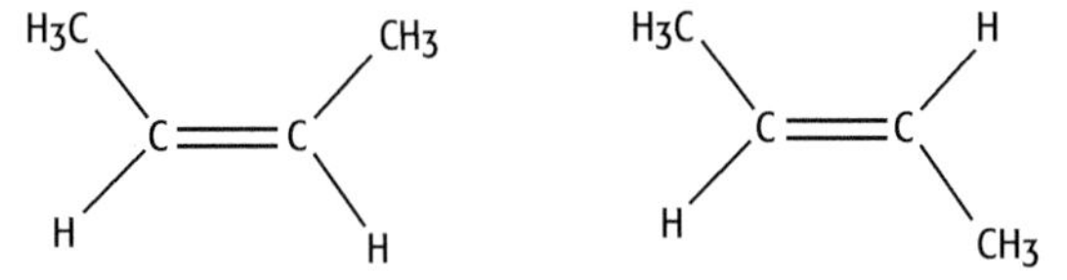

cis-but-2-ene   *trans*-but-2-ene

The isomer with both methyl groups on the same side of the double bond is called the *cis* isomer, and that with the methyl groups on opposite sides is called the *trans* isomer.

The carbon atoms on *both* ends of the double bond must each contain two non-identical groups. The following compounds *do not* exhibit geometrical isomerism because one end, or both ends, of the double bond have two identical atoms or groups attached:

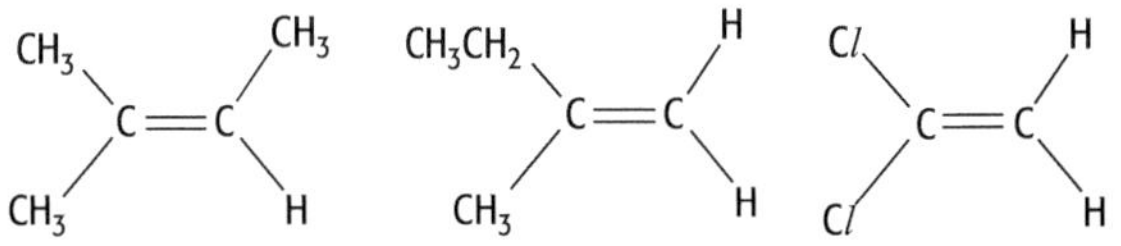

Optical isomerism arises because of the tetrahedral geometry of the sp^3 carbon atom. If a carbon atom is bonded to four different groups, it is called a chiral atom, or **chiral centre**.

There are two different ways of arranging four non-identical groups around a chiral carbon atom, which result in two structures that are non-superimposable mirror images of each other. For example, the amino acid alanine can exist in these two forms:

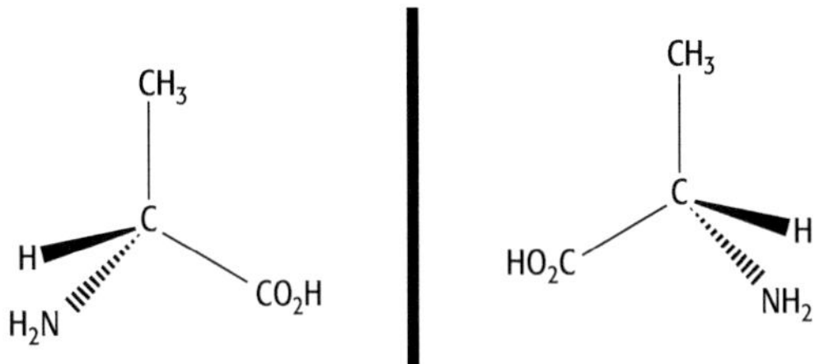

isomer I   mirror   isomer II

Optical isomers have identical chemical reactions (unless they are reacted with other chiral compounds), and all their physical properties are also identical – except for one. If a beam of polarised light is passed through them, one isomer will twist the plane of polarisation in one direction, and the other isomer will twist the plane to exactly the same extent in the opposite direction. Hence the name *optical* isomerism. Some molecules contain more than one chiral centre. Identifying the chiral centres in a complex molecule can be quite difficult, especially if only the skeletal formula is given.

For example, there are three chiral centres in the following molecule:

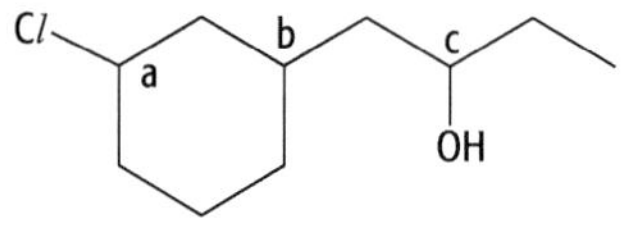

The easiest chiral centre to recognise is carbon atom **c**, attached to the –OH group. This has the four non-identical groups: –H, –OH, –CH$_2$CH$_3$ and –CH$_2$-ring. Carbon atom **b** has –H, –CH$_2$-chain, –CH$_2$-ring, CH$_2$-ring. Note, however, that the two ring CH$_2$ groups are *not* identical – if you extend the chain to the next carbon atom in the ring, one of them is CH$_2$CH$_2$-ring and the other is CH$_2$CHCl-ring. Likewise with carbon atom **a**: this has –H, –Cl, –CH$_2$-ring and –CH$_2$-ring attached to it, but the two ring CH$_2$ groups are *not* identical – extending to the next carbon produces CH$_2$CH$_2$-ring for one and CH$_2$CH(side chain)-ring for the other.

Sometimes stereochemical formulae can be confusing. How many chiral centres can you see in the following molecules?

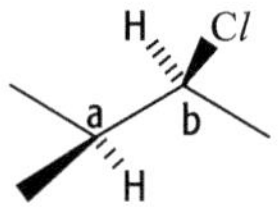

A 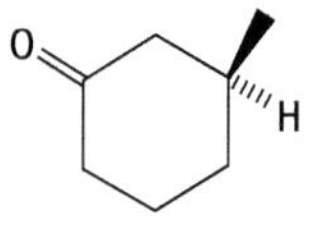   B 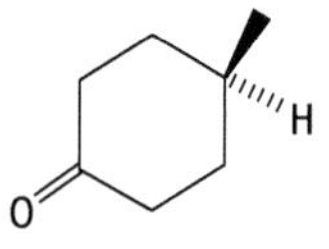   C

Molecule **A** contains just one chiral centre (**b**): carbon atom **a** has two (identical) methyl groups attached to it. Molecule **B** also contains one chiral centre, but molecule **C** contains none: if you follow the carbon atoms in the ring around until they meet, you will find that the two routes to the ketone C=O group are identical.

Worked example 14.01

Draw the structures of all the structural isomers with the formula C$_4$H$_9$OH and name them.

How to get the answer:

Step 1: There are two isomeric alkanes with four carbon atoms: butane and 2-methylpropane.

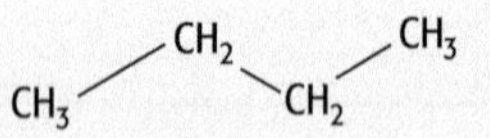

butane 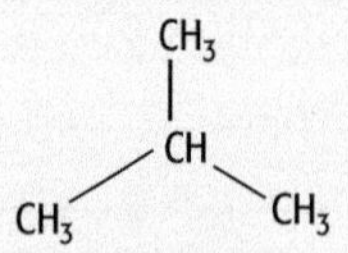   2-methylpropane

Step 2: Each of these contains two different positions where an H can be replaced by an OH, so there are four structural isomers, named as below.

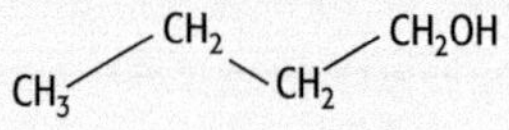

butan-1-ol 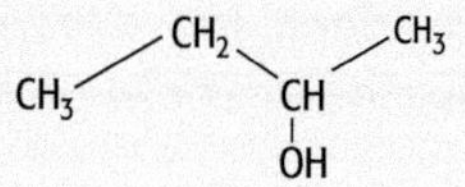   butan-2-ol

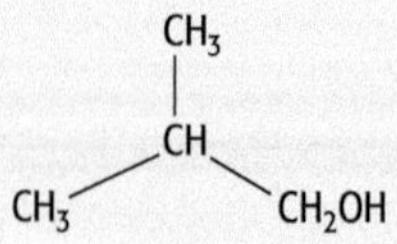

2-methylpropan-1-ol 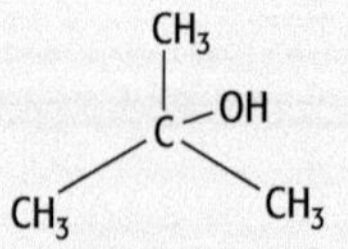   2-methylpropan-2-ol

Progress check 14.03

1. There are five isomers with the molecular formula $C_3H_6Cl_2$. Draw their structures (using stereochemical formulae where necessary), and describe the types of isomerism shown.

2. Draw the structures of all the isomers with the molecular formula C_3H_6BrCl. How many are optical isomers of each other?

3. Draw and name the structures of all the non-cyclic isomers (i.e. not containing a ring of carbon atoms) with the molecular formula C_5H_{10}. How many are geometrical isomers of each other?

14.05 Types of organic reactions

The following units look in detail at the reactions of some functional groups, but it will be useful for reference (and revision) to collect together a list of all the reaction types you need to know (Table 14.07).

Type of reaction	Functional group(s) that undergoes the reaction	Example
electrophilic addition	alkene	$CH_2=CHCH_3 + HBr \rightarrow CH_3CHBrCH_3$
electrophilic substitution	arene	benzene $+ HNO_3 \rightarrow$ nitrobenzene $+ H_2O$
nucleophilic addition	carbonyl (ketone or aldehyde)	$CH_3CH(=O) + HCN \rightarrow CH_3CH(OH)(C\equiv N)$
nucleophilic substitution	bromoalkane	$CH_3Br + OH^- \rightarrow CH_3OH + Br^-$
	alcohol	$CH_3CH_2OH + HBr \rightarrow CH_3CH_2Br + H_2O$
	carboxylic acid	$CH_3CO_2H + SOCl_2 \rightarrow CH_3COCl + SO_2 + HCl$
	acyl chloride	$CH_3COCl + NH_3 \rightarrow CH_3CONH_2 + HCl$
radical substitution	alkane	$CH_3CH_3 + Cl_2 \rightarrow CH_3CH_2Cl + HCl$
elimination	alcohol	$CH_3CH_2CH_2OH \rightarrow CH_3CH=CH_2 + H_2O$
	bromoalkane	$CH_3CH_2Br + OH^- \rightarrow CH_2=CH_2 + H_2O + Br^-$
hydrolysis	ester	$CH_3CO_2CH_2CH_3 + H_2O \rightarrow CH_3CO_2H + CH_3CH_2OH$
	acyl chloride	$CH_3COCl + H_2O \rightarrow CH_3CO_2H + HCl$
	amide	$CH_3CONH_2 + H_2O \rightarrow CH_3CO_2H + NH_3$
	nitrile	$CH_3CH_2C\equiv N + 2H_2O \rightarrow CH_3CH_2CO_2H + NH_3$
	nitroarene	nitrobenzene $+ 6[H] \rightarrow$ aniline $+ 2H_2O$
condensation	carbonyl (ketone or aldehyde)	$CH_3(C_2H_5)C=O + H_2N\text{-}R \rightarrow CH_3(C_2H_5)C=N\text{-}R + H_2O$

Type of reaction	Functional group(s) that undergoes the reaction	Example
condensation	carboxylic acid + alcohol	$CH_3CO_2H + CH_3CH_2OH \rightarrow CH_3CO_2CH_2CH_3 + H_2O$
reduction	alkene	$CH_3CH=CH_2 + H_2 \longrightarrow CH_3CH_2CH_3$
	carbonyl (ketone or aldehyde)	$CH_3CH_2CH=O + 2[H] \longrightarrow CH_3CH_2CH_2OH$
	amide	$CH_3CONH_2 + 4[H] \longrightarrow CH_3CH_2NH_2 + H_2O$
	carboxylic acid	$CH_3CH_2CO_2H + 4[H] \longrightarrow CH_3CH_2CH_2OH + H_2O$
	nitrile	$CH_3CH_2C\equiv N + 4[H] \longrightarrow CH_3CH_2CH_2NH_2$
oxidation	alcohol	$CH_3CH_2OH + [O] \longrightarrow CH_3CHO + H_2O$
	aldehyde	$CH_3CHO + [O] \longrightarrow CH_3CO_2H$

Table 14.07 The reaction types you need to know.

The meanings of some of the terms used in Table 14.07 are given below.

[H] represents hydrogen atoms that have come from a chemical reducing agent (such as $LiAlH_4$ or $Sn + HCl$)

[O] represents oxygen atoms that have come from a chemical oxidising agent (such as $K_2Cr_2O_7 + H_2SO_4$)

An **electrophile** is an atom or group of atoms that reacts with electron-rich centres in other molecules. Electrophiles usually have an empty orbital in their valence shell. They are either positively charged or have a $\delta+$ atom.

A **nucleophile** is an atom or group of atoms that reacts with electron-deficient centres in molecules (such as the $C^{\delta+}$ in $C^{\delta+}-Br^{\delta-}$ compounds). All nucleophiles have a lone pair of electrons and many are anions.

A **radical** is an atom or group of atoms that has an unpaired electron. Radicals (sometimes called **free radicals**) are highly reactive.

An **addition reaction** is one in which an organic molecule (usually containing a double bond) reacts with another molecule to give *only one* product.

A **substitution reaction** is one in which one atom or group replaces another atom or group in a molecule.

An **elimination reaction** is one that forms an alkene by the removal of a molecule of H_2O from an alcohol, or a molecule of HBr from a bromoalkane.

A **hydrolysis reaction** is one where water reacts with a molecule and splits it into two smaller molecules.

A **condensation reaction** is one between two molecules to form a larger molecule, with the removal of a molecule of water.

Reduction is a reaction in which hydrogen is added to, or oxygen is removed from, an organic compound.

Oxidation is a reaction in which oxygen is added to, or hydrogen is removed from, an organic compound.

Sample answer

Question:

Identify the *reaction type* of each of the following reactions.

a cyclohexene → cyclohexanol

b $>=<$ + Br_2 → $>\!\!\!\overset{Br}{\underset{}{<}}$ + HBr

c pyrrolidinone (cyclic amide) → $H_2N\!\!\sim\!\!CO_2H$

d $>=\!\!\!\sim\!\!CHO$ → $>\!\!\!\sim\!\!OH$

Answer:

a is an electrophilic addition. [1 mark]

b is a free radical substitution. [1 mark]

c is a hydrolysis. [1 mark]

d is a reduction. [1 mark]

Revision checklist

Check that you know the following:

- [ ] How to name organic compounds, given their structures (using the prefix-stem-suffix principle).

- [ ] The different type of formulae used to represent organic compounds (molecular, structural, displayed and skeletal), and how to convert one into another.

- [ ] How to work out the structure of an organic compound, given its name and/or its molecular formula.

- [ ] The various types of isomerism – structural (chain, positional and functional group) and stereo (geometrical and optical) – and the molecular features that give rise to these (including chiral centres).

- [ ] Interpret and use the following terminology associated with organic reactions:
  - functional group
  - homolytic and heterolytic fission
  - free radical, initiation, propagation, termination
  - nucleophile, electrophile
  - addition, substitution, elimination, hydrolysis, condensation
  - oxidation and reduction

Exam-style questions

1. For each of the following reactions:
   i. state the *type of reaction* undergone
   ii. give the structure of the organic product
   iii. give the systematic name of the product.

   a. $C_3H_8 + Cl_2 \rightarrow C_3H_7Cl$ [3]

   b. $C_3H_7OH \rightarrow C_3H_6O_2$ [3]

   c. $(CH_3)_3COH \rightarrow C_4H_8$ [3]

   Total: 9

2. Illustrate the following reaction types with a suitable reaction of a straight-chain 4-carbon compound of your choice (e.g. butane, butan-2-ol).

   For each reaction:
   i. write a balanced equation, including the structures of the starting material and the product (you may use [H] to represent a reducing agent adding a hydrogen atom)
   ii. name the starting material and the product.

   a. nucleophilic substitution [3]

   b. nucleophilic addition [3]

   c. electrophilic addition [3]

   d. reduction. [3]

   Total: 12

Unit 15: Hydrocarbons

Learning outcomes

You should be able to:

- ☐ understand the general unreactivity of alkanes, but describe their reactions such as combustion and substitution by chlorine and by bromine
- ☐ describe the mechanism of free-radical substitution with particular reference to the initiation, propagation and termination reactions
- ☐ explain the use of crude oil as a source of both aliphatic and aromatic hydrocarbons
- ☐ suggest how large hydrocarbon molecules are 'cracked' into smaller alkanes and alkenes
- ☐ describe the uses of alkanes as fuels
- ☐ recognise the environmental consequences of carbon monoxide, oxides of nitrogen and unburnt hydrocarbons from car engines, and how they can be removed from exhaust gases using catalytic converters
- ☐ recognise the environmental consequences of gases that contribute to the enhanced greenhouse effect
- ☐ outline the use of infrared spectroscopy in monitoring air pollution
- ☐ describe the following reactions of alkenes:
  - addition of hydrogen, steam, hydrogen halides and halogens
  - oxidation by cold dilute, and hot concentrated, acidified manganate(VII) ions
  - polymerisation
- ☐ describe the mechanism of electrophilic addition in alkenes
- ☐ describe and explain the inductive effects of alkyl groups on the stability of cations
- ☐ describe the characteristics of addition polymerisation
- ☐ deduce the repeat unit of an addition polymer obtained from a given monomer, or identify the monomer(s) present in a given section of a polymer molecule
- ☐ recognise the difficulty of the disposal of poly(alkenes)

There are four homologous series of hydrocarbons you need to know about. They are listed in Table 15.01.

Name of class	General formula	Example compound	Formula	Structure
alkanes	C_nH_{2n+2}	propane	C_3H_8	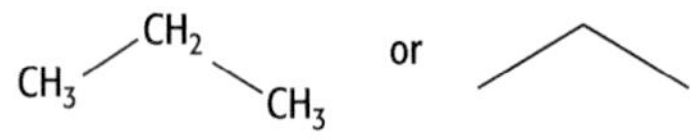 $CH_3-CH_2-CH_3$ or ∧
cycloalkanes	C_nH_{2n}	cyclohexane	C_6H_{12}	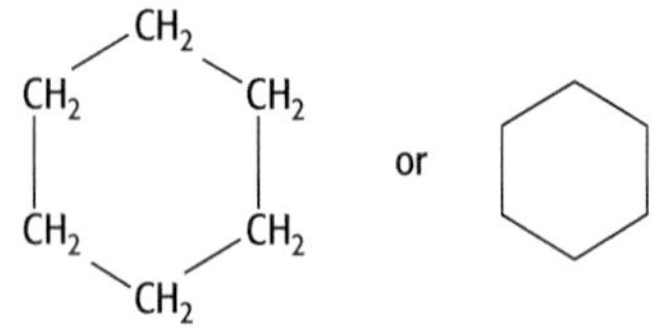 (ring of 6 CH_2) or ⬡
alkenes	C_nH_{2n}	but-1-ene	C_4H_8	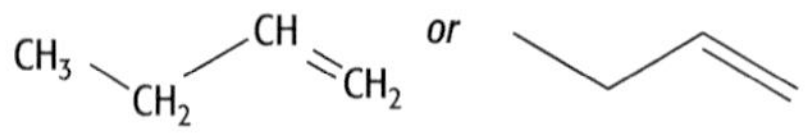 $CH_3-CH_2-CH=CH_2$ or ∕∕═
arenes	C_nH_{2n-6}	methylbenzene	C_7H_8	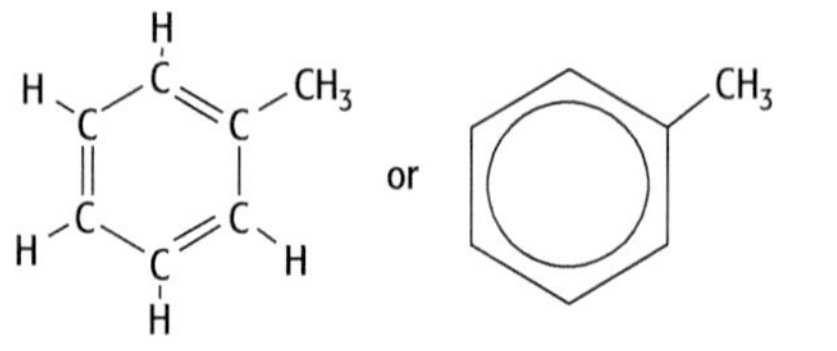 benzene ring with CH_3

Table 15.01 The four classes of hydrocarbon

This unit is concerned with the alkanes, cycloalkanes and the alkenes. Arenes are covered in Unit 25. The reactions of cycloalkanes are very similar to those of alkanes; note, however, their general formula is the same as that of alkenes.

15.01 Physical properties

All alkanes and alkenes are flammable. Those with fewer than five carbon atoms are gases at room temperature; the rest are mostly liquids.

15.02 Isomerism and nomenclature

Alkanes show two types of isomerism. Those with four or more carbon atoms per molecule show *chain isomerism*, whilst those with seven or more carbon atoms can also exist as *optical isomers*.

The stem of their name is worked out from the longest carbon chain, and the positions of side-chains are indicated by numbers, as described in Unit 14.

A

For example, in the compound A, the longest carbon chain has six carbon atoms, with a 2-carbon (= ethyl) side-chain on the third atom of the chain, so its name is 3-ethylhexane.

Alkenes show the same isomerism (chain and optical) as alkanes, but in addition they can show geometrical (*cis–trans*) isomerism (see Unit 14). This is because there is no rotation around the C=C, due to its π bond.

The structural requirements for geometrical isomerism to occur are that there have to be non-identical atoms or groups at *both* ends of the double bond. So structures **B** and **C** have geometrical isomers, whereas structures **D** and **E** do not.

B   C

D   E

Alkenes are named by using the longest carbon chain that includes the C=C as the stem, with the position of the double bond indicated by a number before the stem-end. Two examples are:

trans-2-methylhex-3-ene   cis-3-ethylhex-2-ene

> ## Progress check 15.01
>
> 1   Draw the structural and skeletal formulae of all the isomers with formula C_5H_{12} and name them.
>
> 2   The smallest alkane to show optical isomerism has the molecular formula C_7H_{16}. Draw its structural formula and give its name.
>
> 3   Name, and draw the skeletal formula of, the alkane with the molecular formula C_8H_{18} that contains *two* chiral carbon atoms.

15.03 Crude oil, and how alkanes and alkenes are prepared

Crude oil is a complex mixture of alkanes, cycloalkanes and arenes, whose molecules contain from 1 to 30 or 40 carbon atoms. Fractional distillation separates this mixture into a few fractions, which are still mixtures of many different molecules.

The five most common initial fractions are listed in Table 15.02.

Name of fraction	Number of carbon atoms per molecule	Boiling point range / °C	Percentage in crude oil	Some uses
refinery gas	1–4	<20	2	fuel, petrochemical feedstock
gasoline	5–12	20–180	20	fuel for petrol engines, petrochemical feedstock
kerosene	10–16	180–260	13	aeroplane fuel, paraffin, cracked to give more gasoline
gas oil	12–25	260–380	20	fuel for diesel engines, cracked to give more gasoline
residue	>25	>380	45	lubricating oil, power station fuel, bitumen for roads, cracked to give more gasoline

Table 15.02 The main fractions separated from crude oil

As can be seen from the table, the kerosene and gas oil fractions (the most common in crude oil) are subsequently cracked to produce more gasoline (the fraction most in demand).

By-products of these cracking reactions include the lower alkenes, which are useful starting materials for polymers.

Here is an example:

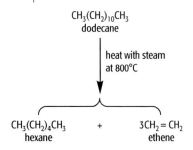

Figure 15.01 A catalytic cracker in an oil refinery

> **TIP**
> The balanced equation for a cracking reaction always includes one or more alkenes, and either one alkane or H_2, but not both an alkane and H_2. Take care to check the cracking equations you write.

Unlike alkenes, alkanes are rarely made in the laboratory. Alkenes are often made by *elimination reactions*, from halogenoalkanes (see Unit 16) or alcohols (see Unit 17).

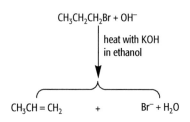

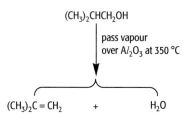

Progress check 15.02

Suggest starting materials and conditions for the synthesis of the following alkenes:

1. $CH_3CH=CHCH_3$
2. 2-methylpent-3-ene
3.

15.04 Reactions of alkanes

Alkanes contain no specific functional group. They are very unreactive. Apart from the cracking of long-chain alkanes, the only two reactions you need to know are combustion and halogenation (reaction with halogens).

a The combustion of alkanes

i   In a plentiful supply of air or oxygen, alkanes burn to give steam and carbon dioxide:

natural gas
$$CH_4(g) + 2O_2(g) \longrightarrow CO_2(g) + 2H_2O(g)$$
$$\Delta H_c^\ominus = -890 \text{ kJ mol}^{-1}$$

petrol/gasoline
$$C_8H_{18}(l) + 12\tfrac{1}{2}O_2(g) \longrightarrow 8CO_2(g) + 9H_2O(g)$$
$$\Delta H_c^\ominus = -5470 \text{ kJ mol}^{-1}$$

The $\Delta H_c^\ominus$ values tell us that each of these is a very exothermic reaction: both methane and octane are important fuels. Most alkanes obtained from the fractional distillation of crude oil are used as fuels.

ii In a limited supply of air, 100% of the hydrogen part of the alkane still gets oxidised to water, but the carbon part does not burn completely. For example:

$$CH_4(g) + 1\tfrac{1}{2}O_2(g) \longrightarrow CO(g) + 2H_2O(g)$$
$$\text{carbon monoxide}$$

$$C_8H_{18}(l) + 6\tfrac{1}{2}O_2(g) \longrightarrow 4CO(g) + 4C(s) + 9H_2O(g)$$
$$\text{carbon monoxide} \quad\quad \text{carbon (soot)}$$

Apart from the poisonous carbon monoxide, and the soot, which can cause lung diseases, other pollutants are formed when hydrocarbon fuels are burned in a car engine. The temperature of combustion can rise high enough to allow the reaction between oxygen and nitrogen (both are gases in the air) to occur:

$$N_2(g) + O_2(g) \longrightarrow 2NO(g)$$
$$2NO(g) + O_2(g) \longrightarrow 2NO_2(g)$$

Both NO (nitric oxide) and NO_2 (nitrogen dioxide) cause smog and acid rain. They are collectively known as NO_x.

The removal of pollutants from vehicle exhausts is achieved by passing the exhaust gases through a catalytic converter. This contains the metals platinum and rhodium, deposited on a honeycomb-shaped grid. The reactions catalysed can be represented by the following equations:

$$CO(g) + NO(g) \longrightarrow CO_2(g) + \tfrac{1}{2}N_2(g)$$
$$CO(g) + \tfrac{1}{2}O_2(g) \longrightarrow CO_2(g)$$
$$\text{unburnt hydrocarbons} + O_2(g) \longrightarrow CO_2(g) + H_2O(g)$$

These reactions remove the main pollutants, but the greenhouse gas CO_2 remains. There is no method yet available to remove it from vehicle exhausts.

> **TIP**
> Although you cannot write a balanced equation for the catalytic removal of unburnt hydrocarbons, you should always make sure that the other equations are balanced in your answers.

Air pollution caused by the combustion of fuels can be monitored by infrared spectroscopy (see Unit 29). The common pollutant gases (CO_2, SO_2, NO_2, NO and CO) all have characteristic absorption bands in their infrared spectra. The intensities of these bands when a sample of air is analysed in a spectrometer can give an indication of their concentrations in ppm (parts per million) in the sample.

b The reaction of alkanes with halogens

In the presence of ultraviolet radiation, or even visible light, alkanes react with chlorine or bromine. During the reaction, one or more of the alkane's hydrogen atoms is *substituted* by a halogen atom.

$$CH_4 + Cl_2 \longrightarrow CH_3Cl + HCl$$
$$CH_3CH_3 + Br_2 \longrightarrow CH_3CH_2Br + HBr$$

The mechanism of the reaction is **free radical substitution**. This is a chain reaction, in which a halogen radical is formed in the first (initiation) step; it reacts with the alkane in the second step; and is re-formed in the third step, to allow the 'chain' to continue.

In this reaction, the radicals are formed by the homolytic breaking of the X–X bond (X = Cl or Br). During homolytic fission, each atom takes one of the two electrons in the X–X bond:

$$:\!\ddot{C}l\!:\!-\!:\!\ddot{C}l\!: \xrightarrow{\text{homolytic fission}} :\!\ddot{C}l\!\cdot + \cdot\ddot{C}l\!:$$

Only the 'odd' unpaired electron in a radical is normally shown, as a dot.

There are three stages to the overall mechanism: initiation, propagation and termination. The propagation stage involves two reactions: the first uses up the initiating chlorine radical, and the second regenerates it.

initiation: $Cl_2 \xrightarrow{light} 2Cl^\bullet$

propagation I: $CH_4 + Cl^\bullet \longrightarrow CH_3^\bullet + HCl$

propagation II: $CH_3^\bullet + Cl_2 \longrightarrow CH_3Cl + Cl^\bullet$

> **TIP**
> Note that free hydrogen atoms, H$^\bullet$, are *never* formed during free radical substitution reactions. Check that you do not write equations that include them.

These two propagation steps can go on and on, as long as there are still methane and chlorine molecules present. Eventually, two radicals collide with each other, producing a stable molecule:

termination: $Cl^\bullet + Cl^\bullet \longrightarrow Cl_2$

or $CH_3^\bullet + Cl^\bullet \longrightarrow CH_3Cl$

or $CH_3^\bullet + CH_3^\bullet \longrightarrow CH_3\text{-}CH_3$ (ethane)

Any one of these three possible termination steps will stop the chain reaction, and a further photon will be needed to start it again. Note that the last termination step produces a 2-carbon alkane, ethane. This could itself undergo free radical chlorination, so an expected byproduct would be CH_3CH_2Cl. This is in fact found, which provides evidence for a mechanism involving $CH_3^\bullet$ as an intermediate.

The main product, CH_3Cl, can also undergo free radical chlorination, so other byproducts include CH_2Cl_2, $CHCl_3$ and CCl_4. Their relative amounts depend on what Cl_2:CH_4 ratio we started with.

Worked example 15.01

1. Write the propagation steps needed to produce CH_2Cl_2.

2. Suggest *four* byproducts, each containing two chlorine atoms per molecule, from the reaction between ethane and chlorine.

3. Suggest starting materials for the preparation of the following compounds by radical substitution reactions:

   a $CH_3CHBrCH_3$

   b (cyclohexane with Cl substituent)

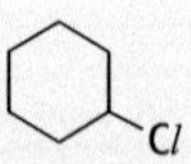

   c (benzene ring with CH_2Br substituent)

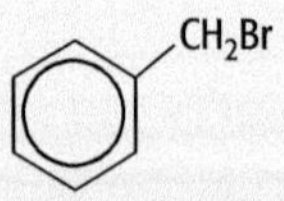

How to get the answer:

1 Step 1: In order to produce a *dichloroalkane*, we need to start with a *monochloroalkane*, in this case chloromethane.

Step 2: $Cl^\bullet + CH_3Cl \longrightarrow CH_2Cl^\bullet + HCl$

$CH_2Cl^\bullet + Cl_2 \longrightarrow CH_2Cl_2 + Cl^\bullet$

2 Step 1: There are only two dichloroethanes, CH_3CHCl_2 and $ClCH_2CH_2Cl$, so the other two compounds that contain two chlorine atoms must be dichlorobutanes (butane is produced by two ethyl radicals combining in a termination step: $2C_2H_5^\bullet \longrightarrow C_4H_{10}$).

Step 2: Any two of the following would fit:

$ClCH_2CH_2CH_2CH_2Cl$ or

$CH_3CHClCH_2CH_2Cl$ or

$CH_3CHClCHClCH_3$ or

$CH_3CH_2CH_2CHCl_2$ or

$CH_3CH_2CCl_2CH_3$

3 Step 1: The starting materials will contain the same carbon skeleton, but with the halogen atom replaced by a hydrogen atom.

Step 2: So the starting materials are:

a propane ($CH_3CH_2CH_3$) + Br_2

b cyclohexane (C_6H_{12}) + Cl_2

c methylbenzene ($C_6H_5\text{–}CH_3$) and Br_2

15.05 Reactions of alkenes

a Electrophilic addition reactions

Unlike alkanes, alkenes do not need light in order to react with bromine. Shaking an alkene with bromine water or bromine in an organic solvent, even in the dark, causes an almost immediate reaction, and the orange colour of the bromine disappears:

$$CH_3CH=CH_2 + Br_2 \longrightarrow CH_3CHBr-CH_2Br$$
$$\text{orange} \qquad \text{colourless}$$

This is a good test for the presence of a C=C double bond in a compound.

Other small molecules also undergo addition reactions with alkenes:

$$CH_3CH=CH_2 + HCl \xrightarrow{\text{room temperature}} CH_3CHClCH_3$$

$$CH_3CH=CH_2 + H_2O \xrightarrow[\text{at 300°C}]{\text{steam + conc.} \ H_2SO_4 \text{ or } H_3PO_4} CH_3CH(OH)CH_3$$

Note that when HCl (or HBr) or H_2O is added to an unsymmetrical alkene such as propene, the hydrogen atom joins onto the carbon with the larger number of hydrogens already, and the Cl, Br or OH joins onto the more substituted carbon. We shall explain why this is later, when we look at the mechanism of these reactions. This is an example of Markovnikov's rule.

b Markovnikov's rule

When HX adds to a double bond:

- the hydrogen atom attaches to the carbon that already has the most hydrogens, or
- the electrophile adds in the orientation that produces the most stable intermediate cation

c The mechanism of electrophilic addition

As we saw in Unit 14 (section 14.05), an electrophile is an atom or group of atoms that reacts with electron-rich centres in other molecules. When a bromine molecule approaches an alkene, the electrons in the Br–Br bond will be repelled by the electron-rich clouds of the π bond, causing an induced dipole to form. Eventually, the electron shift becomes so great as to break the Br–Br bond, forming a bond between the carbon at one end of the C=C bond and the nearest bromine atom, leaving the further one as a bromine anion, having gained the bonding pair of electrons (step 1). This is followed, in step 2, by the bromide ion forming a bond with the intermediate carbocation:

The mechanism is usually represented in a simplified way as follows:

> **TIP**
> When drawing mechanisms, make sure your curly arrows are drawn accurately, starting and finishing in the right places. Also make sure you include *all* charges and $\delta+$ partial charges on the correct atoms, and include *all* relevant lone pairs of electrons.

Note that when the first bromine atom joins onto the double bond, the secondary carbocation, rather than the primary carbocation, is produced:

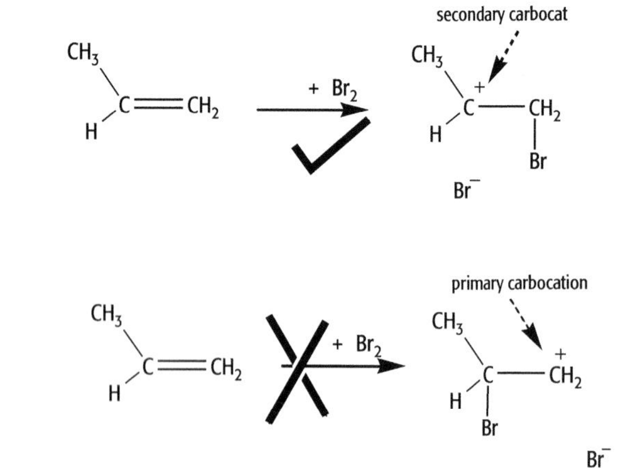

[Mechanism diagrams showing Br₂ addition to propene, with secondary carbocation pathway favoured (✓) over primary carbocation pathway (✗).]

This is because the secondary carbocation has *two* alkyl groups pushing electrons into it via their positive inductive effect, whereas the primary carbocation has only one electron-donating alkyl group on its electron-deficient carbon.

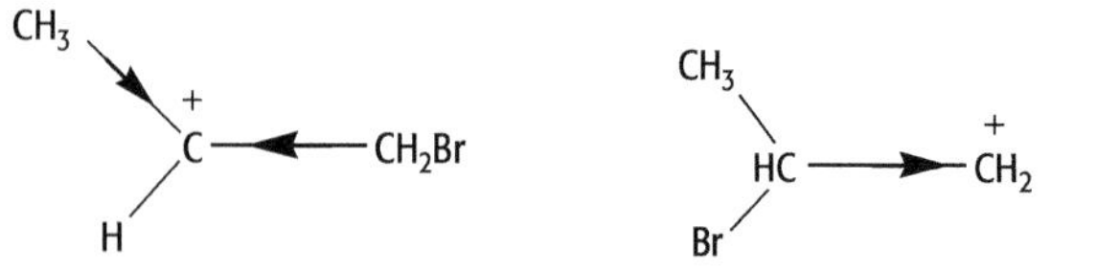

secondary – two electron-donating alkyl groups: more stable

primary – only one electron-donating alkyl group: less stable.

A similar mechanism operates with other electrophiles. Most other electrophiles are unsymmetrical, and already possess a dipole, so one does not have to be induced by the π bond, as in the case of bromine. Because of the inductive effect, Markovnikov addition is usually observed.

With HBr:

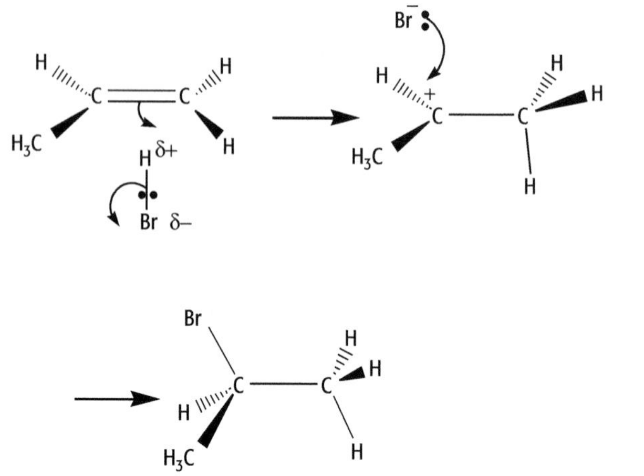

[Mechanism diagram showing HBr addition to propene giving 2-bromopropane.]

With H_2O and acid:

[Mechanism diagram showing acid-catalysed hydration of propene via protonation, water attack on carbocation, and deprotonation to give propan-2-ol.]

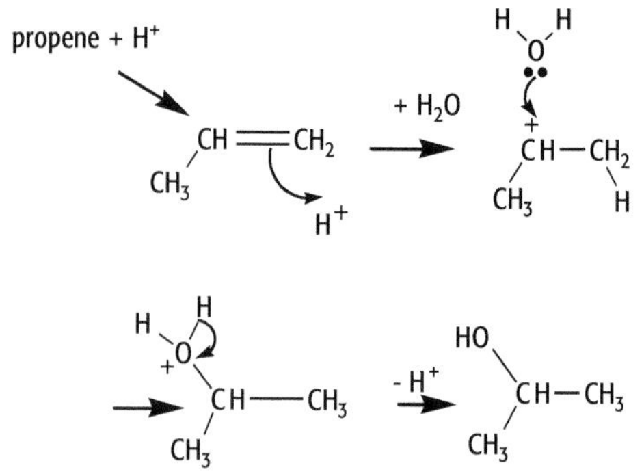

Although reacting alkenes with bromine in an organic solvent (like hexane) produces the dibromide, as outlined above, a significant second product in the reaction between an alkene and bromine water is the bromohydrin, due to the attack of a water molecule on the intermediate carbocation:

$$CH_2=CH_2 + Br_2 + H_2O \longrightarrow CH_2Br-CH_2OH + HBr$$

Progress check 15.03

1. Draw out the mechanism of the reaction between bromine water and propene, giving the bromohydrin, and predict which of the two possible bromohydrins, $CH_3CHBrCH_2OH$ or $CH_3CH(OH)CH_2Br$, will be formed.

2. Draw the structures of the products of the following reactions. Do not forget Markovnikov's rule!

    a $CH_2=CHCH_2CH_3 + HCl \longrightarrow$

    b $CH_2=C(CH_3)_2 + H_2O + H^+ \longrightarrow$

    c [cyclohexene structure] + HBr $\longrightarrow$

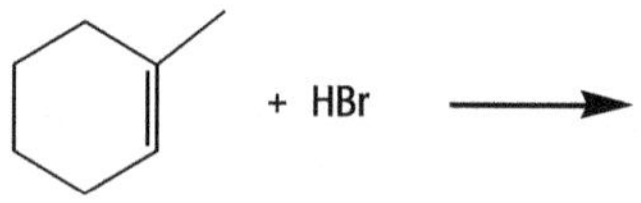

d Catalytic hydrogenation

Alkenes can be reduced to alkanes by reaction with hydrogen over finely divided platinum or nickel.

$$(CH_3)_2C=CHCH_3 + H_2 \xrightarrow{Pt + heat} (CH_3)_2CHCH_2CH_3$$

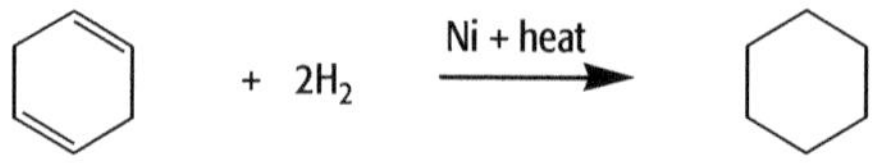

[benzene-like ring] + $2H_2$ $\xrightarrow{Ni + heat}$ [cyclohexane ring]

cyclohexa-1,4-diene
C_6H_8

cyclohexane
C_6H_{12}

e **The oxidation of alkenes**

i Alkenes burn well in oxygen or air:

$$CH_2=CH_2 + 3O_2 \longrightarrow 2CO_2 + 2H_2O$$

ii When shaken with alkenes at room temperature, purple dilute $KMnO_4$(aq) is decolorised. Like the reaction with bromine water, described above, this is an excellent test for the presence of a double bond. The product is the diol:

$$CH_3CH=CH_2 + H_2O + [O]$$
(oxygen atom from $KMnO_4$)
$$\downarrow$$
$$CH_3CH(OH)\text{-}CH_2OH$$
propan-1,2-diol

iii If the potassium manganate(VII) is hot, concentrated and acidified, cleavage of the double bond occurs. The types of compound formed depend on the extent of substitution there is at each end of the double bond:

$$CH_3CH=CH_2 + 5[O] \longrightarrow$$
$$CH_3CO_2H + CO_2 + H_2O$$

$$(CH_3)_2C=CHCH_2CH_3 + 3[O] \longrightarrow$$
$$(CH_3)_2C=O + CH_3CH_2CO_2H$$

Table 15.03 lists the possible products.

If the double bond carbon is ...	Then the oxidised product will be ...
$CH_2=$	CO_2
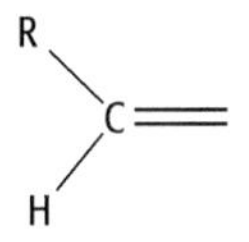 R\C= /H	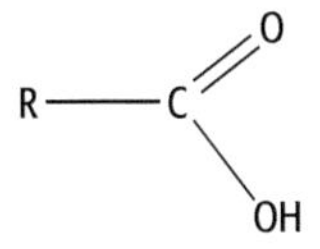 R—C(=O)OH (carboxylic acid)
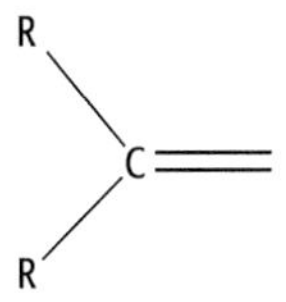 R\C= /R	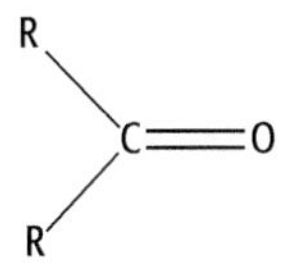 R\C=O /R (ketone)

Table 15.03 The compounds produced from the complete oxidation of C=C bonds

TIP

Note that *neither* aldehydes such as R–CHO, *nor* methanal (CH_2O), *nor* methanoic acid (HCO_2H) are formed. These always get oxidised further. Check that you do not include any of these in the products of $KMnO_4$ oxidations of alkenes.

This reaction is very useful in working out the structure of a compound from its cleavage products. For example, if it is known that the oxidation of compound **X**, C_6H_{12}, with hot $KMnO_4$ gives 3-methylbutanone and ethanoic acid, we can work out that **X** is 3,4-dimethylpent-2-ene:

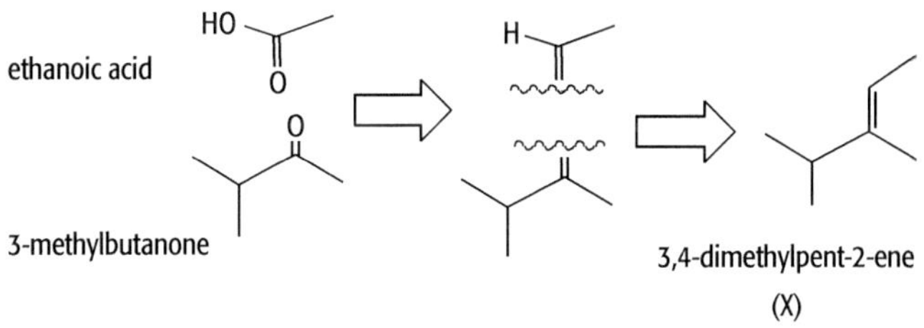

ethanoic acid

3-methylbutanone

3,4-dimethylpent-2-ene (X)

If the double bond is part of a ring, no loss of carbon atoms occurs, although the C=C is still cleaved:

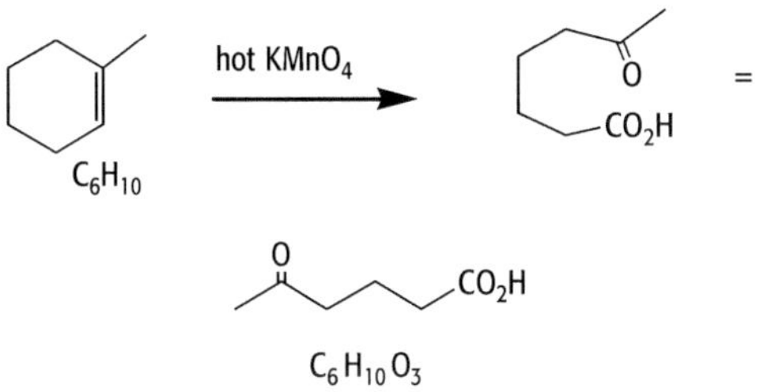

C_6H_{10} → hot $KMnO_4$ →

$C_6H_{10}O_3$

Progress check 15.04

1. Draw the structures of the products of reaction between hot $KMnO_4$ and each of the following isomers of 3,4-dimethylpent-2-ene (**X**):

   a 2,3-dimethylpent-2-ene, $(CH_3)_2C=C(CH_3)CH_2CH_3$

   b 2,3-dimethylpent-1-ene, $CH_3CH_2CH(CH_3)C(CH_3)=CH_2$

   c 2-methylhex-3-ene, $(CH_3)_2CHCH=CHCH_2CH_3$

2. Work out the structures of **Y** and **Z**:

   a Compound **Y**, C_6H_{10}, reacts with hot $KMnO_4$ to give 3-methylpentan-1,5-dioic acid ($HO_2C–CH_2–CH(CH_3)–CH_2–CO_2H$).

   b Compound **Z**, $C_{10}H_{18}$, reacts with hot $KMnO_4$ to give 4-methylcyclohexanone ($C_7H_{12}O$) and propanone (C_3H_6O).

15.06 The polymerisation of alkenes

Industrially, one of the most useful reactions of alkenes is their addition polymerisation.

Polyalkenes are called **addition polymers** because they are made by *adding* monomer units to a growing alkyl chain, and the molecular formula of an addition polymer is a multiple of the molecular formula of the monomer – both have the same empirical formula.

Addition polymerisation is a chain reaction in which the alkene monomer units are added one-by-one to the growing polymer chain. The reaction is initiated by a catalyst – either an organic peroxide or an organo-aluminium compound. If a peroxide is used to initiate the polymerisation of ethene, low density poly(ethene), LDPE, is formed:

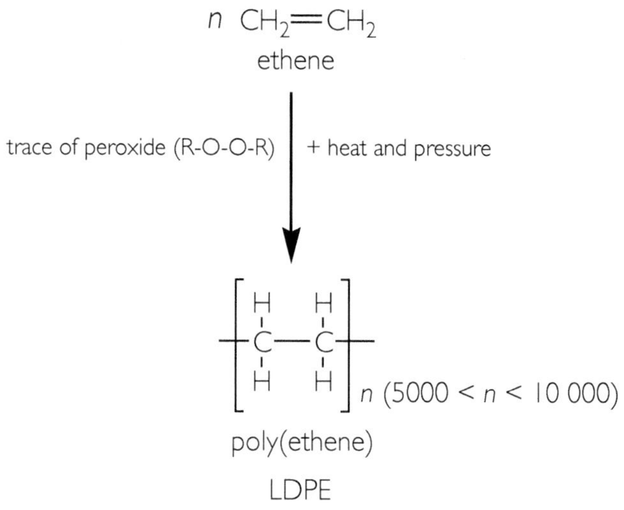

$$n\ CH_2=CH_2 \xrightarrow{\text{trace of peroxide (R-O-O-R)} + \text{heat and pressure}} {\Large[}\!\!\begin{array}{cc} H & H \\ | & | \\ -C-C- \\ | & | \\ H & H \end{array}\!\!{\Large]}_n \quad (5000 < n < 10\,000)$$

poly(ethene)
LDPE

However, if triethylaluminium is the catalyst, the average chain length increases, and high density poly(ethene), HDPE, is formed:

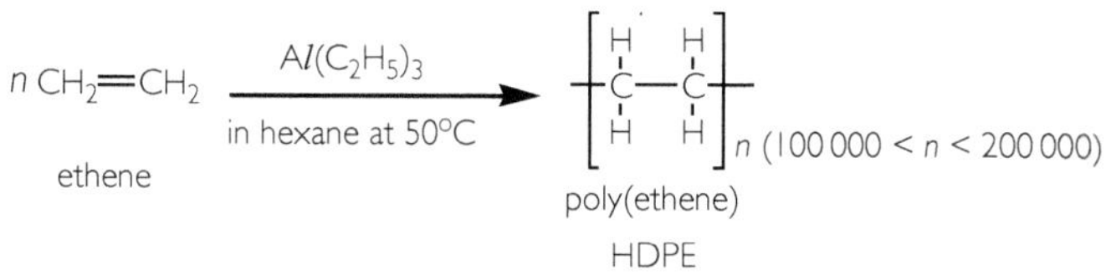

$$n\ CH_2=CH_2 \xrightarrow[\text{in hexane at } 50°C]{Al(C_2H_5)_3} {\Large[}\!\!\begin{array}{cc} H & H \\ | & | \\ -C-C- \\ | & | \\ H & H \end{array}\!\!{\Large]}_n \quad (100\,000 < n < 200\,000)$$

poly(ethene)
HDPE

The van der Waals' forces between the chains in LDPE are less than those in HDPE. This is reflected in their different properties and uses (Table 15.04).

Property / use	Polymer	
	LDPE	**HDPE**
density	low	high
m. pt.	approx. 130 °C	approx. 160 °C
tensile strength	low	higher
flexibility	very flexible	much more rigid
uses	polythene bags, electrical insulation, dustbin liners	bottles, buckets, crates

Table 15.04 Some properties and uses of poly(ethene).

A variety of important polymers can be obtained by polymerising substituted ethenes. Representative polymers are listed in Table 15.05.

Monomer	Structure of polymer (two repeat units shown) and name	Uses
propene, $CH_3-CH=CH_2$	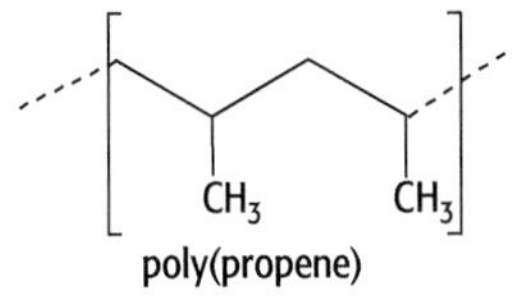 poly(propene)	packaging, ropes, thermal clothing, carpets, car components, containers
chloroethene, $CH_2=CH-Cl$	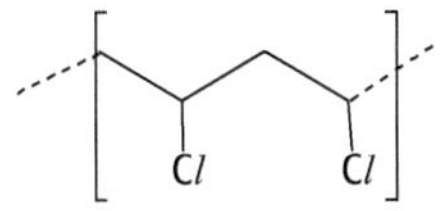 poly(chloroethene) or PVC	water pipes, gutters, upholstery, rubber substitute
phenylethene, 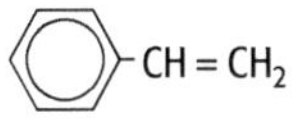 $C_6H_5-CH=CH_2$	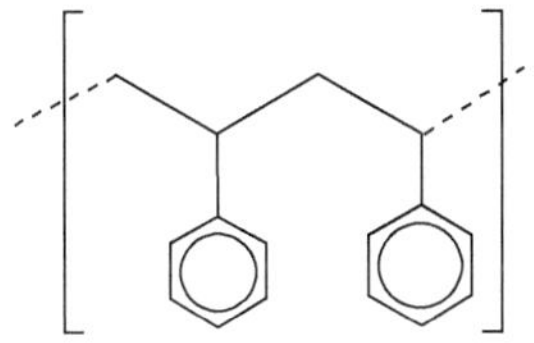 poly(phenylethene), polystyrene	*solid*: disposable cutlery, plastic models, CD and DVD cases *expanded*: packing materials, insulation, foam drinks cups
tetrafluoroethene, $CF_2=CF_2$	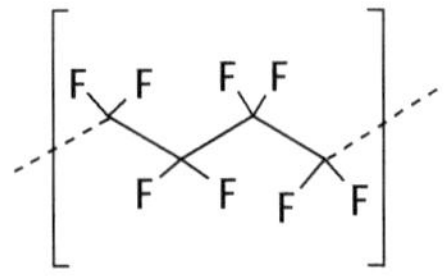 poly(tetrafluoroethene), PTFE	non-stick surfaces for cooking ware; bridge bearings, non-corrosive tubing

Table 15.05 Some other addition polymers and their uses

Worked example 15.02

Draw a section of the polymer chain, showing *two* repeat units, of the polymer formed from each of the following monomers.

a $CH_2=CH-CN$

b $CH_3CH=CHCO_2H$

How to get the answer:

Step 1: Break the double bond, and create two 'free' bonds.

Step 2: Join these free bonds to adjacent monomers.

In general, the polymer will have its side-chains alternating along the chain.

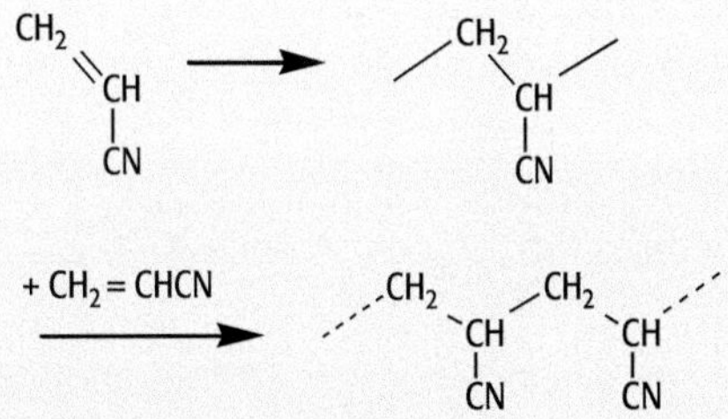

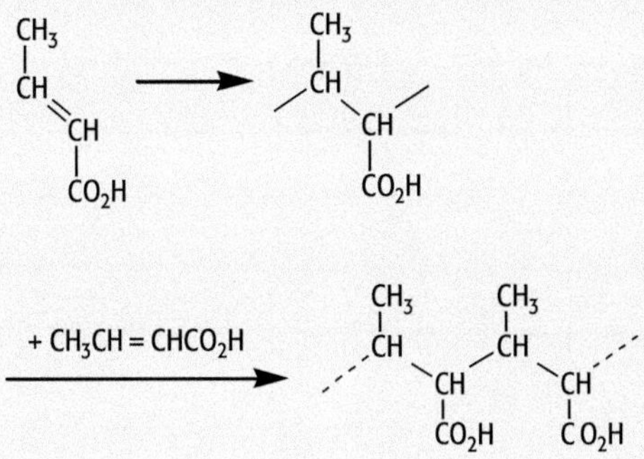

15.07 The disposal of polymers

Polyalkenes are chemically inert and biodegrade only very slowly, so many addition polymers can cause environmental problems when their useful lives are over.

Four methods are available for their safe disposal.

a Incineration
Under the right conditions, incineration can provide useful energy for space heating or power generation. A high temperature of combustion is needed if poisonous combustion products, such as dioxins, are to be avoided. The fumes from the incineration chamber need washing with water to absorb noxious combustion products such as HCl from the burning of PVC.

b Depolymerisation
Heating polymers to high temperatures in the absence of air or oxygen (a process known as pyrolysis) can cause their chains to break up into alkene units. The process is similar to the cracking of long-chain hydrocarbons into alkenes. The monomers produced can be separated by fractional distillation and re-used.

c Recycling
Many polymers can be melted and remoulded.

d Bacterial fermentation
Under the right conditions some polymers can be degraded quite quickly by certain strains of bacteria.

Sample answer

Question:

Hydrocarbon **X**, C_4H_{10}, undergoes the following series of reactions.

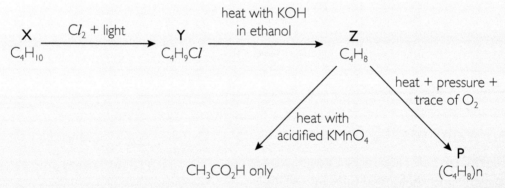

Suggest structures for the compounds **X, Y, Z** and **P**. [4]

Answer:

Z: $CH_3-CH=CH-CH_3$ [1 mark]

Y: $CH_3CH_2CHClCH_3$ [1 mark]

X: $CH_3CH_2CH_2CH_3$ [1 mark]

P: poly(but-2-ene), [structure shown] [1 mark]

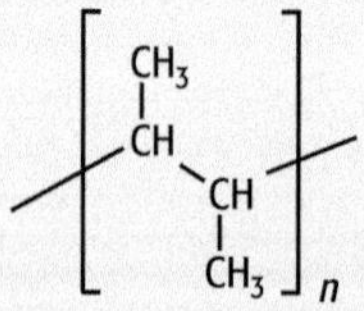

Revision checklist

Check that you know the following:

- [ ] Alkanes are fairly unreactive compounds, and are not affected by common acids, alkalis or oxidising agents.

- [ ] Alkanes undergo three important types of reaction:
  - combustion, giving CO_2 and steam, e.g.
    $C_3H_8 + 5O_2 \longrightarrow 3CO_2 + 4H_2O$
    note that incomplete combustion can produce CO or C(s) (soot)
  - cracking, where long-chain alkanes are broken into shorter chain alkanes and alkenes, e.g. $C_{10}H_{22} \longrightarrow 2C_2H_4 + C_6H_{14}$
  - free-radical substitution with chlorine or bromine, giving, respectively, chloroalkanes or bromoalkanes, e.g.
    $C_2H_6 + Br_2 \longrightarrow C_2H_5Br + HBr$

- [ ] The conditions under which these reactions take place, and the mechanism of free-radical substitution reactions.

- [ ] explain the use of crude oil as a source of hydrocarbons

- [ ] The main reaction of alkenes is electrophilic addition:
  - with Br_2 they give dibromoalkanes, e.g.
    $CH_3CH=CH_2 + Br_2 \longrightarrow CH_3CHBrCH_2Br$
  - with HBr they give monobromoalkanes, e.g.
    $CH_3CH=CH_2 + HBr \longrightarrow CH_3CHBrCH_3$
  - with acidified water they give alcohols, e.g.
    $CH_3CH=CH_2 + H_2O \longrightarrow CH_3CH(OH)CH_3$

- [ ] The conditions under which these reactions take place, the application of Markovnikov's rule, and the mechanism of electrophilic addition reactions.

- [ ] Alkenes are oxidised by $KMnO_4$.
  - In the cold $CH_3CH=CH_2 + H_2O + [O] \longrightarrow CH_3CH(OH)CH_2(OH)$
  - When heated the double bond is completely broken, to give CO_2, ketones or acids, e.g.
    $CH_3CH=CH_2 + 5[O] \longrightarrow CH_3CO_2H + CO_2 + H_2O$

- [ ] Alkenes polymerise to give polyalkenes, e.g.
  $nCH_3CH=CH_2 \longrightarrow [-CH(CH_3)-CH_2-]_n$

- [ ] The various ways of disposing of polymers.

- [ ] alkenes are reduced to alkanes by catalytic hydrogenation using H_2 + Ni or Pt

- [ ] describe and explain the inductive effect of alkyl groups on the stability of cations formed during electrophilic substitution''

- [ ] recognise the difficulty of the disposal of poly(alkenes)

Exam-style questions

1. Crude oil is a mixture of hydrocarbons with molecules containing between 1 and about 40 carbon atoms. The first stage of converting it into useful substances is fractional distillation.

   a. Explain the meaning of the terms *hydrocarbon* and *fractional distillation*. [2]

   b. State the uses of *two* fractions from the fractional distillation of crude oil. For each fraction state the approximate chain length of the hydrocarbons it contains. [2]

   Dodecane, $C_{12}H_{26}$, is an alkane that occurs in crude oil. It can be 'cracked' into smaller alkanes and alkenes.

   c. State the general formulae of alkanes and alkenes. [2]

   d. i State the conditions under which cracking reactions take place.

      ii Write a balanced equation for the cracking of dodecane into hexane and ethene. [2]

e   There are several structural isomers of hexane, C_6H_{14}.

   i   Explain the meaning of the term *structural isomers*.

   ii How many structural isomers are there with the formula C_6H_{14}?

   iii Draw the skeletal formulae of three of these isomers. [4]

Total: 12

2 Alkanes undergo *free-radical substitution* reactions with chlorine.

   a Explain the meaning of the term *free radical*. [1]

   b State the conditions required for free-radical substitution reactions to occur. [1]

   c Write an equation for the reaction between ethane and chlorine. [1]

   d Describe the mechanism of this reaction, labelling all the steps involved. [4]

   e When 2-methylpropane reacts with chlorine, two different compounds with the formula C_3H_7Cl are formed.

      i   Draw structures for the two compounds.

      ii Suggest the ratio of the amounts of the two compounds produced. Explain your answer. [3]

Total: 10

3 a State Markovnikov's rule. [1]

   b Suggest the products of the following reactions of three isomers with the formula C_5H_{10}.

      i $(CH_3)_2CHCH=CH_2$ warmed with dilute sulfuric acid

      ii $(CH_3)_2C=CHCH_3$ treated with HBr(g)

      iii $CH_2=C(CH_3)CH_2CH_3$ treated with Br_2(aq). [3]

   c Some alkenes can show geometrical isomerism. State which, if any, of the above three alkenes show geometrical isomerism. Explain your answer, and draw the structures of any *cis–trans* isomers. [2]

   d The above three isomers can be distinguished by the use of hot acidified $KMnO_4$. State the products that each isomer would give on oxidation by this reagent. [5]

Total: 11

Unit 16

Halogenoalkanes

Learning outcomes

You should be able to:

- write equations to describe the following reactions of halogenoalkanes:
  - nucleophilic substitution, by reacting with $OH^-(aq)$ or water (hydrolysis), NaCN and NH_3
  - elimination, by reacting with ethanolic KOH
- describe the S_N1 and S_N2 mechanisms of nucleophilic substitution in halogenoalkanes, and explain why primary halogenoalkanes react via the S_N2 mechanism but tertiary halogenoalkanes react by the S_N1 mechanism
- explain how the reactivity of a halogenoalkane relates to the strength of the C–Hal bond
- explain some of the uses of fluoroalkanes and how they relate to their lack of chemical reactivity
- recognise the concern about the effect of chlorofluoroalkanes on the ozone layer

16.01 Properties, isomerism and nomenclature

The halogenoalkanes are compounds in which one or more halogen atoms are bonded to an alkyl chain.

The main intermolecular force in halogenoalkanes is the van der Waals' induced dipole force. Their boiling points therefore increase with chain length, just like those of the alkanes.

Halogenoalkanes show chain, positional and optical isomerism. They are named as derivatives of the parent alkanes, by appending the prefix 'chloro', 'bromo', etc. to the alkane name. We can illustrate this by looking at the five isomers with the formula C_4H_9Br, shown below.

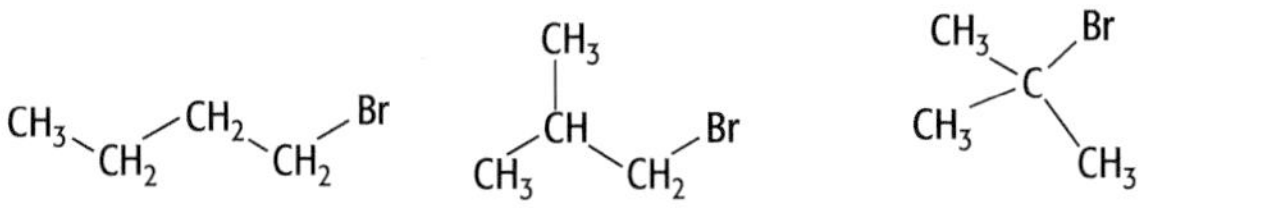

$CH_3CH_2CH_2CH_2Br$ — 1-bromobutane

$(CH_3)_2CHCH_2Br$ — 1-bromo-2-methylpropane

$(CH_3)_3CBr$ — 2-bromo-2-methylpropane

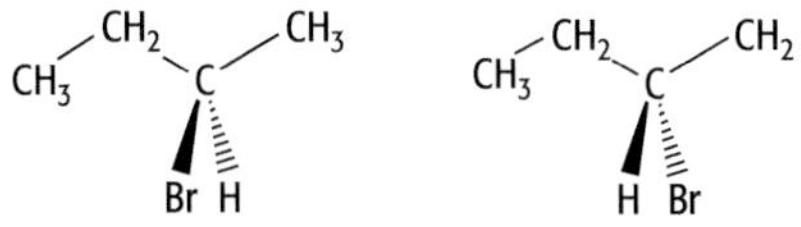

optical isomers of 2-bromobutane (I and II)

We can classify halogenoalkanes as **primary**, **secondary** or **tertiary**, depending on how many alkyl groups are attached to the carbon atom joined to the halogen atom.

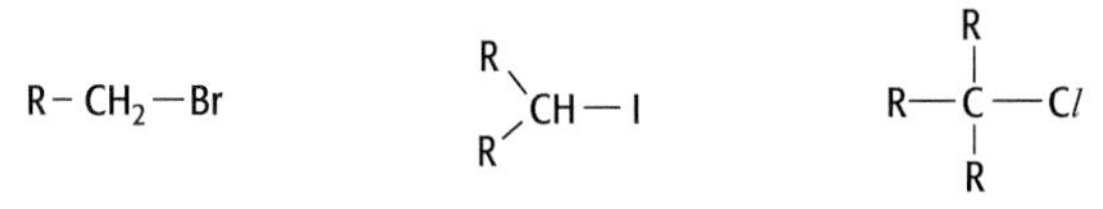

$R-CH_2-Br$ — a primary bromoalkane

R_2CH-I — a secondary iodoalkane

R_3C-Cl — a tertiary chloroalkane

16.02 Methods of preparation

There are three ways of preparing halogenoalkanes.

a **Halogenation of alkanes by free-radical substitution** (see Unit 15)

$C_2H_6 + Br_2 \xrightarrow{light} CH_3CH_2Br + HBr$

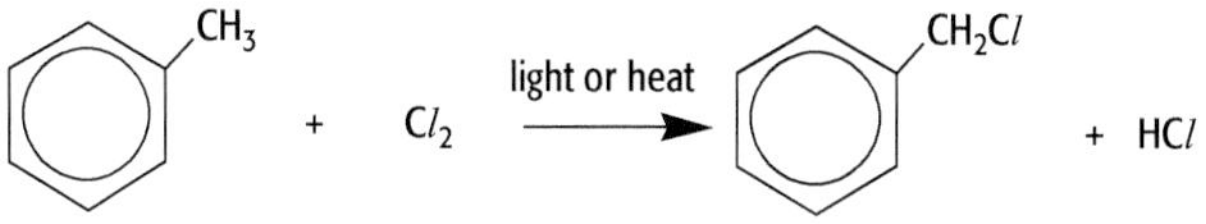

$C_6H_5CH_3 + Cl_2 \xrightarrow{light\ or\ heat} C_6H_5CH_2Cl + HCl$

b **Electrophilic addition to alkenes**

The addition of chlorine or bromine to alkenes gives dihaloalkanes (see Unit 15).

$CH_3CH_2CH=CH_2 + Br_2 \longrightarrow CH_3CH_2CHBrCH_2Br$

The addition of hydrogen halides to alkenes gives monohaloalkanes. The orientation of the halogen follows Markovnikov's rule.

$CH_3CH_2CH=CH_2 + HBr \longrightarrow CH_3CH_2CHBrCH_3$

c Nucleophilic substitution of alcohols

A variety of reagents can be used to make halogenoalkanes from alcohols.

$$CH_3CH_2OH + PCl_5 \xrightarrow{warm} CH_3CH_2Cl + HCl + POCl_3$$

$$CH_3CH_2OH + SOCl_2 \xrightarrow{warm} CH_3CH_2Cl + HCl + SO_2$$

$$CH_3CH_2OH + HBr \xrightarrow{conc.\ H_2SO_4\ +\ NaBr\ +\ heat} CH_3CH_2Br + H_2O$$

16.03 Nucleophilic substitution reactions

Halogens are more electronegative than carbon, and so the C–Br bond in a bromoalkane is strongly polarised $C^{\delta+}$–$Br^{\delta-}$. During the reactions of bromoalkanes, this bond breaks heterolytically to give the bromide ion:

Step 1

The C^+ formed is then liable to attack by nucleophiles (anions or neutral molecules that possess a lone pair of electrons). We shall use hydroxide ion as the nucleophile to describe the mechanism.

Step 2

Sometimes, however, OH^- will attack the $C^{\delta+}$ carbon before the bromide ion has left:

Step 3

TIP
Do not forget to include all relevant electron pairs and dipoles ($\delta+$ and $\delta-$) when drawing reaction mechanisms. Your 'curly arrows' should always be drawn accurately, showing exactly where the electron pairs are coming from and going to.

Whether the nucleophile attacks before or after the bromide ion has left depends on whether the halogenoalkane is primary, secondary or tertiary. We can explain this as follows.

Steps 1 and 2 together describe the S_N1 mechanism. The 'S' stands for substitution; the '$_N$' stands for nucleophilic; and the '1' means that only one molecule is involved in the slow step of the reaction. Step 1 is likely to be much slower than step 2, since it involves the breaking of a fairly strong bond and the separation of opposite charges – both of which are endothermic processes.

Its rate therefore depends only on how much bromoalkane there is, and not how much OH^-. If $[OH^-]$ is increased, the rate of step 2 would increase, but the rate of the overall reaction would not increase, as step 1 would still be the slower step.

Reaction 3, on the other hand, involves a single step that includes both RBr and OH^-, so the rate of reaction depends on how much there is of each reactant. This is called the S_N2 mechanism. The 'S' and the '$_N$' mean the same as they did for the S_N1 reaction, but the '2' now means that two species are involved in the slow (and only) step of the reaction.

There are two factors that determine whether a bromoalkane reacts via the S_N1 or the S_N2 mechanism, and both work in the same direction, encouraging primary bromoalkanes to react via the S_N2 mechanism, and tertiary bromoalkanes to react via the S_N1 mechanism. (Secondary bromoalkanes usually react via a mixture of the two mechanisms.)

As we saw in Unit 15, when we looked at the Markovnikov addition of HBr to alkenes, alkyl groups are electron-donating, and so their presence stabilises **carbocations** by spreading out the positive charge. Since the slow step in an S_N1 reaction is the formation of a carbocation, this mechanism is favoured in the case of tertiary bromides.

3 electron-donating alkyl groups stabilise the carbocation

On the other hand, the 'half-way-house' (called the *transition state*) in an S_N2 reaction involves squeezing five groups around the central carbon atom. This is less favoured with tertiary bromoalkanes, which have three alkyl groups on the central carbon, compared to primary bromoalkanes, which have only one alkyl group.

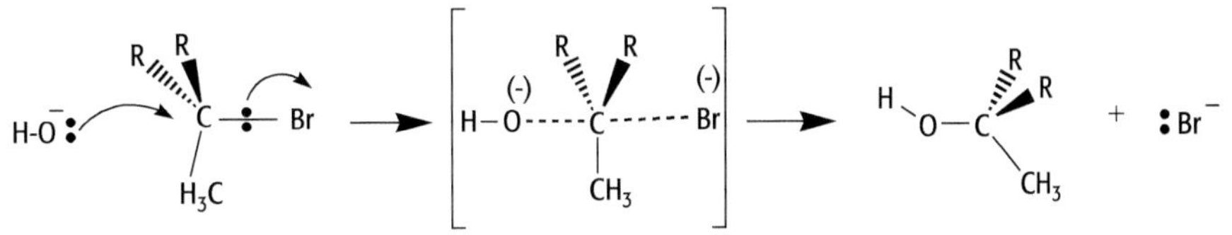

If R = alkyl, this transition state is more hindered, so is less stable, than if R = H.

These conclusions can be summarised in Table 16.01.

Type of bromoalkane	Mechanism	Reasons
primary	S_N2	Less steric hindrance in 5-coordinated transition state. Little stabilisation of carbocation by just one alkyl group.
secondary	mix of S_N1 and S_N2	A fair amount of steric hindrance in 5-coordinated transition state. Only poor stabilisation of carbocation by two alkyl groups.
tertiary	S_N1	Large amount of steric hindrance in 5-coordinated transition state. Good stabilisation of carbocation by three electron-donating alkyl groups.

Table 16.01 The relationship between mechanism and structure of halogenoalkanes

Progress check 16.01

Predict whether each of the following compounds will react with OH^- by an S_N1 or S_N2 mechanism, or a mixture of the two.

1. 2-chloropentane
2. 3-bromopentane
3. 1-bromobutane
4. 3-chloro-3-methylpentane.

A number of different nucleophiles undergo nucleophilic substitution with bromoalkanes. Some of them are listed in Table 16.02.

Nucleophile		Reagents and conditions	Products when reacted with bromoethane
Name	Formula		
water	H_2O:	heat in water	$CH_3CH_2OH + HBr$
hydroxide ion	$HO:^-$	heat with NaOH(aq)	$CH_3CH_2OH + Br^-$
ammonia	$:NH_3$	heat under pressure in a sealed tube with NH_3 in ethanol	$CH_3CH_2NH_2 + HBr$
cyanide ion	$^-:C\equiv N$	heat in ethanol with NaCN	$CH_3CH_2CN + Br^-$

Table 16.02 The various nucleophilic substitution reactions of halogenoalkanes

> **TIP**
>
> Be especially careful to remember the exact reagents and conditions for these reactions. Under different conditions, hydroxide ions can cause elimination reactions to occur with bromoalkanes (see later in this unit); and cyanide ions are also used as a homogeneous catalyst in nucleophilic addition reactions with carbonyl compounds (see unit 18).

Worked example 16.01

For each of the following compounds, suggest a suitable starting bromoalkane, and give the reagents and conditions needed to make the compound.

a $(CH_3)_2CHCH_2NH_2$

b $CH_3CH_2CH(OH)CH_3$

How to get the answer:

a **Step 1:** To make a primary amine, R–NH_2, we need a bromoalkane with the same carbon skeleton, so the starting material will be $(CH_3)_2CHCH_2Br$.

   Step 2: The reagent and conditions we need are: *heat with NH_3 under pressure in ethanol* (pressure is needed because NH_3 is a gas at atmospheric pressure, so would escape from the heated solution before it had time to react).

b **Step 1:** The starting material will have the same carbon skeleton as the alcohol product: $CH_3CH_2CHBrCH_3$.

   Step 2: We need to treat this 2-bromobutane with *hot NaOH(aq)*.

Progress check 16.02

Write balanced equations, including the structures of the products, for the reactions between:

1. 1-bromopropane and sodium hydroxide
2. 2-bromobutane and sodium cyanide
3. 2-chloropropane and ammonia.

The product from the reaction of a halogenoalkane and cyanide ion is called a **nitrile**.

Nitriles are important in organic synthesis, as they provide either carboxylic acids on hydrolysis or primary amines on reduction:

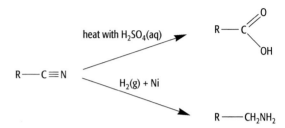

The relative rates of hydrolysis of chloro-, bromo- and iodoalkanes can be compared by their reaction with silver nitrate in an ethanol–water mixed solvent. The halogenoalkane slowly hydrolyses, releasing the halide ion, which reacts with Ag^+ ions to form a silver halide precipitate:

$$R-X + H_2O + Ag^+ \longrightarrow R-OH + H^+ + AgX(s)$$

The results are shown in Table 16.03.

Compound	Observation on reaction with $AgNO_3 + H_2O$ in ethanol	Colour of precipitate
$(CH_3)_2CH-Cl$	slight cloudiness after 1 hour	white
$(CH_3)_2CH-Br$	cloudiness appears after 10–20 minutes	pale cream
$(CH_3)_2CH-I$	thick precipitate appears within a minute	pale yellow

Table 16.03 How silver nitrate is used to differentiate between halogenoalkanes

This trend in reactivity corresponds to the strengths of the C–X bond (Table 16.04). The weaker the bond (i.e. the smaller the bond enthalpy), the easier it is to break, so the faster the reaction.

Bond	Bond enthalpy / kJ mol^{-1}
C–F	485
C–Cl	338
C–Br	285
C–I	213

Table 16.04 C–halogen bond enthalpies

16.04 Elimination reactions

In section 16.03 above, we saw hydroxide ions acting as a nucleophile towards a $\delta+$ carbon atom. Hydroxide ions can also act as bases, taking protons away from other compounds, for example:

$$HO^- + HCl \longrightarrow H_2O + Cl^-$$

Under the right conditions, hydroxide ions can take a proton off a bromoalkane, which causes an elimination reaction to occur, forming an alkene.

$$OH^- + CH_3CH_2Br \longrightarrow H_2O + CH_2=CH_2 + Br^-$$

The mechanism is as follows:

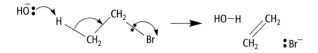

In general:

- elimination reactions have higher activation energies, so are favoured by higher temperatures
- they are less prone to steric hindrance, so are favoured with tertiary bromoalkanes
- they involve a greater spreading-out of charge in the transition state than either the S_N1 or S_N2 reactions, so are favoured by less polar solvents such as ethanol

The usual conditions for an elimination reaction are to *heat RBr under reflux with KOH in ethanol* (KOH is more soluble in ethanol than is NaOH).

Progress check 16.03

Suggest suitable bromoalkanes, reagents and reaction conditions for the synthesis of the following compounds.

1. $CH_3CH_2C(CH_3)CH_2CN$
2. $CH_3C(CH_3)CHCH_2NH_2$
3. $CH_3CH=CHCH_3$
4. $CH_2=CHCH_2CH_3$.

Sample answer

Question:

Suggest structures for compounds **A–D** in the following scheme, and state the reagents and conditions for reactions 2 and 3. [6]

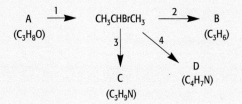

Answer:

A is $CH_3CH(OH)CH_3$ [1 mark]

B is $CH_3CH=CH_2$ [1 mark]

C is $(CH_3)_2CH-NH_2$ [1 mark]

D is $(CH_3)_2CH-CN$ [1 mark]

Reagent and conditions for reaction 2: heat with KOH in ethanol [1 mark]

Reagent and conditions for reaction 3: heat with NH_3 in ethanol under pressure [1 mark]

16.05 Useful halogeno compounds and their environmental impact

Halothane ($CF_3CHBrCl$), isoflurane ($CF_3CHCl-O-CHF_2$) and sevoflurane ($(CF_3)_2CH-O-CH_2F$) are important anaesthetics, used in surgery the world over.

The halons, such as $CBrClF_2$, are useful fire extinguishers. All of these contain many fluorine atoms: the C–F bond is very strong, and so is quite unreactive.

Many polychloroalkanes such as 1,1,1-trichloroethane (CH_3CCl_3) are used as solvents.

Polychloroethene (PVC) is a very useful plastic, with a variety of uses (see Unit 15). Once finished with, it has to be disposed of with care: incineration under the wrong conditions can produce irritant HCl(g), or the highly poisonous tetrachlorodioxins.

CFCs (chlorofluorocarbons) are inert, volatile liquids that at one time found favour as refrigerator fluids and aerosol propellants. However, it was found that, once released into the atmosphere, they diffused into the stratosphere, where ultraviolet radiation from the Sun decomposed them into chlorine atoms. These free radicals catalysed the destruction of the ozone layer, which is important in filtering out the harmful UV radiation from the Sun.

$$CF_3Cl \xrightarrow{\text{ultraviolet light}} CF_3^\bullet + Cl^\bullet$$

Ozone is formed from oxygen molecules by the action of UV light:

$$O_2 \longrightarrow 2O$$
$$O + O_2 \longrightarrow O_3$$

The chlorine atoms produced from CFCs act as homogeneous catalysts (see Unit 9), decomposing ozone, O_3, into oxygen, O_2, by a free radical chain reaction:

$$Cl^\bullet + O_3 \longrightarrow ClO^\bullet + O_2$$
$$ClO^\bullet + O \longrightarrow Cl^\bullet + O_2$$

It has been estimated that one chlorine atom can destroy over 10^5 ozone molecules before it eventually diffuses back into the lower atmosphere. There it can react with water vapour to produce hydrogen chloride, which can be flushed out by rain as dilute hydrochloric acid.

Other stratospheric pollutants such as nitric oxide from high-flying aircraft also destroy ozone.

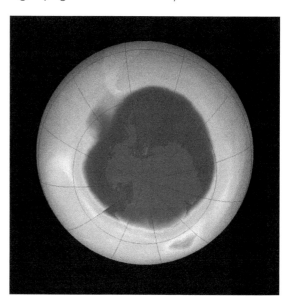

Figure 16.01 The depleted ozone layer over the Antarctic

Revision checklist

Check that you know the following:

- How to work out what structural isomers are possible for a given halogenoalkane molecular formula, and how to name them.
- How to make halogenoalkanes from alkanes, alkenes and alcohols.
- How halogenoalkanes react with hydroxide ions (two ways), ammonia and cyanide ions.
- Describe the mechanisms of the S_N1 and the S_N2 reactions, and which types of halogenoalkane react by each mechanism.
- How $AgNO_3(aq)$ can be used to distinguish between the halogenoalkanes RCl, RBr and RI.
- Some uses of halogenoalkanes.
- How CFCs destroy the ozone layer.

Exam-style questions

1. Study the following scheme showing some reactions of 2-bromobutane and answer the questions below.

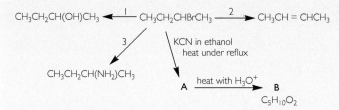

   a Name the type of reaction undergone in reactions 1 and 2. [2]
   b State the reagents and the conditions for reactions 1, 2 and 3. [6]
   c Suggest the structures of compounds **A** and **B**. [2]

   Total: 10

2. a Chloroethane, bromoethane and iodoethane differ in their reactivities in nucleophilic substitution reactions.

   i State the trend in reactivity, and explain the difference. [2]
   ii Describe how $AgNO_3$(aq) can be used to investigate this difference in reactivity. You should describe the observations you would make with each halogenoalkane. [3]

   b Chlorofluorocarbons (CFCs), such as dichlorodifluoromethane, CCl_2F_2, were once used as aerosol propellants.

   i State two reasons why CFCs were used as aerosol propellants. [2]
   ii Explain what happens to CFCs in the upper atmosphere in the presence of ultraviolet light, and explain why they are harmful to the ozone layer. Use CCl_2F_2 as an example of a CFC in any equations you write. [3]

   Total: 10

Alcohols, esters and carboxylic acids

Learning outcomes

You should be able to:

- use and interpret the general, structural, displayed and skeletal formulae of alcohols, esters and carboxylic acids

- recall the chemistry of alcohols in the following reactions: combustion, substitution to give halogenoalkanes, reaction with sodium, oxidation to give carbonyl compounds and carboxylic acids, dehydration to alkenes, formation of esters

- describe how alcohols are classified into primary, secondary and tertiary, and how these three classes of alcohol can be distinguished

- deduce the presence of a $CH_3CH(OH)-$ group in an alcohol from its reaction with alkaline aqueous iodine to form tri-iodomethane

- describe the formation of carboxylic acids from alcohols, aldehydes and nitriles

- describe the reactions of carboxylic acids in the formation of salts, esters and alcohols

- state the major commercial uses of esters as solvents, perfumes, flavourings

- Describe the acid and base hydrolysis of esters

17.01 Formulae and properties

The alcohols form a homologous series with the general formula $C_nH_{2n+1}OH$. Carboxylic acids have the general formula $C_nH_{2n+1}CO_2H$.

Alcohols with small M_r values are totally miscible with water, due to hydrogen bonding between the alcohol molecules and water molecules:

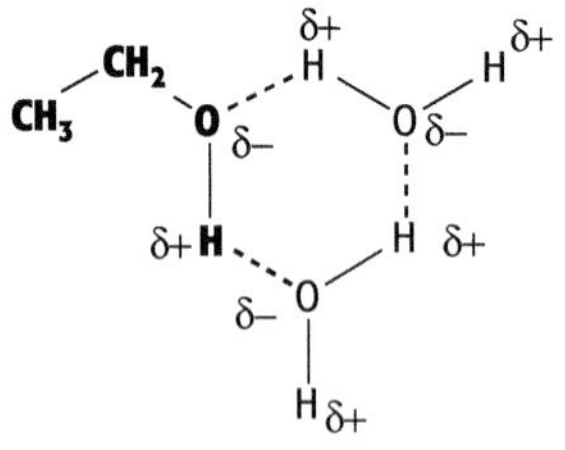

The boiling points of the alcohols are all much higher than alkanes with similar M_r values, due to strong intermolecular hydrogen bonding:

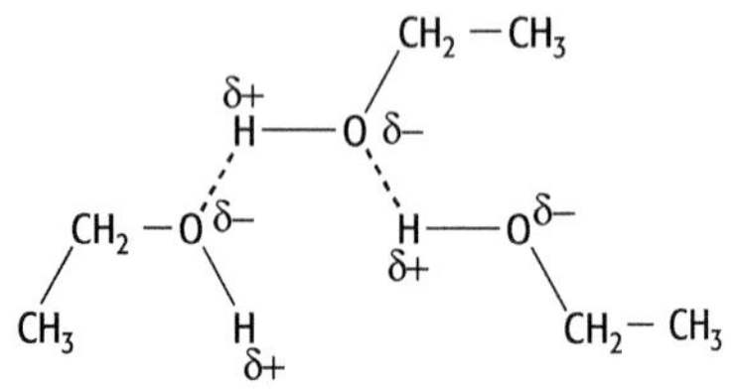

Two carboxylic acid molecules readily form dimers:

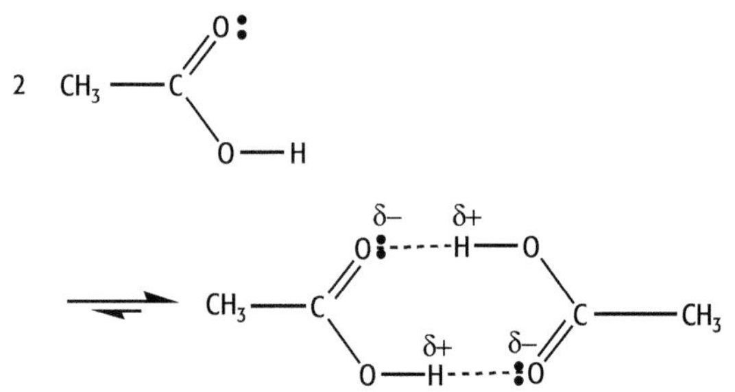

Carboxylic acids contain three sites for hydrogen bonding to water molecules:

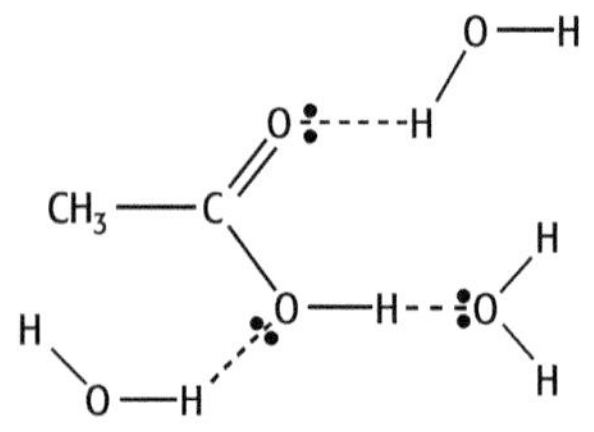

As a result of the electron-donating properties of alkyl chains, alcohols are slightly less acidic than water, because the resulting alkoxide ion is less stable than is the hydroxide ion.

As their name suggests, carboxylic acids are acidic, though they are generally weak acids, incompletely ionised in aqueous solution:

$CH_3CO_2H(aq) \rightleftharpoons CH_3CO_2^-(aq) + H^+(aq)$

In a $1.0 \, mol \, dm^{-3}$ solution of ethanoic acid, only about 1 molecule in 1000 (0.1%) is ionised.

Nevertheless, carboxylic acids are about 10^{11} times more acidic than alcohols. This is due to the ability of the carbonyl group to delocalise the negative charge in the anion, thus making the anion much more stable than the alkoxide ion from alcohols.

alcohol: R—O—H → R—O:⁻ + H⁺
anion not stabilised by delocalisation

carboxylic acid: R—C(=O)(O—H) → H⁺ + R—C(=O)(O⁻) ↔ R—C(O⁻)(=O)

or

R—C(O(½⁻))(O(½⁻))

anion stabilised by delocalisation

17.02 Isomerism and nomenclature

Alcohols are named by adding the suffix -ol to the name-stem of the alkane that has the same number of carbon atoms. A number is added before the '-ol' to show where on the chain the –OH group is situated. If there are side-chains in the molecule, their position is indicated by a number added before the side-chain prefix.

Alcohols show three types of isomerism. The isomers of C_4H_9OH can be used to illustrate these, and also their nomenclature.

Alcohols with the formula C_4H_9OH also illustrate the three *types* of alcohol.

a **Chain isomerism**

$CH_3CH_2CH_2CH_2OH$ butan-1-ol
$(CH_3)_2CHCH_2OH$ 2-methylpropan-1-ol

b **Positional isomerism**

$CH_3CH_2CH_2CH_2OH$ butan-1-ol
$CH_3CH_2CH(OH)CH_3$ butan-2-ol
$(CH_3)_2CHCH_2OH$ 2-methylpropan-1-ol
$(CH_3)_3COH$ 2-methylpropan-2-ol

c **Optical isomerism**

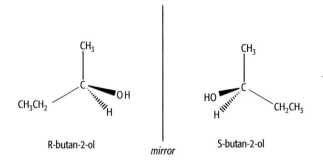

R-butan-2-ol   mirror   S-butan-2-ol

The two non-superimposable mirror images of butan-2-ol.

d **Primary alcohols**

These contain the —CH_2OH group.

Examples are:

butan-1-ol $CH_3CH_2CH_2CH_2OH$

2-methylpropan-1-ol $(CH_3)_2CHCH_2OH$

e **Secondary alcohols**

These contain the >CHOH group.

An example is:

butan-2-ol $CH_3CH_2CH(OH)CH_3$

f **Tertiary alcohols**

These contain the >C—OH group.

An example is:

2-methylpropan-2-ol $(CH_3)_3COH$

g **Functional group isomerism**

Both carboxylic acids and esters contain the carboxyl group, $-CO_2-$, and those with the same molecular formula are **functional group isomers** of each other. The following three compounds are isomers with the molecular formula $C_3H_6O_2$. The first is a carboxylic acid, and the other two are esters.

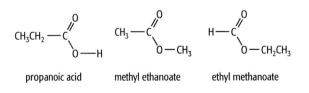

propanoic acid   methyl ethanoate   ethyl methanoate

Progress check 17.01

1. a Draw the skeletal formulae, and write down the names, of all the alcohols with the molecular formula $C_5H_{11}OH$.

   b Identify which of these are primary, secondary and tertiary alcohols.

   c How many of these alcohols contain chiral carbon atoms?

2. Draw the skeletal formulae, and write down the names, of all the compounds with the molecular formula $C_4H_8O_2$ that contain the carboxyl group, $-CO_2-$.

17.03 How are alcohols made?

a Nucleophilic substitution

Heating a halogenoalkane with aqueous sodium hydroxide produces an alcohol.

$$CH_3CH_2CH_2Br + NaOH(aq) \longrightarrow CH_3CH_2CH_2OH + NaBr(aq)$$

b Electrophilic addition

Ethene undergoes a hydration reaction when it is passed with steam over a phosphoric acid catalyst at an elevated temperature and pressure. This is the main industrial method of making ethanol:

$$CH_2=CH_2 + H_2O \xrightarrow{H_3PO_4 \text{ at } 300°C \text{ and } 70\,atm} CH_3CH_2OH$$

The hydration can also be carried out in the laboratory by absorbing the alkene in concentrated sulfuric acid, and then diluting with water:

$$CH_2=CH_2 + H_2SO_4 \longrightarrow CH_3CH_2-OSO_3H$$

$$CH_3CH_2-OSO_3H + H_2O \longrightarrow CH_3CH_2OH + H_2SO_4$$

If the alkene is not symmetrical, the addition follows Markovnikov's rule (see Unit 15), and the OH joins to the more substituted end of the double bond.

$$CH_3CH=CH_2 + H_2O \xrightarrow{H^+} CH_3CH(OH)CH_3$$

propene   propan-2-ol

c Reduction

Aldehydes and ketones can be reduced to alcohols by a variety of reducing agents:

- hydrogen gas over a nickel catalyst
- sodium borohydride, $NaBH_4$, in water
- lithium aluminium hydride ($LiAlH_4$) in dry ether

Aldehydes give primary alcohols:

$$(CH_3)_2CHCHO \xrightarrow{NaBH_4} (CH_3)_2CHCH_2OH$$

Ketones give secondary alcohols:

$$CH_3CH_2COCH_3 \xrightarrow{H_2 + Ni} CH_3CH_2CH(OH)CH_3$$

Carboxylic acids can be reduced to primary alcohols. The stronger reducing agent, $LiAlH_4$, is required:

$$CH_3CH_2CO_2H \xrightarrow{LiAlH_4} CH_3CH_2CH_2OH$$

Progress check 17.02

Give the structural formulae and the names of the alcohols produced in the following reactions.

1. $CH_3I + KOH(aq) \longrightarrow$
2. $(CH_3)_2C=CH_2 + H_2O\ (+ H^+) \longrightarrow$
3. $CH_3COC(CH_3)_2CHO + NaBH_4 \longrightarrow$

17.04 How are carboxylic acids made?

a Oxidation of primary alcohols or aldehydes

$$CH_3CH_2OH + 2[O] \xrightarrow{Na_2Cr_2O_7 + H_2SO_4(aq) + heat} CH_3CO_2H + H_2O$$

$$CH_3CHO + [O] \xrightarrow{Na_2Cr_2O_7 + H_2SO_4(aq) + heat} CH_3CO_2H$$

b Hydrolysis of nitriles

$$CH_3-C\equiv N \xrightarrow{heat\ with\ H_2SO_4(aq)} CH_3CO_2H$$

> **TIP**
>
> You may be asked for the other product of this reaction. It is the ammonium ion, NH_4^+. The full balanced equation is:
>
> $R-CN + H^+ + 2H_2O \longrightarrow R-CO_2H + NH_4^+$
>
> Write ionic rather than molecular equations whenever you can (but do not forget the charges). The molecular equation for this reaction is a little more complicated!
>
> $2R-CN + H_2SO_4 + 4H_2O \longrightarrow 2R-CO_2H + (NH_4)_2SO_4$

c **Hydrolysis of acid derivatives such as esters or acyl chlorides**

This is not a generally useful method, as acid derivatives are usually made from the carboxylic acids in the first place.

Progress check 17.03

Suggest the structures of the carboxylic acids formed by the following reactions, identifying the intermediate compounds **X** and **Y**.

1. $(CH_3)_2CHCH_2OH + Na_2Cr_2O_7$ and $H^+(aq) \longrightarrow W$
2. $(CH_3)_2CHCH_2OH + HBr \longrightarrow X$
3. $X + KCN \longrightarrow Y$
4. $Y + H_3O^+(aq)$, and heat $\longrightarrow Z$

17.05 The reactions of alcohols

a **Oxidation**

All alcohols burn well in air or oxygen, producing carbon dioxide and water vapour:

$CH_3CH_2OH + 3O_2 \longrightarrow 2CO_2 + 3H_2O$

Ethanol is an important fuel in its own right (e.g. in camping stoves) and in some countries it is also a component of petrol (gasoline) for motor transport.

Under more controlled conditions in the laboratory, primary alcohols can be oxidised to aldehydes or carboxylic acids, depending on the conditions used; and secondary alcohols are oxidised to ketones.

The most commonly used oxidising agent is acidified potassium dichromate, which turns from orange to green ($Cr^{3+}(aq)$) as it is reduced. Another oxidising agent that can be used is hot $KMnO_4 + H^+(aq)$.

i Secondary alcohols:

$(CH_3)_2CH(OH) + [O] \xrightarrow[\text{dil } H_2SO_4]{\text{heat under reflux with } K_2Cr_2O_7 +} (CH_3)_2C=O + H_2O$
propan-2-ol $\qquad\qquad\qquad\qquad\qquad$ propanone (a ketone)

ii Primary alcohols, with an excess of oxidising agent:

$CH_3CH_2CH_2OH + 2[O] \xrightarrow[\text{dil } H_2SO_4]{\text{heat under reflux with } K_2Cr_2O_7 +} CH_3CH_2CO_2H + H_2O$
propan-1-ol $\qquad\qquad\qquad\qquad\qquad$ propanoic acid

iii Primary alcohols, adding the oxidising agent to the alcohol, and distilling off the aldehyde as soon as it is formed (see Unit 18):

$CH_3CH_2CH_2OH + [O] \xrightarrow[\text{as it is formed}]{\text{heat with } K_2Cr_2O_7 + \text{dil } H_2SO_4, \text{distil product}} CH_3CH_2CH=O + H_2O$
propan-1-ol (b. pt. 97 °C) $\qquad\qquad\qquad$ propanal (b. pt. 49 °C)

iv Tertiary alcohols are resistant to oxidation.

$(CH_3)_3C-OH \xrightarrow{K_2Cr_2O_7 + H_2SO_4(aq)}$ no reaction
2-methylpropan-2-ol

b **Nucleophilic substitution**

There are several ways of converting alcohols into halogenoalkanes.

i With an inorganic chloride:

$CH_3CH_2OH + SOCl_2 \xrightarrow{\text{gentle warming}} CH_3CH_2Cl + SO_2 + HCl$

$(CH_3)_2CHCH_2OH + PCl_5 \xrightarrow{\text{gentle warming}} (CH_3)_2CHCH_2Cl + POCl_3 + HCl$

ii With concentrated HBr(aq) or a mixture of $KBr + H_2SO_4(aq)$

$(CH_3)_2CH(OH) + HBr \xrightarrow{\text{heat under reflux}} (CH_3)_2CH-Br + H_2O$

iii Tertiary alcohols are so reactive they can be converted into chlorides by shaking with concentrated HCl at room temperature.

$$(CH_3)_3C\text{–}OH + HCl \longrightarrow (CH_3)_3C\text{–}Cl + H_2O$$

c **Elimination**

Alcohols can be dehydrated to alkenes by heating with concentrated sulfuric acid or phosphoric acid, or passing their vapour over heated aluminium oxide:

$$CH_3CH_2CH_2OH \xrightarrow{\text{heat at } 180\,°C \text{ with } H_2SO_4 \text{ or } H_3PO_4} CH_3CH=CH_2 + H_2O$$

$$CH_3CH_2OH \xrightarrow{\text{pass vapour over } Al_2O_3 \text{ at } 350\,°C} CH_2=CH_2 + H_2O$$

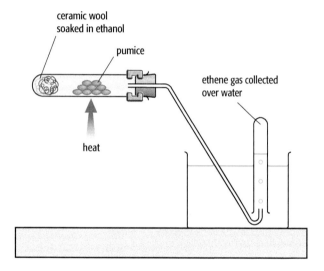

Figure 17.01 Dehydration of ethanol to ethene

d **Reactions of the –O–H bond**

i Reaction with sodium

When a small piece of sodium metal is added to a pure alcohol, fizzing takes place as hydrogen gas is evolved:

$$CH_3CH_2OH + Na \rightarrow CH_3CH_2O^-Na^+ + \tfrac{1}{2}H_2(g)$$

ii Esterification

When an alcohol is heated under reflux with a carboxylic acid and a catalytic quantity of concentrated sulfuric acid, an ester is formed:

$$CH_3CH_2CH_2OH + CH_3C\underset{OH}{\overset{O}{\diagup\diagdown}} \xrightarrow{\text{heat with } H_2SO_4 \text{ catalyst}} CH_3C\underset{O-CH_2CH_2CH_3}{\overset{O}{\diagup\diagdown}} + H_2O$$

e **The triiodomethane reaction**

This is a specific reaction undergone by alcohols containing the –CH(OH)CH$_3$ grouping. When warmed with alkaline aqueous iodine, these alcohols produce a pale yellow precipitate of **triiodomethane**. This is a good test for the presence of a methyl secondary alcohol group in a molecule. The reaction is also undergone by methyl ketones (see Unit 18).

$$CH_3CH(OH)CH_3 + 6OH^-(aq) + 4I_2(aq) \longrightarrow CH_3CO_2^- + CHI_3(s) + 5I^- + 5H_2O$$

> **TIP**
> All alcohols that give a positive triiodomethane test are *secondary* alcohols except for one: the primary alcohol ethanol (CH$_3$CH$_2$OH) also gives a positive result.

17.06 Reactions of carboxylic acids

a **Reaction with reactive metals**

Since they contain an –OH group, carboxylic acids react with sodium metal, giving off hydrogen. With ethanoic acid, the salt, sodium ethanoate, is formed:

$$CH_3CO_2H + Na \longrightarrow CH_3CO_2^-Na^+ + \tfrac{1}{2}H_2(g)$$

Carboxylic acids also evolve hydrogen with other reactive metals, such as lithium, magnesium and calcium.

b **Reaction with sodium hydroxide**

Carboxylic acids undergo typical acid–base reactions, and their solutions can be titrated with an alkali:

$$CH_3CO_2H + NaOH \longrightarrow CH_3CO_2Na + H_2O$$

c **Reaction with carbonates**

Carboxylic acids are acidic enough to liberate CO$_2$ from carbonates.

$$2CH_3CO_2H + Na_2CO_3 \longrightarrow 2CH_3CO_2Na + H_2O + CO_2(g)$$

d **Reaction with alcohols: formation of esters**

Esters are produced when a carboxylic acid and an alcohol are heated under reflux with a small amount of concentrated sulfuric acid. The sulfuric acid is a homogeneous catalyst for the reaction, and can drive the equilibrium over to the ester side by removing the water produced:

$CH_3CO_2H + CH_3CH_2OH \rightleftharpoons CH_3CO_2CH_2CH_3 + H_2O$
ethyl ethanoate

$H_2O + H_2SO_4 \longrightarrow H_3O^+ + HSO_4^-$

e **Reduction to alcohols**

When carboxylic acids are treated with lithium aluminium hydride in dry ether, they are reduced to primary alcohols (see section 17.3).

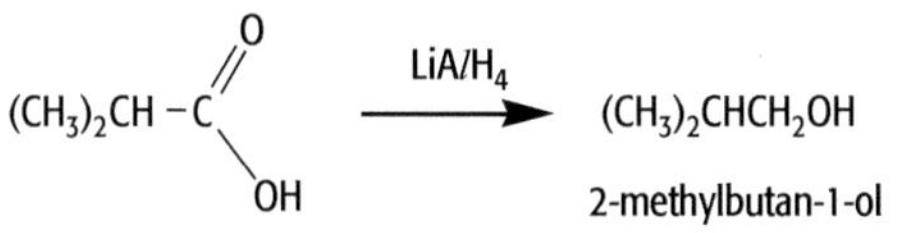

$(CH_3)_2CH-C(=O)OH \xrightarrow{LiAlH_4} (CH_3)_2CHCH_2OH$
2-methylbutan-1-ol

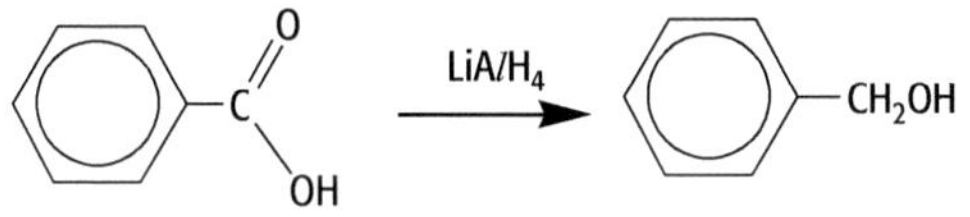

C₆H₅–C(=O)OH $\xrightarrow{LiAlH_4}$ C₆H₅–CH₂OH
phenylmethanol

> **TIP**
>
> LiAlH₄ is a very strong reducing agent that will reduce carboxylic acids, esters, aldehydes and ketones to alcohols. The weaker reducing agent NaBH₄ will only reduce aldehydes and ketones, and so can be used to selectively reduce these groups in a multi-functional compound:
>
> $HO_2C-CH_2-CHO \xrightarrow{LiAlH_4} HOCH_2-CH_2-CH_2OH$
>
> $HO_2C-CH_2-CHO \xrightarrow{NaBH_4} HO_2C-CH_2-CH_2OH$

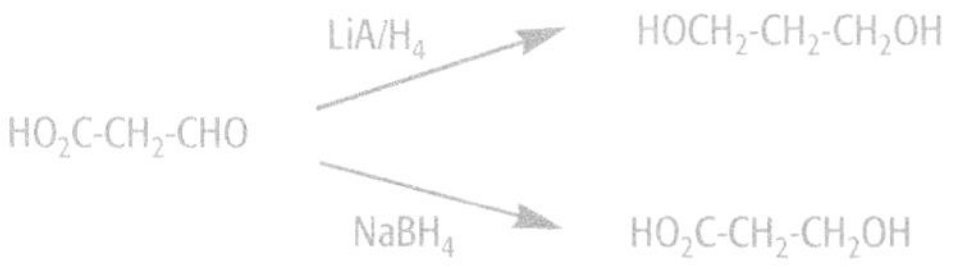

Sample answer

Compound **A**, $C_6H_8O_3$, decolorises $Br_2(aq)$, effervesces with $Na_2CO_3(aq)$, gives an orange precipitate with 2,4-DNPH but does not produce a silver mirror when warmed with Tollens' reagent.

Compound **A** undergoes the following series of reactions.

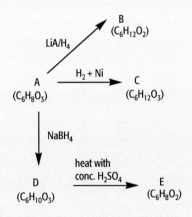

Question:

Suggest structures for compounds **A–E** and explain the reactions involved. [9]

Answer:

Note that the question asks for explanations, in addition to the structures of **A–E**.

The results of the tests show that **A** contains a –CO₂H group, a C=C and a ketone group. [1 mark]

LiAlH₄ will reduce both the –CO₂H and the >C=O groups, but leave the C=C group. [1 mark]

H₂ + Ni will reduce the C=C and the >C=O groups, but not the –CO₂H group. [1 mark]

NaBH₄ will reduce the >C=O to a >CH(OH) group, but not the other two groups. [1 mark]

The reaction from **D** to **E** looks like an 'internal' esterification between the –CO₂H and the >CH(OH) group in **D**.

Possible structures are:

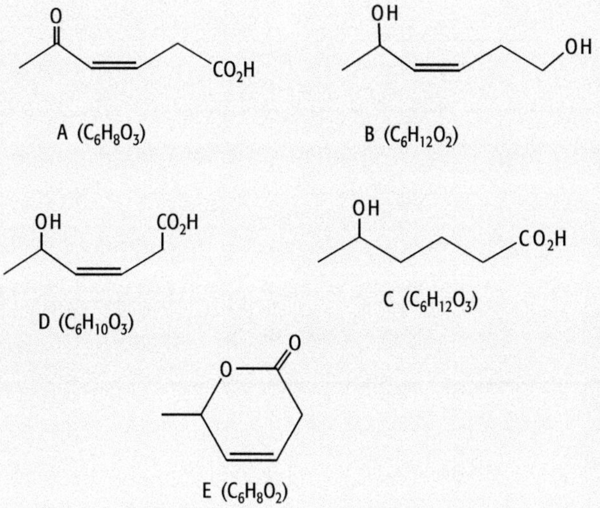

A (C₆H₈O₃): CH₃–CO–CH=CH–CH₂–CO₂H

B (C₆H₁₂O₂): CH₃–CH(OH)–CH=CH–CH₂–CH₂OH

C (C₆H₁₂O₃): CH₃–CH(OH)–CH₂–CH₂–CH₂–CO₂H

D (C₆H₁₀O₃): CH₃–CH(OH)–CH=CH–CH₂–CO₂H

E (C₆H₈O₂): six-membered lactone (δ-lactone with C=C)

[5]

17.07 The hydrolysis of esters

If an ester is heated under reflux with dilute aqueous sulfuric acid, it is hydrolysed to a carboxylic acid and an alcohol. This is the reverse of the equilibrium reaction described above, which makes an ester from a carboxylic acid and an alcohol. The equilibrium is shifted in the hydrolysis direction by the large amount of water that is present.

$$CH_3CO_2CH_2CH_3 + H_2O \underset{}{\overset{H_3O^+/heat}{\rightleftharpoons}} CH_3CO_2H + CH_3CH_2OH$$

The reaction proceeds more quickly, and with a better yield, if hydroxide is used as the catalyst instead of an acid. The formation of the salt of the acid drives the equilibrium well over to the hydrolysis side:

$$CH_3CO_2CH_2CH_3 + NaOH \underset{}{\overset{NaOH(aq) + heat}{\rightleftharpoons}} CH_3CO_2^-Na^+ + CH_3CH_2OH$$

Worked example 17.01

Suggest the products of the following reactions.

a $CH_3CH_2CO_2CH_3 + H_3O^+ + heat \longrightarrow A + B$

b $HCO_2CH_2CH_3 + NaOH(aq) + heat \longrightarrow C + D$

How to get the answer:

a Step 1: Splitting the molecule at the C–O bond in the carboxyl group produces the two fragments CH_3CH_2CO and OCH_3.

   Step 2: As these pick up water, the products will be **A**: $CH_3CH_2CO_2H$ (propanoic acid) and **B**: CH_3OH (methanol).

b Step 1: Splitting the molecule at the C–O bond in the carboxyl group produces the two fragments HCO and OCH_2CH_3.

   Step 2: Hence **C** is HCO_2H (methanoic acid) and **D** is CH_3CH_2OH (ethanol).

17.08 Tests for alcohols and carboxylic acids

Alcohols and carboxylic acids fizz with sodium metal (giving off $H_2(g)$) and with PCl_5 (giving off steamy fumes of $HCl(g)$). Unlike carboxylic acids, alcohols are neutral, so universal indicator stays green when added to an alcohol, whereas with acids it turns red.

Primary, secondary and tertiary alcohols can be distinguished by treating them with acidified dichromate, with results as in Table 17.01.

Alcohol type	Observation on warming with reagent	Effect of the distillate on universal indicator
R_3C–OH, tertiary	stays orange	neutral (stays green): water only produced
R_2CH–OH, secondary	turns green	neutral (stays green): ketone produced
RCH_2–OH, primary	turns green	acidic (goes red): carboxylic acid produced

Table 17.01 How primary, secondary and tertiary alcohols can be distinguished

Progress check 17.04

- Three compounds **A**, **B** and **C** are isomers with the molecular formula $C_5H_{12}O$.
- All three react with sodium metal.
- Only compounds **B** and **C** react with acidified potassium dichromate when heated under reflux: compound **B** gives an acidic product, whereas compound **C** gives a neutral product.
- On heating with concentrated sulfuric acid, compounds **A** and **C** give 2-methylbut-2-ene, whereas compound B gives pent-1-ene.

Explain the reactions described above, suggest structures for compounds **A**, **B** and **C**, and state which, if any, contain a chiral carbon atom.

17.09 Uses of alcohols, acids and esters

Both methanol and ethanol are used as additives for automobile fuels ('gasohol' is petrol (gasoline) containing 10–20% ethanol). Ethanol is also used as a solvent, and a skin disinfectant in hospitals.

Esters with low M_r find many uses as solvents for paints. Natural food flavours are often esters, and many of these are made artificially, to flavour fruit drinks, sweets and desserts. Some of the naturally occurring esters now made synthetically and used as artificial flavours in foodstuffs are as follows (their flavours are in brackets after their names):

2-methylpropyl methanoate (raspberry)

pentyl ethanoate (pear)

methyl butanoate (apple)

ethyl butanoate (pineapple)

Ethanoic acid is an important solvent for various plastics and glues. It is also found in vinegar. Salts of carboxylic acids, such as sodium benzoate, are used as food preservatives, and acid+salt mixtures, such as citric acid and sodium citrate, are used as buffering agents to keep the pH of foodstuffs constant.

Revision checklist

Check that you know the following:

- The different types of alcohol (primary, secondary and tertiary) and their different reactions, e.g. mild oxidation.
- The various methods used to make alcohols (from alkenes and halogenoalkanes), carboxylic acids (from alcohols, aldehydes and nitriles) and esters (from acids and alcohols).
- The reactions of alcohols: oxidation, substitution, elimination, esterification and reaction with sodium.
- The reactions of acids: with sodium metal, hydroxides, carbonates, alcohol (to make esters). Reduction with LiAlH$_4$.
- The use of the triiodomethane reaction to test for the CH$_3$CH(OH)– group.

Exam-style questions

1 Malic acid (HO$_2$CCH$_2$CH(OH)CO$_2$H) occurs in apples and is used as a food additive.

   a The following is an outline of a 4-step synthesis of malic acid starting from hydroxyethanal.

   HOCH$_2$CHO $\xrightarrow{\text{step 1}}$ A $\xrightarrow{\text{step 2}}$ B $\xrightarrow{\text{step 3}}$

   C $\xrightarrow{\text{step 4}}$ HO$_2$CCH$_2$CH(OH)CO$_2$H

   i Suggest possible structures for the intermediate compounds **A**, **B** and **C**. [3]

   ii Suggest suitable reagents and conditions for steps 1–4. [4]

   b Explain whether or not malic acid exhibits optical isomerism. [1]

   c When malic acid is heated with concentrated H$_3$PO$_4$, a mixture of two isomeric compounds **D** and **E** with the molecular formula C$_4$H$_4$O$_4$ is formed. Suggest the structures of **D** and **E** and explain the isomerism they demonstrate. [3]

   Total: 11

2 The ester **F** is used for flavouring fruit drinks.

   CH$_3$CH$_2$CO$_2$CH(CH$_3$)$_2$

   F

When **F** is treated with hot aqueous acid, two compounds (**G** and **H**) are formed.

a i What *type of reaction* is occurring here? [1]

   ii Write a balanced equation for the reaction, showing the structures of the two products **G** and **H**. [3]

b Describe a test (reagents and observations) by which samples of **G** and **H** could be distinguished. [2]

c Draw the structures of the following isomers of **F**, all of which are esters.

   i   Isomer **J** can be made from a tertiary alcohol.

   ii Isomer **K** can be made from a primary alcohol that has a chiral carbon atom.

   iii Isomer **L** can be made from an acid whose molecules contain a chiral centre. [3]

d Describe how the alcohols that make up esters **F**, **J** and **L** could be distinguished from each other. [3]

Total: 12

Carbonyl compounds

Learning outcomes

You should be able to:

- describe the formation of aldehydes and ketones from primary and secondary alcohols respectively using $Cr_2O_7^{2-} + H^+$

- describe the reduction of aldehydes and ketones by $NaBH_4$ or $LiAlH_4$

- describe the reaction of aldehydes and ketones with HCN and NaCN

- describe the mechanism of the nucleophilic addition reactions of hydrogen cyanide with aldehydes and ketones

- describe the use of 2,4-dinitrophenylhydrazine (2,4-DNPH) reagent to detect the presence of carbonyl compounds

- deduce the nature (aldehyde or ketone) of an unknown carbonyl compound from the results of simple tests (Fehling's and Tollens' reagents)

- describe the reaction of CH_3CO- compounds with alkaline aqueous iodine to give triiodomethane

18.01 Properties, isomerism and nomenclature

The carbonyl group, >C=O, is a subunit of many other functional groups, such as carboxylic acids, amides and esters, but the term **carbonyl compound** is reserved for those compounds in which it appears on its own.

There are two types of carbonyl compounds: **aldehydes** and **ketones**.

The properties of aldehydes and ketones are very similar to each other – almost all the reactions of ketones are also shown by aldehydes. But aldehydes show additional reactions associated with their lone hydrogen atom.

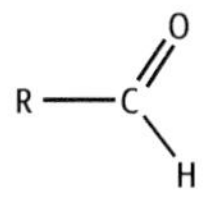

general formula of aldehydes 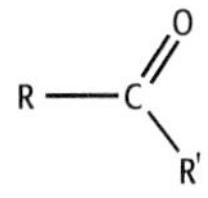   general formula of ketones

Aldehydes are named by adding the suffix -al to the stem, whereas ketones have the suffix -one (pronounced 'own'). Aldehydes and ketones are positional isomers of each other. They can also show structural and optical isomerism.

Worked example 18.01

There are eight carbonyl compound isomers with the molecular formula $C_5H_{10}O$. What are their structures and their systematic names?

How to get the answer:

Step 1: Let us start with a straight chain of five carbon atoms (pentane). The carbonyl group can be on C-1, C-2 or C-3.

Step 2: If we consider the 2-methylbutane skeleton, the carbonyl group can be on C-1, C-3 or C-4, the first of which can exist as optical isomers.

Step 3: Lastly, there is one carbonyl compound based on the 2,2-dimethylpropane skeleton.

The structures are as follows. Notice how they are named.

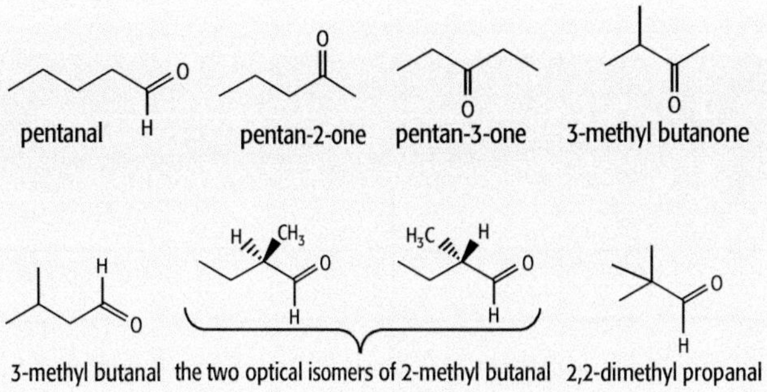

pentanal   pentan-2-one   pentan-3-one   3-methyl butanone

3-methyl butanal   the two optical isomers of 2-methyl butanal   2,2-dimethyl propanal

Because of the electronegativity of oxygen, the carbonyl group is highly polar:

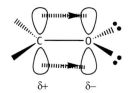

The dipole moment of the carbonyl group is sometimes represented by the following **mesomeric** pair of formulae.

This polarisation dominates the reactivity of carbonyl compounds, but also has an effect on their physical properties. Their molecules attract each other by dipole–dipole forces, so they have higher boiling points than alkanes of similar M_r. In addition, the lone pairs on oxygen can hydrogen bond with water molecules, so carbonyl compounds with low M_r are soluble in water.

18.02 Methods of preparation

Carbonyl compounds are usually made in the laboratory by oxidising alcohols. Either potassium manganate(VII) or potassium dichromate(VI) can be used, although the latter reagent is the more usual. Both reagents need to be in acidified solution, and heated.

If you are asked to write a balanced equation for these oxidation reactions, do not panic! You can use the symbol [O] for oxygen atoms coming from oxidising agents, so the last equation can be written:

$$CH_3CH(OH)CH_3 + [O] \longrightarrow CH_3COCH_3 + H_2O$$

A special procedure needs to be carried out to ensure that the oxidation of primary alcohols stops at the aldehyde stage, since aldehydes are easily oxidised further to carboxylic acids. The conditions used are as follows:

- set up a flask for simple distillation, and add the alcohol to the flask
- slowly drip the solution of acidified dichromate(VI) into the flask containing the alcohol, so that the alcohol is always the reagent in excess
- keep the temperature of the flask just below the boiling point of the alcohol

If this procedure is followed, the aldehyde that is formed will immediately distil off, and can be collected after it has been liquefied in the condenser. This technique relies on the fact that aldehydes are always more volatile than the corresponding alcohols, because the alcohol molecules hydrogen bond to each other, whereas the aldehyde molecules do not. Table 18.01 shows two examples.

Alcohol		Aldehyde		Difference / °C
formula	b. pt. / °C	formula	b. pt. / °C	
CH_3CH_2OH	78	CH_3CHO	20	58
$CH_3CH_2CH_2OH$	97	CH_3CH_2CHO	49	48

Table 18.01 Comparing the boiling points of aldehydes and alcohols

18.03 Reactions common to aldehydes and ketones

A mentioned above, the carbon atom in the carbonyl group is slightly positively charged. Carbonyl compounds are therefore attacked by nucleophiles. The initial reaction is an addition to the C=O double bond, but this can be followed by an elimination.

a Nucleophilic addition of HCN

In the presence of a small amount of sodium cyanide (or sodium hydroxide, which reacts with HCN to form sodium cyanide), hydrogen cyanide adds on to carbonyl compounds:

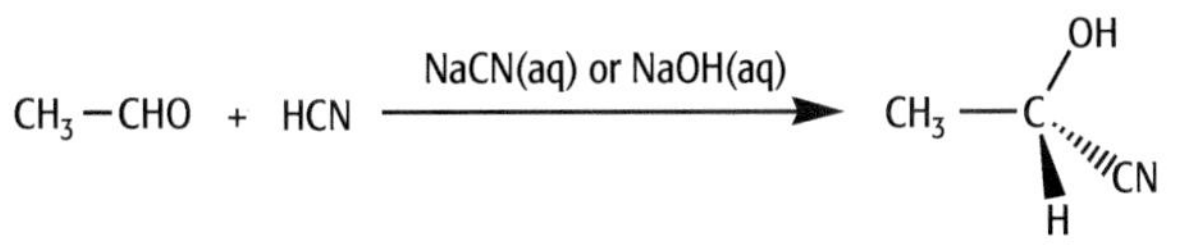

$$CH_3-CHO + HCN \xrightarrow{NaCN(aq) \text{ or } NaOH(aq)} CH_3-\underset{H}{\overset{OH}{C}}-CN$$

2-hydroxypropanenitrile

The reaction is catalysed by cyanide ions from the sodium cyanide. Like the reaction between sodium cyanide and a halogenoalkane, the reaction is an effective method of adding a carbon atom to a chain. Cyanohydrins are useful compounds: they can be hydrolysed to 2-hydroxycarboxylic acids by heating with dilute sulfuric acid, and can also be reduced to 2-hydroxyamines:

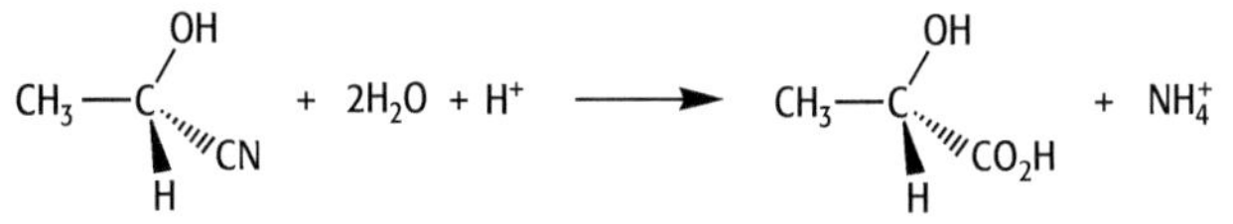

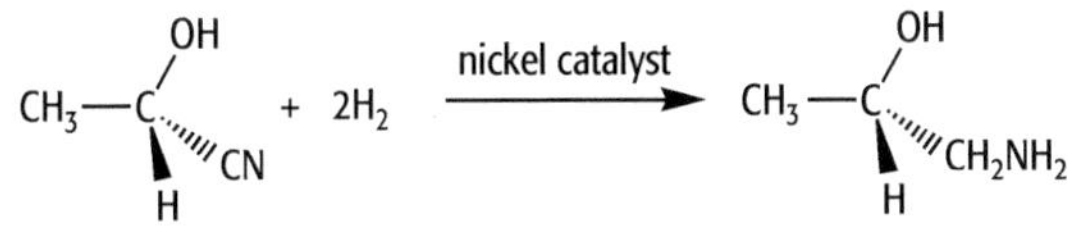

$$CH_3-\underset{H}{\overset{OH}{C}}-CN + 2H_2O + H^+ \longrightarrow CH_3-\underset{H}{\overset{OH}{C}}-CO_2H + NH_4^+$$

$$CH_3-\underset{H}{\overset{OH}{C}}-CN + 2H_2 \xrightarrow{\text{nickel catalyst}} CH_3-\underset{H}{\overset{OH}{C}}-CH_2NH_2$$

> **Progress check 18.01**
>
> Suggest two-stage routes to the following two compounds, starting in each case with a carbonyl compound.
>
> 1 $CH_3CH_2C(OH)(CH_3)CH_2NH_2$
> 2 $(CH_3)_2C(OH)CO_2H$.

b The mechanism of the reaction

The first step is the attack on the $\delta+$ carbon atom by the nucleophilic cyanide ion, ^-CN:

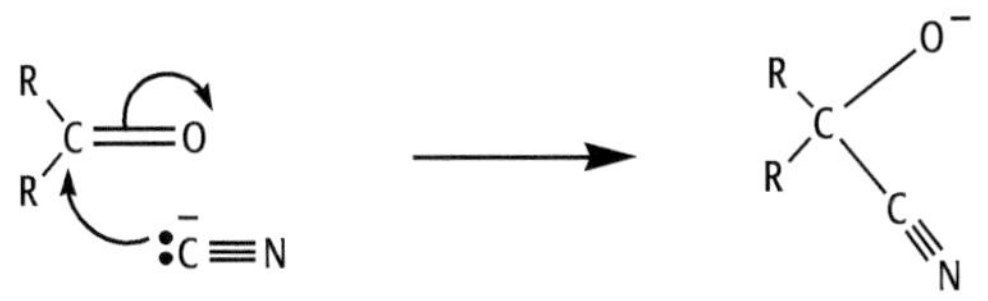

The intermediate anion is a strong base. It can take a proton off an un-ionised molecule of hydrogen cyanide:

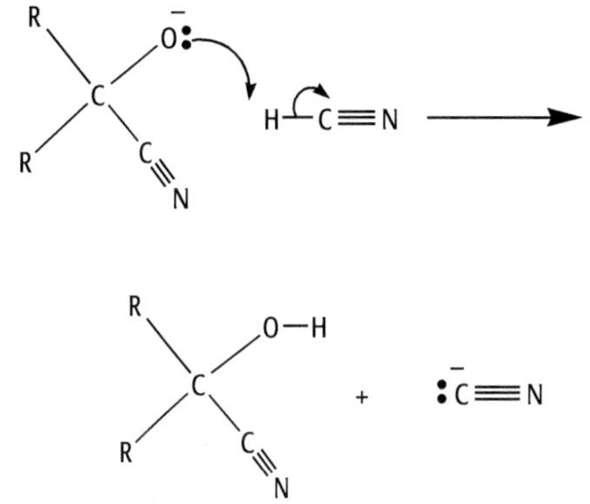

The cyanide ion that was used up in the first step is re-formed in the second step. It is acting as a **homogeneous catalyst**, so only a small amount of NaCN is required.

c Reduction of the carbonyl group

Carbonyl compounds can be reduced to alcohols using either complex metal hydrides, or using hydrogen gas in the presence of nickel or platinum.

Aldehydes give primary alcohols, whilst ketones give secondary alcohols:

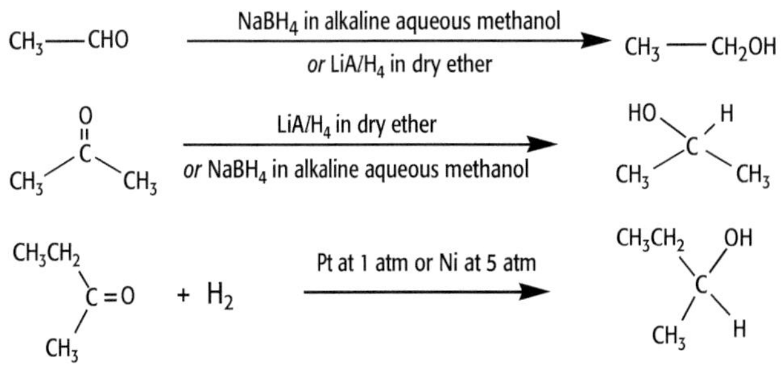

$$CH_3-CHO \xrightarrow[\text{or } LiAlH_4 \text{ in dry ether}]{NaBH_4 \text{ in alkaline aqueous methanol}} CH_3-CH_2OH$$

$$CH_3-\underset{CH_3}{\overset{O}{\underset{\|}{C}}}-CH_3 \xrightarrow[\text{or } NaBH_4 \text{ in alkaline aqueous methanol}]{LiAlH_4 \text{ in dry ether}} \underset{CH_3}{\overset{HO}{\underset{|}{C}}}\overset{H}{\underset{|}{}}CH_3$$

$$\underset{CH_3}{\overset{CH_3CH_2}{C}}=O + H_2 \xrightarrow{\text{Pt at 1 atm or Ni at 5 atm}} \underset{CH_3}{\overset{CH_3CH_2}{\underset{|}{C}}}\overset{OH}{\underset{H}{}}$$

> **TIP**
>
> Although reducing agents for organic reactions are usually denoted by the single-atom symbol [H] in equations, you should still make sure that any equation you write is atom balanced.

d Condensation reactions

Carbonyl compounds react with primary amines (RNH_2), hydrazines ($RNHNH_2$) and hydroxylamine ($HONH_2$) to give compounds containing a C=N double bond. The reaction involves an initial nucleophilic addition to the C=O group (like the reaction with HCN discussed above) and subsequent loss of water:

$$R_2C=O + NH_2\text{-}R' \longrightarrow \underset{R}{\overset{R}{\underset{|}{C}}}\!\!\begin{smallmatrix}OH\\ \\NH\text{-}R'\end{smallmatrix} \xrightarrow{-H_2O} \underset{R}{\overset{R}{C}}\!\!=\!\!N_{\diagdown R}$$

The most important class of compounds formed in this way are the 2,4-dinitrophenylhydrazones. These are crystalline orange solids whose formation as precipitates is a definitive test for the presence of a carbonyl compound:

[Reaction scheme showing R₂C=O + H₂N—NH-C₆H₃(NO₂)₂ (2,4-dinitrophenylhydrazine, 2,4-DNPH) → in methanol, + H⁺ → R₂C=N—NH-C₆H₃(NO₂)₂ (a 2,4-dinitrophenylhydraz**one**, orange precipitate) + H₂O]

The 2,4-dinitrophenylhydrazones of many aldehydes and ketones have distinctive melting points. Taking the melting point of the recrystallised orange precipitate enables us to identify *which* carbonyl compound is present.

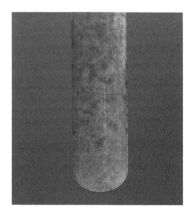

Figure 18.01 The orange precipitate formed by the reaction between 2,4-DNPH and a carbonyl compound

18.04 Oxidation reactions of aldehydes

a With acidified dichromate(VI)

Unlike ketones, aldehydes can be oxidised to carboxylic acids. The reaction takes place on gentle warming, and the colour of the reagent changes from orange to green.

$$CH_3\text{-}CHO + [O] \xrightarrow[\text{heat}]{K_2Cr_2O_7 + H_2SO_4(aq)} CH_3\text{-}COOH$$

b With Fehling's solution

The bright blue **Fehling's solution** is a solution of Cu^{2+} ions, complexed with salts of tartaric acid, in an aqueous alkaline solution. When warmed with an aldehyde, the Cu^{2+} ions are reduced to Cu^+ ions, which in the alkaline solution form a dull red precipitate of copper(I) oxide. The aldehyde is oxidised to the salt of the corresponding carboxylic acid:

$$CH_3CHO(aq) + 2Cu^{2+}(aq) + 5OH^-(aq) \longrightarrow$$
$$\text{blue solution}$$
$$CH_3CO_2^-(aq) + Cu_2O(s) + 3H_2O(l)$$
$$\text{red precipitate}$$

c With Tollens' reagent

Tollens' reagent contains silver ions, complexed with ammonia, in an aqueous alkaline solution.

These silver ions are readily reduced to silver metal on gentle warming with an aldehyde. The metal will often silver-plate the inside of the test tube. Once again, the aldehyde is oxidised to the salt of the corresponding carboxylic acid:

$$CH_3CH_2CHO + 2Ag^+ + 3OH^- \longrightarrow$$
$$CH_3CH_2CO_2^- + 2Ag(s) + 2H_2O$$
$$\text{silver mirror}$$

Either the formation of a red precipitate on warming with Fehling's solution, or the formation of a silver mirror on warming with Tollens' reagent, is a good way to distinguish an aldehyde from a ketone.

Figure 18.02 The silver mirror formed when Tollens' reagent is warmed with an aldehyde.

> **TIP**
>
> You should be aware that there is one other compound that reacts with Fehling's solution or Tollens' reagent. This is methanoic acid (HCO_2H), which contains a –CHO group.

Progress check 18.02

Three isomers, **A**, **B** and **C**, have the molecular formula C_4H_8O.

- **A** and **B** form orange precipitates with 2,4-DNPH whereas **C** does not.
- **C** decolorises bromine water, but **A** and **B** do not.
- **A** gives a silver mirror when warmed with Tollens' reagent, but **B** and **C** do not.
- On treatment with hydrogen and nickel, both **B** and **C** give the same compound, butan-2-ol, whereas **A** gives butan-1-ol.

Suggest structures for **A**, **B** and **C**, and explain all the reactions described.

18.05 The triiodomethane reaction

Ketones that contain the group –$COCH_3$ (methyl ketones) form a yellow precipitate of **triiodomethane** (iodoform) when added to an aqueous alkaline solution of iodine. The reaction is also shown by methyl alcohols, containing the group –$CH(OH)CH_3$.

$$\underset{CH_3}{\overset{R}{>}}C=O + 4OH^- + 3I_2 \longrightarrow \underset{^-O}{\overset{R}{>}}C=O + CHI_3 + 3H_2O + 3I^-$$

iodoform: pale yellow precipitate

Sample answer

Question:

Three ketones, **D**, **E** and **F**, have the molecular formula $C_5H_{10}O$. **D** and **E** undergo the triiodomethane reaction, but **F** does not. On treatment with hydrogen and nickel, followed by heating with concentrated H_2SO_4, compounds **E** and **F** produce the same alkene, **G**, with molecular formula C_5H_{10}, whereas compound **D** produces an isomeric alkene **H**.

Suggest the structures of **D**, **E** and **F**, and the structures of the alkenes **G** and **H**. Explain your answers. [7]

Answer:

Note that explanations need to be given, not just the structures for **D**–**H**.

If compounds **D** and **E** undergo the iodoform reaction, they must contain the –$COCH_3$ group. [1 mark]

Taking this away from the molecular formula leaves C_3H_7, so they must be $CH_3CH_2CH_2COCH_3$ and $(CH_3)_2CHCOCH_3$ (we do not know which is which yet).

If **F** does not undergo the iodoform reaction, it must be the ketone $CH_3CH_2COCH_2CH_3$. [1 mark]

Compounds **E** and **F** contain the same carbon chain, because on reduction and dehydration they produce the same alkene. [1 mark]

Compound **E** is $CH_3CH_2CH_2COCH_3$. [1 mark]

This leaves compound **D** to be $(CH_3)_2CHCOCH_3$. [1 mark]

Alkene **G** can only be $CH_3CH_2CH=CHCH_3$, [1 mark] but alkene **H** could be either $(CH_3)_2CHCH=CH_2$ or $(CH_3)_2C=CHCH_3$. [1 mark]

The following scheme shows the reactions involved.

Revision checklist

Check that you know the following:

- [ ] The reagents and conditions needed to oxidise alcohols to carbonyl compounds, and to reduce carbonyl compounds to alcohols.

- [ ] The products of, and the mechanism for, the reaction between carbonyl compounds and HCN.

- [ ] How 2,4-DNPH is used to test for the presence of a carbonyl compound.

- [ ] How Fehling's solution or Tollens' reagent is used to test for the presence of an aldehyde.

- [ ] How the triiodomethane reaction tests for the presence of the CH_3CO- group.

Exam-style questions

1. The compound 4-hydroxynonenal is produced by the oxidation of lipids in the body.

   [structure of 4-hydroxynonenal]

   4-hydroxynonenal

   a i Name the three functional groups in the molecule of nonenal. [3]

   ii Describe how you could test for the presence of each of these functional groups. For each test you should state the reagent used, and the observation you would make. [6]

   b Calculate the empirical formula of 4-hydroxynonenal. [1]

   c 4-hydroxynonenal shows two types of stereoisomerism. Explain this statement. [2]

   d Suggest the structures of the products of the following reactions.

   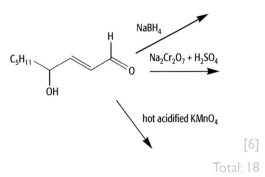

   [6]

   Total: 18

2. Compound **A** is a component of some suntan creams. **A** has an M_r of 90 and its empirical formula is CH_2O. Compound **A** gives an orange precipitate with 2,4-DNPH but does not react with Fehling's solution. **A** also effervesces with sodium metal, but does not effervesce with $Na_2CO_3(aq)$.

   a Suggest a structure for **A**. Explain your reasoning. [3]

   b Suggest structures for organic products **B**, **C**, **D** and **E**, of the following reactions.

   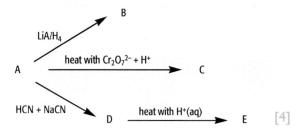

   [4]

   c State the molecular formula of compound **E**, and calculate the volume of $H_2(g)$ that would be evolved (measured at 25 °C and 1 atm) if 1.00 g of **E** were treated with an excess of sodium metal. [4]

   d Show the mechanism of the reaction **A** ⟶ **D**. In your answer you should use curly arrows to show the movement of electrons, and include relevant lone pairs and charges. [3]

   Total: 14

Unit 19

Lattice energy

Learning outcomes

You should be able to:

- explain and use *lattice energy* as an exothermic enthalpy change and explain the factors affecting its magnitude,
- apply Hess' Law to construct Born Haber cycles,
- interpret and explain the trends in the thermal stability of Group 2 nitrates and carbonates
- apply Hess' Law to find enthalpy changes for ionic solids and solutions
- interpret and explain the trend in solubility of Group 2 sulfates and hydroxides.

19.01 Lattice energy

In ionic substances, the ions are held together by non-directional electrostatic attractions and the energy changes occur when making or breaking an ionic lattice.

The lattice energy is the energy released when gaseous ions are brought to their places in an ionic lattice to make 1 mole of ionic solid:

$$Na^+(g) + Cl^-(g) \longrightarrow NaCl(s)$$
$$\Delta H_{latt}[NaCl] = -780 \, kJ\,mol^{-1}$$
$$Mg^{2+}(g) + O^{2-}(g) \longrightarrow MgO(s)$$
$$\Delta H_{latt}[MgO] = -3791 \, kJ\,mol^{-1}$$
$$Ca^{2+}(g) + 2Br^-(g) \longrightarrow CaBr_2(s)$$
$$\Delta H_{latt}[CaBr_2] = -2176 \, kJ\,mol^{-1}$$

Lattice energies are always negative as energy is released when an ionic lattice is formed.

Ions with higher charges have a greater electrostatic attraction and correspondingly more negative lattice energies; larger ions cannot get so close to each other and so less energy is released on forming the lattice:

- The lattice energy gets less negative for a larger Group 1 ion with the same anion or a larger halide ion with the same cation.
- Ca^{2+} and Na^+ ions are approximately the same size but the lattice energy of $CaCl_2$ is much more negative than the lattice energy of $NaCl$.
- Mg^{2+} and Li^+ ions are of similar size and S^{2-} and Cl^- ions are of similar size but the lattice energy of MgS is much more negative than the lattice energy of LiCl.

Worked example 19.01

Suggest the relative sizes of lattice energy for $CaCl_2$, CaS and Na_2S.

How to get the answer:

Step 1: State the charges on the ions: Ca^{2+}, Cl^-, S^{2-}, Na^+.

Step 2: State the relative sizes of the ions: Cl^- and S^{2-} are about the same size; Ca^{2+} and Na^+ are about the same size.

Step 3: Lattices with the highest ionic charges have the most negative lattice energy: CaS.

Step 4: With similar sizes and similar charges, Na_2S and $CaCl_2$ have similar lattice energies.

19.02 Born–Haber cycles

It is possible to calculate a lattice energy from the charges on the ions and their positions in the lattice but we usually calculate them from experimental data using a Hess cycle – see Figure 19.01.

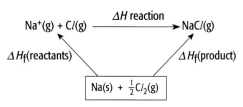

Figure 19.01 Forming the NaCl lattice

In Figure 19.01:

$\Delta H_{reaction}$ = the lattice energy of NaCl

ΔH_f[product] = the enthalpy of formation of NaCl(s) = $-411\,kJ\,mol^{-1}$

It is less clear how to evaluate ΔH_f[reactants] as it is made up of a number of components (see Unit 6 for enthalpy changes).

Solid sodium needs to be vaporised and ionised:

$$Na(s) \longrightarrow Na(g)$$
The enthalpy change of atomisation, $\Delta H_{at} = +107\,kJ\,mol^{-1}$

$$Na(g) \longrightarrow Na^+(g)$$
The first ionisation energy, $\Delta H_{ion} = +496\,kJ\,mol^{-1}$

The chlorine molecules need to be atomised and then made into negative ions:

$$½Cl_2(g) \longrightarrow Cl(g)$$
The enthalpy change of atomisation, $\Delta H_{at} = +121\,kJ\,mol^{-1}$

$$Cl(g) \longrightarrow Cl^-(g)$$
The electron affinity, $\Delta H_{ea} = -349\,kJ\,mol^{-1}$

The **electron affinity** is the energy change when each atom in 1 mole of gaseous atoms gains one electron to become a negative ion.

These energy changes are put into a Hess cycle known as a Born–Haber cycle (Figure 19.02).

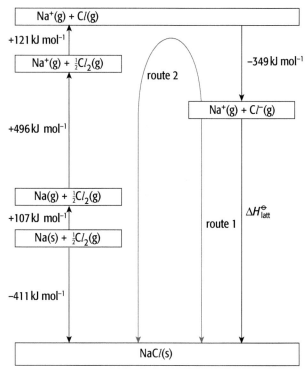

Figure 19.02 The Born–Haber cycle for NaCl

In Figure 19.02, most values are known and so the lattice energy can be calculated.

$$\Delta H_{latt} = -\Delta H_{at}[Na(s)] - \Delta H_{ion}[Na(g)] - \Delta H_{at}[Cl(g)] \\ - \Delta H_{ea}[Cl(g)] + \Delta H_f[NaCl]$$

$$= -[(+107) + (+496) + (+121) + (-349)] + \\ (-411) = -786\,kJ\,mol^{-1}$$

 If the route direction opposes the enthalpy change arrow, multiply the enthalpy change by -1.

Worked example 19.02

Calculate the lattice energy of $CaCl_2$ from the enthalpy values given.

$\Delta H_{at}[Ca(s)] = +178\,kJ\,mol^{-1}$

$\Delta H_{ion}[Ca(g)] = +590\,kJ\,mol^{-1}$

$\Delta H_{ion}[Ca^+(g)] = +1145\,kJ\,mol^{-1}$

$\Delta H_f[CaCl_2] = -796\,kJ\,mol^{-1}$

$\Delta H_{ea}[Cl(g)] = -349\,kJ\,mol^{-1}$

How to get the answer:

Step 1: The equation for the lattice energy:
$$Ca^{2+}(g) + 2Cl^-(g) \longrightarrow CaCl_2(s)$$

Step 2: The equation for the formation:
$$Ca(s) + Cl_2(g) \longrightarrow CaCl_2(s)$$

Step 3: What energy changes are needed for Ca(s) ⟶ Ca²⁺(g)? Atomisation and ionisation: Ca(s) ⟶ Ca(g) ⟶ Ca²⁺(g)

Step 4: How many ionisation energies are required? First and second.

Step 5: What energy changes are needed for $Cl_2(g)$ ⟶ $2Cl^-(g)$? Atomisation and electron affinity: $Cl_2(g)$ ⟶ $2Cl(g)$ ⟶ $2Cl^-(g)$

Step 6: How many electron affinities are needed? Two for $2Cl(g)$.

Step 7: Write the equation using the Born–Haber cycle: $\Delta H_{latt} = -\Delta H_{at}[Ca(s)] - \Delta H_{ion}[Ca(g)] - \Delta H_{ion}[Ca^+(g)] - 2\Delta H_{at}[Cl(g)] - 2\Delta H_{ea}[Cl(g)] + \Delta H_f[CaCl_2]$

Step 8: Insert the values: $\Delta H_{latt} = -[(+178) + (+590) + (+1145) + 2(+121) + 2(-349)] + (-796) = -2253 \text{ kJ mol}^{-1}$

To calculate the lattice energy of calcium oxide, you need to form an oxide ion as well as including the first two ionisation energies for calcium. The first electron added to the gaseous O atom releases energy but energy is needed to add a second electron to an already negative ion:

$\Delta H_{at}[O(g)] = +249 \text{ kJ mol}^{-1}$

$\Delta H_f[CaO(s)] = -635 \text{ kJ mol}^{-1}$

$\Delta H_{ea}[O(g)] = -141 \text{ kJ mol}^{-1}$

$\Delta H_{ea}[O^-(g)] = +798 \text{ kJ mol}^{-1}$

Progress check 19.01

1. Identify each of the energy changes in Figure 19.02 with the values in the equation to calculate the lattice energy.

2. For each of the following pairs of ionic solids, explain which would have the most negative lattice enthalpy and why:
   a. potassium bromide and potassium iodide
   b. potassium bromide and sodium oxide
   c. magnesium oxide and calcium sulfide.

3. Using values from the text above, show that the lattice enthalpy for CaO = $-3454 \text{ kJ mol}^{-1}$.

19.03 Thermal stability of Group 2 nitrates and carbonates

a Group 2 carbonates

We saw in Unit 11 that Group 2 carbonates are decomposed by heat but that they become more thermally stable down the group:

$MCO_3(s) \longrightarrow MO(s) + CO_2(g)$

The temperature of decomposition is lowest for $MgCO_3$ at 623 K and highest for $BaCO_3$ at 1123 K. The reason lies in the different ion sizes. The cationic radius increases down the group and so its polarising power decreases. The anion is relatively large (Figure 19.03).

Figure 19.03 Ion polarisation: a small highly charged cation can distort the shape of the anion but a larger cation cannot

As shown in Figure 19.03, as the cationic radius increases its polarising power decreases and so the polarisation of the large carbonate ion decreases down the group.

The carbonate ion is large; the electrons are easily pulled towards a small positive charge, distorting the electron cloud. Mg^{2+} can polarise the carbonate electron cloud the most but the distortion gets less as the positive ion gets larger. The carbonate in $MgCO_3$ is most distorted, breaking a carbon–oxygen bond and releasing $CO_2(g)$. As the distortion decreases down the group, a higher and higher temperature is needed to decompose the carbonate.

b Group 2 nitrates

All the Group 2 nitrates decompose:

$M(NO_3)_2(s) \longrightarrow MO(s) + 2NO_2(g) + \frac{1}{2}O_2(g)$

For similar reasons to the carbonates, the decomposition temperature is higher going down the group.

c Lattice energy and decomposition

The decomposition of both the carbonates and the nitrates produces the oxide. The O^{2-} ion is smaller than the NO_3^- or CO_3^{2-} ions and so the lattice energies of the oxides are much more negative than the lattice energies of the carbonates or nitrates. The difference in lattice energy is a driving force for the decomposition.

> ## Progress check 19.02
>
> A white solid is a mixture of barium carbonate and barium nitrate. State the observations when the following four tests are carried out on separate portions of the mixture and explain what has happened.
>
> 1. Shake a portion with water.
> 2. Heat a portion of the mixture and apply a glowing splint to any gas produced.
> 3. Heat a portion of the mixture with a Bunsen burner and then shake the residue with dilute hydrochloric acid.
> 4. Heat a portion of the mixture and then shake the residue with water and add universal indicator.

19.04 Formation of ionic solids and solutions

Above we discussed gaseous ions forming an ionic solid and giving out the lattice energy.

When gaseous ions dissolve in water, they produce hydrated ions. The positive and negative ions are surrounded by water molecules and kept in solution by ion–dipole attractive forces (Figure 19.04).

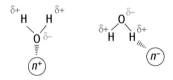

Figure 19.04 Ions surrounded by water molecules experience ion–dipole attractive forces

The $\delta-$ O atoms of H_2O molecules form ion–dipole attractive forces with the positive ions; the $\delta+$ H atoms form ion–dipole attractive forces with the negative ions.

These forces keep the ions in solution and the ions are hydrated. Ions getting hydrated release energy – the enthalpy change of hydration.

When ionic solids dissolve in water, energy is needed to pull the ions from their positions in the lattice and energy is then released when the ions are hydrated. The overall enthalpy change is the enthalpy change of solution.

19.05 The solubility of Group 2 sulfates and hydroxides

Group 2 sulfates become less soluble going down the group.

The energy change when an ionic solid dissolves in water is represented in Figure 19.05.

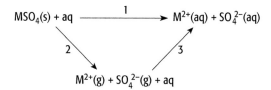

Figure 19.05 Solids dissolving can be thought of as the sum of two processes

- (1) is the enthalpy change of solution
- (2) is the reverse of the lattice energy
- (3) is the enthalpy of hydration.

(2) depends on the size of the ion and becomes less positive as the M^{2+} ion gets larger going down the group.

(3) depends on the charge density of the ions, which does not change for the sulfate ion. As the Group 2 ions get larger, their charge density gets smaller and the ion–dipole attraction between ion and water molecules gets weaker. The hydration enthalpy gets less negative.

From the Hess cycle:

the enthalpy change of solution = −lattice energy + hydration enthalpy

(− lattice energy is energy absorbed; hydration enthalpy is energy released)

The sulfate ion is so large, it masks changes in the small positive Group 2 ion when we consider the lattice energy so lattice energy changes are fairly small. However, when going into solution, the ions are being hydrated separately; the radius of the positive ion changes greatly and so the changes in hydration enthalpy are much larger.

TIP Both the lattice energy and the hydration enthalpy get smaller going down Group 2, but the hydration enthalpy changes more.

Thus the enthalpy change of solution becomes more positive going down the group and so the sulfates get less soluble.

The hydroxide ion is much smaller than the sulfate ion and for Group 2 hydroxides, the lattice energy decreases more than the hydration enthalpies as the group is descended. The enthalpy change of solution becomes less positive and the hydroxides become more soluble.

Sample answer

Question:

Explain why $CuSO_4$ is soluble but $BaSO_4$ is insoluble in water.

Answer:

The solubility depends on ΔH_{sol} [1 mark] which is equal to − lattice energy + hydration enthalpy [1 mark].

Cu^{2+} is much smaller than Ba^{2+} [1 mark], so $\Delta H_{latt}[CuSO_4]$ is greater than $\Delta H_{latt}[BaSO_4]$ [1 mark].

$\Delta H_{hyd}[Cu^{2+}]$ is greater than $\Delta H_{hyd}[Ba^{2+}]$ [1 mark].

SO_4^{2-} is large and masks changes in ΔH_{latt} [1 mark].

ΔH_{hyd} changes more than ΔH_{latt} [1 mark], ΔH_{sol} is less +ve for $CuSO_4$ [1 mark].

Progress check 19.03

The ionic radii of some positive ions in crystal lattices are given below:

Ba^{2+} 0.135 nm

Co^{2+} 0.065 nm

Mg^{2+} 0.072 nm

1. Is $CoSO_4$ soluble in water? Explain your answer.
2. Is $Co(OH)_2$ soluble in water? Explain your answer.

Revision checklist

Check that you know the following:

- [ ] explain and use the terms *electron affinity* and *lattice energy*
- [ ] explain the effect of ionic charge and radius on the numerical magnitude of a lattice energy
- [ ] apply Hess' Law to construct a Born-Haber cycle and use it to calculate enthalpy changes
- [ ] how to apply Hess' Law to find enthalpy changes applying to ionic solids and aqueous solutions
- [ ] interpret and explain the trend in the thermal stability of the nitrates and carbonates of Group 2 elements
- [ ] interpret and explain the variation in solubility of the Group 2 sulfates and hydroxides

Exam-style questions

1. a Define *lattice energy*. [2]

   b i A student was asked to calculate $\Delta H_{latt}[BeO]$. He looked up the following values:

   $\Delta H_f[BeO(s)]$; $\Delta H_{at}[Be(s)]$; first ionisation energy of Be(g); electron affinity of O(g)

   What other enthalpy values does he need? [3]

   ii Consider your answers in part i and, for each of the enthalpy values in your answer in turn, state whether the lattice energy calculated without including that enthalpy value will be more or less negative than the proper value. [3]

   c How would you expect $\Delta H_{latt}[BeO]$ to compare with $\Delta H_{latt}[MgS]$? [1]

   Total: 9

2. Explain why:

   a $MgSO_4$ is soluble but $BaSO_4$ is insoluble. [6]

   b $Mg(OH)_2$ is insoluble but $Ba(OH)_2$ is soluble. [4]

   Total: 10

Electrochemistry

Learning outcomes

You should be able to:

- predict which substance is liberated during electrolysis and calculate the mass or volume produced
- state and apply $F = Le$ and describe how to determine the Avogadro constant
- describe the standard hydrogen electrode and its use in measuring standard electrode potentials
- define *standard electrode potential* and *standard cell potential* and combine electrode potentials to calculate cell potentials
- deduce which way the electrons flow in an electrochemical cell and predict reaction feasibility
- apply knowledge of electrode potentials to the halogens
- predict how the electrode potential varies with concentration and use the Nernst equation
- state advantages of fuel cells and rechargeable batteries

20.01 $F = Le$

Electrolysis is the decomposition of an ionic substance (an **electrolyte**) by the passage of an electric current. The ions need to be free to move, so electrolysis can occur in molten or aqueous salts, bases and aqueous acids. Electrodes immersed in the electrolyte allow electric current to pass. Positive ions attracted to the negatively charged cathode are discharged by gaining electrons; electrons flow from the anode to the cathode (Figure 20.01).

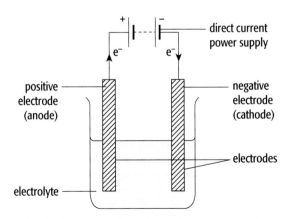

Figure 20.01 Electrons flow from the anode to the cathode

With $CuSO_4(aq)$ electrolyte, at the cathode:

$$Cu^{2+} + 2e^- \longrightarrow Cu(s)$$

1 mole of copper is deposited by 2 moles of electrons.

The charge on 2 moles of electrons = no. of electrons × charge on 1 electron = $2\,Le$

(L = Avogadro's number; e = charge on 1 electron)

The quantity of charge on a mole of electrons, Le, is the Faraday, F.

$F = Le = 6.022 \times 10^{23} \times 1.602 \times 10^{-19}$
$= 96\,500\,C\,mol^{-1}$

20.02 Substance liberated during electrolysis

a Molten lead bromide

In molten lead bromide, $PbBr_2$, the ions present are Pb^{2+} and Br^-.

At the cathode: $Pb^{2+} + 2e^- \longrightarrow Pb(l)$

At the anode: $2Br^- \longrightarrow Br_2(g) + 2e^-$

Lead and bromine are produced if the applied voltage overcomes the ionic attractions.

b Ions discharged from a mixture

The ion discharged from a salt mixture is the one which takes least energy. A molten mixture of lead bromide and lead chloride produces bromine at the anode as Br^- is more easily oxidised than Cl^-.

> Species are oxidised at the anode and reduced at the cathode.

c In aqueous solution

In aqueous solution, small amounts of H^+ and OH^- from the ionisation of water may be discharged.

d Other conditions

i Material of the electrodes

$CuSO_4$(aq) electrolysed with inert electrodes of platinum or carbon produces copper at the cathode and oxygen gas at the anode:

$$2H_2O(l) \longrightarrow O_2(g) + 4H^+(aq) + 4e^-$$

Using copper electrodes, Cu is oxidised at the anode and goes into solution as Cu^{2+}; the anode dissolves.

Similarly, $AgNO_3$(aq) using platinum electrodes produces O_2(g) at the anode but using silver electrodes, Ag goes into solution as Ag^+.

ii Concentration effects

The high activation energy for the discharge of hydrogen or oxygen may prevent their discharge. In the electrolysis of sodium chloride solutions, H^+ is discharged to form hydrogen gas at the cathode but Cl^- is preferentially discharged at the anode. Only at very low Cl^- concentrations is OH^- discharged.

20.03 Electrolysis calculations

An ammeter included in the circuit of Figure 20.01 can measure the current flowing through the cell; a variable voltage prevents the current changing as the solutions warm up. The amount of charge flowing is given by:

$$Q = I \times t$$

I = current in amps; t = time in seconds; Q = quantity of charge in coulombs.

Worked example 20.01

If a current of 0.5 A is passed for 2 hours, what mass of copper is deposited on the cathode?

How to get the answer:

Step 1: Work out the quantity of electricity
$= 0.5 \times 2 \times 60 \times 60$ C

Step 2: Quantity of electricity $= nF$; n = no. of moles of electrons, F = Faraday

Step 3: Work out n: $nF = 0.5 \times 2 \times 60 \times 60$
$n = \dfrac{(0.5 \times 2 \times 60 \times 60)}{96\,500} = 0.0373$ mol

Step 4: Write the equation:
$Cu^{2+}(aq) + 2e^- \longrightarrow Cu(s)$

Step 5: The number of moles of Cu
$= \frac{1}{2} \times 0.0373 = 0.0187$ mol

Step 6: Mass of Cu deposited
$= 0.0187 \times 63.5$ g $= 1.18$ g

Sample answer

Question:

0.65 A is passed through H_2SO_4(aq) for 1.5 hours. What volumes of hydrogen and oxygen, measured at 25 °C and 1 atm, are produced? [5]

Answer:

Cathode: $2H^+(aq) + 2e^- \longrightarrow H_2(g)$

Anode: $2H_2O(l) \longrightarrow O_2(g) + 4H^+(aq) + 4e^-$
[1 mark]

quantity of electricity $= 0.65 \times 1.5 \times 60 \times 60$ C
$= 3510$ C [1 mark] $= nF$

$n = \dfrac{3510}{96\,500}$ moles $= 0.0364$ mol [1 mark]

no. of moles of hydrogen $= \frac{1}{2} \times 0.0364$
$= 0.0182$ mol

volume of hydrogen $= 0.0182 \times 24$ dm³ $= 436$ cm³

volume of oxygen $= \frac{1}{4} \times 0.0364 \times 24$ dm³
$= 218$ cm³ [1 mark]

In practice, as oxygen is slightly soluble in water, a smaller volume of oxygen than this is collected [1 mark].

20.04 Determination of the Avogadro constant

An apparatus similar to Figure 20.01 with copper electrodes dipping in $CuSO_4(aq)$ and including an ammeter and a variable voltage supply can be used to determine the Avogadro constant.

The clean, dry cathode is weighed before being used. A current of 0.5 A is kept constant for 30 minutes. The cathode is then removed, dried and reweighed.

mass of cathode at start = m_1 g

mass of cathode at end = m_2 g

mass of copper produced = $(m_2 - m_1)$ g

quantity of electricity passed = $0.5 \times 30 \times 60$ C = nF = nLe

$A_r[Cu] = 63.5$, no. of moles Cu = $\dfrac{m_2 - m_1}{63.5}$

$Cu^{2+}(aq) + 2e^- \longrightarrow Cu(s)$

no. of moles electrons = $\dfrac{2(m_2 - m_1)}{63.5}$

charge on the electron = 1.602×10^{-19} C

$0.5 \times 30 \times 60 = \dfrac{2(m_2 - m_1)}{63.5} \times 1.602 \times 10^{-19} \times L$

Progress check 20.01

1. What volume of hydrogen is produced at the cathode when a current of 0.75 A is passed through concentrated sodium chloride solution for 25 minutes?

2. 123 cm³ of another gas is produced at the anode of the cell in question 1. What is the gas? Suggest why the volume produced is 123 cm³.

3. In the electrolytic production of aluminium, the current passed through molten alumina is high. What mass of aluminium is produced by passing a current of 100 000 A through molten alumina for 1 hour?

4. When a current of 0.80 A is passed through molten lead bromide for 45 minutes, the mass of the cathode increases by 4.33 g. Find a value for the Avogadro constant (charge on an electron = 1.602×10^{-19} C; A_r for lead = 207).

20.05 Electrode potentials and cell potentials

A copper rod immersed in water sets up an equilibrium (Figure 20.02); some copper ions enter the water, leaving their electrons on the rod:

$$Cu^{2+}(aq) + 2e^- \rightleftharpoons Cu(s)$$

A potential difference between the rod and the water is set up.

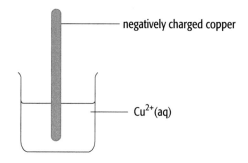

Figure 20.02 A piece of copper immersed in water

If copper is immersed in $CuSO_4(aq)$, fewer Cu^{2+} can go into solution so the potential difference between rod and solution will be smaller. Silver immersed in silver nitrate solution becomes less negative than copper as silver ions are less good at leaving.

Redox reactions are an electron competition. Copper immersed in silver nitrate solution disappears:

$$Cu(s) + 2Ag^+(aq) \longrightarrow Cu^{2+}(aq) + 2Ag(s)$$

Writing the oxidation and reduction reactions separately:

$$Cu(s) \longrightarrow Cu^{2+}(aq) + 2e^-$$

$$Ag^+(aq) + e^- \longrightarrow Ag(s)$$

We cannot measure the potential difference between a piece of metal and a solution of its ions but we can compare redox equilibria by carrying out the oxidation and reduction reactions separately (Figure 20.03).

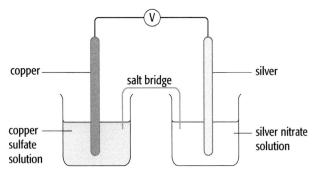

Figure 20.03 A silver half-cell combined with a copper half-cell

Both oxidised and reduced form of each redox reaction are needed so that they can move in either direction; in Figure 20.03, each beaker has a metal rod dipping into a solution of its ion. The high-resistance voltmeter allows little current to flow (otherwise, the potential difference falls). The solutions are in electrical contact using a salt bridge – a piece of filter paper soaked in potassium nitrate solution or a glass tube filled with potassium nitrate solution held in place with porous plugs. The salt bridge allows ion movement but keeps the solutions separate.

The potential of the cell is the voltage read from the voltmeter when no current flows in the circuit. It is the difference between the two electrode potentials. Electrode potentials are measured against a **standard hydrogen electrode**; which is assigned an electrode potential of 0.0 V.

20.06 The standard hydrogen electrode

The electrode equilibrium:

$$H^+(aq) + e^- \rightleftharpoons \tfrac{1}{2}H_2(g)$$

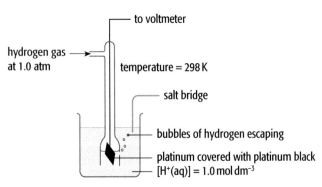

Figure 20.04 The standard hydrogen electrode

Hydrogen gas is bubbled into HCl(aq) over an inert platinum electrode – see Figure 20.04 for the conditions. The equilibrium is achieved slowly and so a powdered platinum catalyst (called platinum black) is used. The electrode potential of this half-cell is given the value 0.0 V; all other electrode potentials are measured against the standard hydrogen electrode.

20.07 Measuring standard electrode potentials

The **standard electrode potential**, $E^\ominus$, is the voltage produced when a half-cell is connected to a standard hydrogen electrode under standard conditions.

The standard conditions are:

- pressure of gases = 1.0 atm
- concentration of solutions = 1.0 mol dm^{-3}
- T = a defined temperature, usually 298 K

a **A metal dipping into a solution of one of its ions**

See Figure 20.05 for the apparatus used to measure the standard electrode potential between $Cu^{2+}(aq)$ and $Cu(s)$.

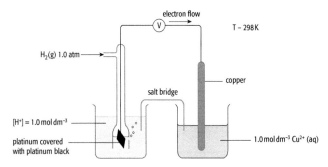

Figure 20.05 Measuring the standard electrode potential of the $Cu^{2+},Cu(s)$ electrode

Both the voltage and polarity of the electrodes are noted. Here, the voltage is 0.34 V and copper is positive; electrons flow towards the copper:

$$Cu^{2+}(aq) + 2e^- \rightleftharpoons Cu(s) \qquad E^\ominus = +0.34\,V$$

The reactions that actually occur:

reduction: $Cu^{2+}(aq) + 2e^- \longrightarrow Cu(s)$

oxidation: $H_2(g) \longrightarrow 2H^+(aq) + 2e^-$

The overall reaction is: $Cu^{2+}(aq) + H_2(g) \longrightarrow Cu(s) + 2H^+(aq)$

b Two different oxidation states in solution

Examples of redox equilibria are:

$$Cl_2(aq) + 2e^- \rightleftharpoons 2Cl^-(aq)$$
$$MnO_4^-(aq) + 8H^+(aq) + 5e^- \rightleftharpoons Mn^{2+}(aq) + 4H_2O(l)$$

An inert electrode, usually bright platinum, is needed to allow electrons to flow (Figure 20.06).

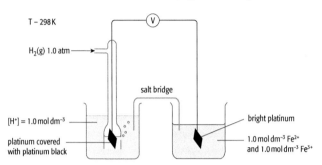

Figure 20.06 Measuring the standard electrode potential of two aqueous oxidation states.

Both the oxidised and reduced forms (Fe^{3+} and Fe^{2+}) have to be in the beaker so the reaction is reversible.

The right-hand electrode is positive and $E^\ominus[Fe^{3+}(aq) + e^-, Fe^{2+}(aq)] = +0.77\,V$:

The '+' sign shows that $Fe^{3+}(aq)$ is reduced more readily than $H^+(aq)$. A more positive electrode potential shows the oxidised form of an electrode is a stronger oxidising agent.

$E^\ominus[MnO_4^-(aq) + 8H^+(aq) + 5e^-, Mn^{2+}(aq) + 4H_2O(l)] = +1.51\,V$ shows that acidified $MnO_4^-(aq)$ is a much better oxidising agent than $Fe^{3+}(aq)$.

Standard electrode potentials are quoted for *reduction* electrode equilibria.

$Ag^+(aq) + e \rightleftharpoons Ag(s)$ $E^\ominus = +0.80\,V$

$Zn^{2+}(aq) + 2e \rightleftharpoons$ $E^\ominus = -0.76\,V$

See a Data Booklet for other values of electrode potentials.

> **TIP**
> When measuring a standard electrode potential, a half-cell must contain both the oxidised and the reduced form in standard concentrations.

20.08 Calculating standard cell potentials

The standard cell potential, $E^\ominus_{cell}$, is the voltage across two connected electrodes under standard conditions when no current is taken from the cell. It is also called the electromotive force, e.m.f.

$E^\ominus_{cell} = E^\ominus(\text{right-hand electrode}) - E^\ominus(\text{left-hand electrode})$

> ### Worked example 20.02
> What is the $E^\ominus_{cell}$ for $Ag^+(aq), Ag(s)$ connected to $Zn^{2+}(aq), Zn(s)$ and what is the cell reaction?
>
> How to get the answer:
>
> Step 1: Assume the silver electrode is at the left:
> $E^\ominus_{cell} = -0.76 - +0.8 = -1.56\,V$
>
> Step 2: The right-hand electrode has the sign of $E^\ominus_{cell}$: zinc is negative.
>
> Step 3: Oxidation occurs at the negative electrode: $Zn(s) \longrightarrow Zn^{2+}(aq) + 2e^-$
>
> Step 4: The silver electrode is positive:
> $Ag^+(aq) + e^- \longrightarrow Ag(s)$
>
> Step 5: The cell reaction is the sum of the oxidation and the reduction reactions in the right proportions to cancel the electrons:
> $Zn(s) + 2Ag^+(aq) \longrightarrow Zn^{2+}(aq) + 2Ag(s)$

20.09 Electron flow and feasibility of a reaction

The sign of $E^\ominus_{cell}$ is the sign of the right-hand electrode.

The positive electrode lacks electrons because they have been used in a reduction reaction. The negative electrode has excess electrons which have been left there after an oxidation reaction.

Electrons from the negative electrode move through the voltmeter to the positive electrode.

Using these ideas, you can decide whether a reaction is feasible.

Worked example 20.03

Does chlorine oxidise cobalt to $CoCl_2$ or $CoCl_3$?

How to get the answer:

Step 1: Find the electrode potentials:

$Cl_2(aq), 2Cl^-(aq)$ $E^\ominus = +1.36\,V$

$Co^{2+}(aq), Co(s)$ $E^\ominus = -0.28\,V$

$Co^{3+}(aq), Co^{2+}(aq)$ $E^\ominus = +1.82\,V$

Step 2: Assuming cobalt is at the left, $E^\ominus_{cell}$ + 1.36 − (−0.28) = +1.64 V

Step 3: From the polarity, the right-hand electrode is positive and where reduction occurs.

Step 4: Write the half-reaction equations:

$Cl_2(aq) + 2e^- \longrightarrow 2Cl^-(aq)$

$Co(s) \longrightarrow Co^{2+}(aq) + 2e^-$

Step 5: Put the half-reactions together:

$Cl_2(aq) + Co(s) \longrightarrow 2Cl^-(aq) + Co^{2+}(aq)$

Step 6: Repeat the process for $CoCl_2$ to $CoCl_3$: $E^\ominus_{cell} = +1.36 - 1.82 = -0.46\,V$

Step 7: The chlorine electrode is negative, where oxidation occurs.

Step 8: Write the equations:

$2Cl^-(aq) \longrightarrow Cl_2(aq) + 2e^-$

$Co^{3+}(aq) + e^- \longrightarrow Co^{2+}(aq)$

Step 9: Put the half-reactions together and sum up:

$2Cl^-(aq) + 2Co^{3+}(aq) \longrightarrow Cl_2(aq) + 2Co^{2+}(aq)$

Chlorine cannot oxidise $CoCl_2$ to $CoCl_3$ but instead $CoCl_3$ oxidises chloride to chlorine.

Progress check 20.02

1. Using standard electrode potentials from the text, what is the $E^\ominus_{cell}$, which is the positive electrode and where does oxidation take place for:

   a. a silver electrode connected to a copper electrode?

   b. a hydrogen electrode connected to a chlorine electrode?

2. Can $AlCl_3$ oxidise chromium to $CrCl_2$ or $CrCl_3$?

   $Al^{3+}(aq), Al(s)$ $E^\ominus = -1.66\,V$

   $Cr^{2+}(aq), Cr(s)$ $E^\ominus = -0.91\,V$

   $Cr^{3+}(aq), Cr(s)$ $E^\ominus = -0.74\,V$

3. Can $FeCl_3(aq)$ oxidise $Fe(s)$ to $FeCl_2(aq)$?

   $Fe^{2+}(aq), Fe(s)$ $E^\ominus = -0.44\,V$

   $Fe^{3+}(aq), Fe^{2+}(aq)$ $E^\ominus = +0.77\,V$

20.10 Balancing redox equations

In Unit 7, you balanced redox equations using oxidation numbers. Here we balance redox equations from half-reactions. The oxidation and reduction equations are combined in the right proportions to balance the electrons.

For the oxidation of $Fe^{2+}(aq)$ by acidified chromate(VI), the two half-reactions are:

$$Fe^{3+}(aq) + e^- \rightleftharpoons Fe^{2+}(aq)$$

$$Cr_2O_7^{2-}(aq) + 14H^+(aq) + 6e^- \rightleftharpoons 2Cr^{3+}(aq) + 7H_2O(l)$$

Write the first equation as an oxidation ×6 and the second as a reduction and then add them:

$$6Fe^{2+}(aq) + Cr_2O_7^{2-}(aq) + 14H^+(aq) \rightleftharpoons 6Fe^{3+}(aq) + 2Cr^{3+}(aq) + 7H_2O(l)$$

Check the balance in numbers of the different types of atoms and in charge.

20.11 Electrode potentials and the reactivity of the halogens

The standard electrode potentials for $X_2(aq) + 2e^- \rightleftharpoons 2X^-(aq)$ become less positive down Group 17 and so the halogens become weaker oxidising agents.

20.12 Variation of electrode potential with concentration

Ion concentrations affect redox equilibria.

$Cu^{2+}(aq) + 2e^- \rightleftharpoons Cu(s)$ moves to the left if $[Cu^{2+}(aq)]$ is lowered and more copper goes into solution as Cu^{2+} ions, leaving more electrons on the copper. The (reduction) electrode potential becomes less positive, which affects E_{cell}.

20.13 The Nernst equation

The quantitative effect of concentration changes on electrode potentials is given by the Nernst equation.

a For a metal dipping into a solution of its ions

electrode potential = $E = E^\ominus + \dfrac{2.3RT}{zF} \log_{10}[\text{ion}]$

z = number of electrons involved in the electrode reaction

2.3 is a conversion factor

R = molar gas constant

T = temperature (in K)

F = Faraday constant

At a constant temperature of 298 K, the factor $(2.3RT/F) = 0.059$ V and the equation becomes:

$E = E^\ominus + \dfrac{0.059}{z} \log_{10}[\text{ion}]$

For copper dipping into:

- $[Cu^{2+}] = 1.0 \text{ mol dm}^{-3}, \log_{10}(1.0) = 0$

  $E = E^\ominus = 0.34$ V

- $[Cu^{2+}] = 0.5 \text{ mol dm}^{-3}, \log_{10}(0.5) = -0.30$

  $E = E^\ominus - \dfrac{0.059}{2} \times 0.30 = 0.33$ V

- $[Cu^{2+}] = (1.0 \times 10^{-5}) \text{ mol dm}^{-3}$,

  $\log_{10}(1.0 \times 10^{-5}) = -5.0$

  $E = E^\ominus - \dfrac{0.059}{2} \times 5.0 = 0.19$ V

The effect is small but significant at very low concentrations.

b If the electrode system is two ions

$Fe^{3+}(aq) + e^- \rightleftharpoons Fe^{2+}(aq)$

The Nernst equation becomes:

$E = E^\ominus + \dfrac{0.059}{z} \log_{10} \dfrac{[\text{oxidised form}]}{[\text{reduced form}]}$

z = number of electrons transferred (here $z = 1$)

c For complex electrodes

$VO_2^+(aq) + 2H^+(aq) + e^- \rightleftharpoons VO^{2+}(aq) + H_2O(l)$:

$E = E^\ominus + \dfrac{0.059}{z} \log_{10} \dfrac{[VO_2^+(aq)][H^+(aq)]^2}{[VO^{2+}(aq)]}$

$[H^+]$ is included although $[H_2O]$ changes can be ignored.

d Combining Nernst equations

Cell potentials for two complex electrodes are found by combining the Nernst equations for the two electrodes.

> ### Progress check 20.03
>
> 1. What is the cell potential between copper dipping into 0.5 mol dm^{-3} $CuSO_4$ and zinc dipping into 0.5 mol dm^{-3} $ZnSO_4$?
>
> 2. What is the cell potential between copper dipping into $1.0 \times 10^{-3} \text{ mol dm}^{-3}$ $CuSO_4$ and zinc dipping into 1.0 mol dm^{-3} $ZnSO_4$?
>
> 3. What is the electrode potential of $MnO_4^-(aq) + 8H^+(aq) + 5e^- \rightleftharpoons Mn^{2+}(aq) + 4H_2O(l)$ if the acid is only 0.5 mol dm^{-3} although all the other species are 1.0 mol dm^{-3}?

TIP: Do not forget to include the number of transferred electrons in the Nernst equation.

20.14 The fuel cell and batteries

a Fuel cells

Burning fuels to get power is inefficient. A fuel cell converts fuel directly into electrical energy, is much more efficient and only water is produced.

Typically $H_2(g)$ (the fuel) loses electrons at the negative electrode and H^+ moves through the electrolyte to react with $O_2(g)$ at the positive electrode. The cells are stacked to produce higher voltages but they are light. Hydrogen, however, is explosive and difficult to store.

b **Batteries**

i Nickel / metal hydride batteries rely on the metal hydride producing hydrogen which then loses electrons, as in the fuel cell. The electrolyte is usually KOH. The cell can be recharged and keeps a constant voltage for a long time.

ii Lithium ion batteries rely on Li losing electrons. They are light, rechargeable and need low maintenance but provide a high voltage.

Revision checklist

Check that you know the following:

- [ ] state and apply $F = Le$
- [ ] predict the identity of the substance liberated during electrolysis
- [ ] carry out electrolysis calculations
- [ ] describe a method to find the Avogadro constant
- [ ] describe the standard hydrogen electrode and how to use it
- [ ] define 'standard electrode potential' and 'standard cell potential'
- [ ] describe how to measure standard electrode potentials
- [ ] combine two standard electrode potentials to calculate a standard cell potential and their use in predicting the direction of electron flow and feasibility of reaction
- [ ] use half reactions to construct redox equations
- [ ] use electrode potentials to determine the reactivity of the halogens
- [ ] predict how the value of an electrode potential varies with the ion concentration
- [ ] use the Nernst equation
- [ ] state the advantages of fuel cells and improved batteries

Exam-style questions

1. a Draw a labelled diagram of the apparatus needed to measure the standard electrode potential of $Fe^{3+}(aq) + e^- \rightleftharpoons Fe^{2+}(aq)$. [5]

   b The following two electrodes are combined in an electrochemical cell:

   $I_2(aq) + 2e^- \rightleftharpoons 2I^-(aq)$ $E^\ominus = +0.54\,V$

   $Fe^{3+}(aq) + e^- \rightleftharpoons Fe^{2+}(aq)$ $E^\ominus = +0.77\,V$

   i What is the $E^\ominus_{cell}$?

   ii Write the equation for the reaction which occurs. [3]

   c What is the electrode potential for the Fe^{3+}, Fe^{2+} electrode if both $[Fe^{3+}(aq)]$ and $[Fe^{2+}(aq)]$ are $0.50\,mol\,dm^{-3}$? [1]

   Total: 9

2. A coin is an alloy of 84% by moles of copper and the rest is nickel. It is the anode of a cell with $CuSO_4(aq)$ electrolyte. A current of 0.5 A is passed for 1 hour. What are the changes in mass of the cathode and anode? [6]

   Total: 6

Unit 21: Further aspects of equilibria

Learning outcomes

You should be able to:

- explain pH, K_a, pK_a and K_w and use them in calculations
- describe and explain the changes in pH during titration reactions
- choose a suitable acid base indicator for a titration
- explain how a buffer solution controls pH and calculate the pH of buffer solutions
- understand and use the solubility product in calculations
- state what is meant by partition coefficient and use a partition coefficient in calculations

21.01 pH, K_a, pK_a, K_w

The lower the pH, the more acidic is the solution; [H⁺(aq)] ranges from about $1.0\,\text{mol}\,\text{dm}^{-3}$ to $1.0 \times 10^{-14}\,\text{mol}\,\text{dm}^{-3}$. The pH scale is an easier way to represent acidity.

$$\text{pH} = -\log_{10}[\text{H}^+(\text{aq})]$$

Strong acids are fully ionised:

$$\text{HC}l(\text{aq}) \longrightarrow \text{H}^+(\text{aq}) + \text{C}l^-(\text{aq})$$

[H⁺(aq)] equals the concentration of the acid put into the solution.

Weak acids are only partially ionised and [H⁺(aq)] is far lower than the concentration of the acid put into solution. An equilibrium is set up:

$$\text{CH}_3\text{COOH}(\text{aq}) + \text{H}_2\text{O}(l) \rightleftharpoons \text{CH}_3\text{COO}^-(\text{aq}) + \text{H}_3\text{O}^+(\text{aq})$$

$$K_c = \frac{[\text{CH}_3\text{COO}^-(\text{aq})][\text{H}_3\text{O}^+(\text{aq})]}{[\text{CH}_3\text{COOH}(\text{aq})][\text{H}_2\text{O}(l)]}$$

The concentration of water in pure water $= \frac{1000}{18} = 55.6\,\text{mol}\,\text{dm}^{-3}$ and in a dilute solution changes very little from this value, so it can be taken as constant. A new constant, K_a, is defined:

$$K_a = \frac{[\text{CH}_3\text{COO}^-(\text{aq})][\text{H}_3\text{O}^+(\text{aq})]}{[\text{CH}_3\text{COOH}(\text{aq})]}$$

K_a is the acid dissociation constant and indicates how much ionisation of the acid takes place. Values of K_a range over many powers of 10 so pK_a is often used:

$$\text{p}K_a = -\log_{10} K_a$$

Water is not only the solvent but is central to the ionisation of acids and bases as it is slightly ionised:

$$\text{H}_2\text{O}(l) + \text{H}_2\text{O}(l) \rightleftharpoons \text{H}_3\text{O}^+(\text{aq}) + \text{OH}^-(\text{aq})$$

The equilibrium constant for this ionisation:

$$K_c = [\text{H}_3\text{O}^+(\text{aq})][\text{OH}^-(\text{aq})] / [\text{H}_2\text{O}(l)]^2$$

but as the concentration of water changes little, a new constant, K_w, the ionic product of water, is defined.

$$K_w = [\text{H}_3\text{O}^+(\text{aq})][\text{OH}^-(\text{aq})]$$

This important equilibrium constant links [H₃O⁺] and [OH⁻]; if one increases, then the other decreases.

At room temperature, $K_w = 1.0 \times 10^{-14}\,\text{mol}^2\,\text{dm}^{-6}$.

TIP Remember the value of K_w for use in calculations.

In pure water, [H₃O⁺(aq)] = [OH⁻(aq)] and so $1.0 \times 10^{-14} = [\text{H}_3\text{O}^+(\text{aq})]^2$

Thus [H₃O⁺(aq)] = $1.0 \times 10^{-7}\,\text{mol}\,\text{dm}^{-3}$ and pH = 7

21.02 Calculating H⁺ concentration and pH

a **Strong monobasic acids**

[H₃O⁺(aq)] is the same as the acid concentration in the solution.

pH of $0.50\,\text{mol}\,\text{dm}^{-3}$ HCl = 0.30

b **Weak acids**

K_a describes the amount of ionisation.

> ## Worked example 21.01
>
> Calculate the pH of a $0.50\,\text{mol}\,\text{dm}^{-3}$ CH_3COOH solution.
>
> How to get the answer:
>
> Step 1: Write the equilibrium:
> $$CH_3COOH(aq) + H_2O(l) \rightleftharpoons CH_3COO^-(aq) + H_3O^+(aq)$$
>
> Step 2: Write the K_a expression:
> $$K_a = [CH_3COO^-(aq)][H_3O^+(aq)] / [CH_3COOH(aq)]$$
>
> Step 3: Find the K_a value (from a data book):
> $$K_a = 1.7 \times 10^{-5}\,\text{mol}\,\text{dm}^{-3}$$
>
> Step 4: From the equilibrium equation: $[H_3O^+(aq)] = [CH_3COO^-(aq)]$
>
> Step 5: Consider $[CH_3COOH]$: the amount of ionisation is very small so the un-ionised acid concentration is little changed from the original concentration added to the solution $= 0.50\,\text{mol}\,\text{dm}^{-3}$.
>
> Step 6: Put the values in the K_a expression:
> $$1.7 \times 10^{-5} = \frac{[H_3O^+(aq)]^2}{0.50}$$
>
> Step 7: Work out $[H_3O^+(aq)] = 2.92 \times 10^{-3}\,\text{mol}\,\text{dm}^{-3}$
>
> Step 8: Work out $pH = -\log_{10} 2.92 \times 10^{-3} = 2.54$
>
> Generally, $[H_3O^+] = \sqrt{c \times K_a}$ where c = concentration of the acid.

c **Strong bases**

These are completely ionised in solution.

$[OH^-(aq)]$ in $0.5\,\text{mol}\,\text{dm}^{-3}$ NaOH $= 0.5\,\text{mol}\,\text{dm}^{-3}$

To calculate the $[H_3O^+(aq)]$, use $K_w = 1.0 \times 10^{-14}$
$= [H_3O^+(aq)] \times 0.5$

$$[H_3O^+(aq)] = \frac{1.0 \times 10^{-14}}{0.5} = 2.0 \times 10^{-14}\,\text{mol}\,\text{dm}^{-3}$$

$pH = -\log_{10}(2.0 \times 10^{-14}) = 13.70$

> ## Progress check 21.01
>
> 1. Calculate the $[H_3O^+(aq)]$ and pH of
>   a. $0.001\,\text{mol}\,\text{dm}^{-3}$ HCl
>   b. $0.05\,\text{mol}\,\text{dm}^{-3}$ HNO_3.
>
> 2. Using the K_a values given, calculate the $[H_3O^+(aq)]$ and pH of
>   a. $0.01\,\text{mol}\,\text{dm}^{-3}$ HCOOH, $K_a = 1.6 \times 10^{-4}\,\text{mol}\,\text{dm}^{-3}$
>   b. $0.005\,\text{mol}\,\text{dm}^{-3}$ C_6H_5OH, $K_a = 1.28 \times 10^{-10}\,\text{mol}\,\text{dm}^{-3}$.
>
> 3. Using the value of K_w given previously, calculate the $[H_3O^+(aq)]$ and pH of:
>   a. $0.005\,\text{mol}\,\text{dm}^{-3}$ KOH
>   b. $0.001\,\text{mol}\,\text{dm}^{-3}$ NaOH.

21.03 Changes of pH during titrations

When an acid is added to a base, the pH of the mixture, as measured with a pH meter, decreases. The curve of pH against the volume of acid added depends on whether the acid and base are strong or weak.

Although the reaction is called neutralisation, it is rare that the final mixture is neutral with a pH of 7, as we shall see.

An **acid–base indicator** changes colour at the **end point**, providing a visual method of knowing when the right amount of acid has been added.

The **equivalence point** is when the correct amount of acid has been added according to the equation.

The pH of the equivalence point and the end point should be the same; for this to happen, we need to choose the right indicator for the titration.

a **Titrating a strong acid against a strong base**

See Figure 21.01.

- $20\,\text{cm}^3$ of $0.1\,\text{mol}\,\text{dm}^{-3}$ NaOH is placed in the titration flask. $pH = 13.0$.
- $0.1\,\text{mol}\,\text{dm}^{-3}$ HCl is added from the burette. The pH falls gradually.

- Adding 19.5 cm³ of HCl, pH = 11.1; adding 20.1 cm³ of HCl, pH = 3.6. The pH drops abruptly around the equivalence point.

The equivalence point is at the centre of the abrupt pH drop at pH = 7. An indicator needs to change colour at a pH lying on the abrupt pH drop.

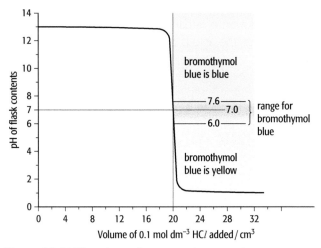

Figure 21.01 Titration of a strong base against a strong acid

b Titrating a strong acid against a weak base

See Figure 21.02.

- 20 cm³ of 0.1 mol dm⁻³ ammonia solution is placed in the titration flask. This time, the pH, approximately 11.1, is lower than for the NaOH(aq).
- As 0.1 mol dm⁻³ HCl is added, the pH falls rapidly and then the rate of fall slows as a mixture of the buffer solution (see section 21.05), ammonia and ammonium chloride, is formed.
- The pH drops at the equivalence point from approximately 7.5 to 3.5 with the equivalence point at approximately pH = 5.
- As a weak base is only partially ionised, the onset of the pH drop is at a much lower value than for a strong base.
- Any indicator needs to change colour in the pH range 7.5 to 3.5.
- The equivalence point when a strong acid is titrated with a weak base is below 7.

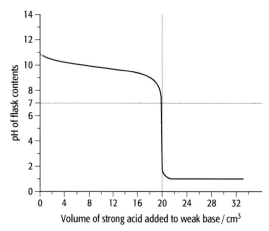

Figure 21.02 Titration of a strong acid against a weak base.

c Titrating a weak acid against a strong base

See Figure 21.03.

- 25 cm³ of 0.1 mol dm⁻³ NaOH, with a pH = 13, is placed in the flask and 0.1 mol dm⁻³ CH₃COOH is added from a burette.
- The main pH drop is from approximately pH = 12 to 7.0.
- The equivalence point is around pH = 9.5.
- The pH does not fall as far as for a strong acid.
- Any indicator needs to change colour within the range 7.0 to 11.9.

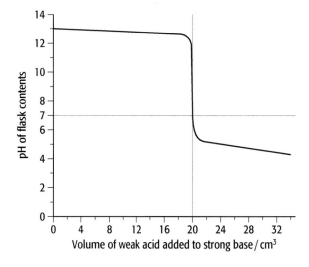

Figure 21.03 Titration of a weak acid against a strong base.

d Titration of a weak acid with a weak base

There is no clear pH drop and therefore no suitable indicator can be chosen.

21.04 Choice of indicators

Indicators show one colour at low pH and another at high pH. They are weak acids:

$$HIn(aq) + H_2O(l) \rightleftharpoons H_3O^+(aq) + In^-(aq) \quad [Eqn\ 1]$$

The ionisation constant: $K_{In} = \dfrac{[H_3O^+(aq)][In^-(aq)]}{[HIn]}$

$$pK_{In} = pH + \log_{10}\dfrac{[HIn(aq)]}{[In^-(aq)]} \quad [Eqn\ 2]$$

In acid solution, eqn 1 is pushed to the left; the indicator is mostly HIn.

In alkaline solution, $H_3O^+(aq)$ is removed and eqn 1 moves to the right; the indicator is mostly In^-.

HIn gives the colour in acid and In^- the colour in alkali.

At the end point, $[HIn(aq)] = [In^-(aq)]$ so $pK_{In} = pH$.

Different indicators have different pK_{In} values and change colour at different pH values – see Table 21.01.

Indicator	pK_{In}	Colour change from low to high pH
methyl orange	3.7	red to yellow
bromophenol blue	4.0	yellow to blue
methyl red	5.1	red to yellow
bromothymol blue	7.0	yellow to blue
phenolphthalein	9.3	colourless to red
alizarin yellow R	12.5	yellow to orange

Table 21.01 pK_{In} values for some indicators

A suitable indicator for a titration has a pK_{In} value lying within the sudden pH drop.

Progress check 21.02

Suggest indicators from Table 21.01 which would show the end point for the following titrations.

1. $0.1\ mol\,dm^{-3}$ HBr against $0.5\ mol\,dm^{-3}$ KOH
2. $0.1\ mol\,dm^{-3}$ HCl against $0.1\ mol\,dm^{-3}$ CH_3NH_2
3. $0.1\ mol\,dm^{-3}$ CCl_3COOH against $0.1\ mol\,dm^{-3}$ NaOH
4. $0.5\ mol\,dm^{-3}$ CH_3COOH against $0.1\ mol\,dm^{-3}$ NaOH.

21.05 Buffer solutions

a **Definition**

A buffer solution is one which minimises any change in pH when small amounts of acid or base are added. It is a solution of a weak acid with one of its salts or a weak base with one of its salts.

For a solution of ethanoic acid and sodium ethanoate, the acid is partially ionised:

$$K_a = 1.7 \times 10^{-5} = \dfrac{[H_3O^+(aq)][CH_3COO^-(aq)]}{[CH_3COOH(aq)]}\ mol\,dm^{-3}$$

Sodium ethanoate is completely ionised:

$$CH_3COO^-Na^+(s) + aq \longrightarrow CH_3COO^-(aq) + Na^+(aq)$$

In this mixture, there are appreciable concentrations of both un-ionised acid (CH_3COOH) and its conjugate base (CH_3COO^-) and for this buffer, $[H_3O^+(aq)] \neq [CH_3COO^-(aq)]$.

b **Finding the pH of a buffer solution**

The un-ionised acid concentration is the concentration of the acid at the start less a small amount which has ionised. For weak acids, this amount is very small and can be ignored.

The concentration of the conjugate base is the concentration of the salt plus a small amount from the ionisation of the acid. Again, this very small amount can be ignored.

The expression for K_a becomes:

$$K_a = \dfrac{[H_3O^+(aq)][salt]}{[acid]}$$

As K_a and the two starting concentrations are known, the $[H_3O^+(aq)]$ and pH can be calculated.

c **How does a buffer solution work?**

A buffer solution needs high concentrations of both the acid and its conjugate base in the solution.

$$CH_3COOH(aq) + H_2O(l) \rightleftharpoons CH_3COO^-(aq) + H_3O^+(aq)$$

If acid is added, the equilibrium moves to the left to remove it.

If base is added, it reacts with $H_3O^+(aq)$ to form H_2O and the equilibrium moves to the right.

Worked example 21.02

What is the pH of a buffer solution made by adding 50 cm³ of 0.1 mol dm⁻³ CH_3COOH to 50 cm³ 0.1 mol dm⁻³ CH_3COONa?
$K_a[CH_3COOH] = 1.7 \times 10^{-5}$ mol dm⁻³.

How to get the answer:

Step 1: Look for the initial concentrations:
$[CH_3COOH] = [CH_3COONa] = 0.05$ mol dm⁻³

Step 2: Write down the K_a expression for a buffer solution:
$$K_a = \frac{[H_3O^+(aq)][\text{salt}]}{[\text{acid}]}$$

Step 3: Use the value of K_a:
$$1.7 \times 10^{-5} = \frac{[H_3O^+(aq)](0.05)}{(0.05)}$$

Step 4: Calculate $[H_3O^+(aq)] = 1.7 \times 10^{-5}$ mol dm⁻³

Step 5: Find pH = $-\log_{10}[H_3O^+(aq)] = 4.77$.

TIP The best buffering action is when the acid and its salt have equal concentrations.

d Controlling the pH in blood

For proteins to remain active, a pH in the range 6.8 to 7.8 has to be maintained. This is achieved by a hydrogencarbonate buffering system coupled with breathing out carbon dioxide:

$$CO_2(g) + H_2O(l) \rightleftharpoons H_2CO_3(aq) \rightleftharpoons HCO_3^-(aq) + H^+(aq)$$

After exercise, acid enters the blood and reacts with HCO_3^- – pushing the equilibria to the left and finally expelling CO_2. Thus the pH is maintained.

e Other uses of buffer solutions

- Cosmetics are buffered to prevent ingredients such as sodium hydroxide irritating the skin.
- Baby shampoo is buffered at a pH of 6.0.
- Paints are buffered to prevent any caustic action from their use.

Progress check 21.03

1 Fill in the blanks in the following table.

Acid	K_a / mol dm⁻³	Concentration of acid in solution / mol dm⁻³	Concentration of sodium salt in solution	$[H_3O^+(aq)]$ / mol dm⁻³	pH
CH_3COOH	1.7×10^{-5}	0.1	0.5 mol dm⁻³		
HCOOH	1.6×10^{-4}	1.0			3.5
C_6H_5OH		0.1	0.2 mol dm⁻³	6.4×10^{-11}	
CH_3COOH	1.7×10^{-5}	0.5	2.5 g in 100 cm³		

2 What is the pH of the buffer solution made by dissolving 0.34 g sodium methanoate, HCOONa, in 100 cm³ of 0.1 mol dm⁻³ methanoic acid? The K_a(HCOOH) = 1.6×10^{-4} mol dm⁻³.

3 Using an acid from the table, suggest how you could make a buffer solution of pH = 4.8.

21.06 Solubility product and common-ion effect

a Solubility product

Many insoluble salts have a small solubility, producing a few ions in solution and setting up an equilibrium which lies well over to the left:

$$AgCl(s) \rightleftharpoons Ag^+(aq) + Cl^-(aq)$$

$$K_c = \frac{[Ag^+(aq)][Cl^-(aq)]}{[AgCl(s)]}$$

[AgCl(s)] depends on density, which is constant, so the **solubility product**, K_{sp}, is defined as:

$$K_c \times [AgCl(s)] = K_{sp} = [Ag^+(aq)][Cl^-(aq)]$$

For $Fe(OH)_2$, $Fe(OH)_2(s) \rightleftharpoons Fe^{2+}(aq) + 2OH^-(aq)$

$$K_{sp} = [Fe^{2+}(aq)][OH^-(aq)]^2$$

b Calculating a concentration from a solubility product

For AgI: $K_{sp} = [Ag^+(aq)][I^-(aq)]$
$= 8.0 \times 10^{-17}$ mol^2 dm^{-6}

But $[Ag^+(aq)] = [I^-(aq)]$,
so $[Ag^+(aq)]^2 = 8.0 \times 10^{-17}$

$[Ag^+(aq)] = 8.9 \times 10^{-9}$ mol dm^{-3}

To find the units of the solubility product, insert the concentration units into the expression.

As the [AgI] in the solution is the same as [Ag$^+$], the solubility of AgI is 8.9×10^{-9} mol dm^{-3}.

c Calculating a solubility product from the solubility

For $ZnCO_3$, $K_{sp} = [Zn^{2+}(aq)][CO_3^{2-}(aq)]$ and $[Zn^{2+}(aq)] = [CO_3^{2-}(aq)]$.

The solubility of $ZnCO_3$ is 3.74×10^{-6} mol dm^{-3} = concentration of $Zn^{2+}(aq)$.

$K_{sp} = [Zn^{2+}(aq)]^2 = (3.74 \times 10^{-6})^2$
$= 1.4 \times 10^{-11}$ mol^2 dm^{-6}

d The common-ion effect

AgCl(s) in pure water: $AgCl(s) \rightleftharpoons Ag^+(aq) + Cl^-(aq)$

If Cl^- ions are added, the equilibrium is driven to the left and less AgCl dissolves. Any common ion added will lower the solubility.

K_{sp} for AgCl $= 2.0 \times 10^{-10}$ mol^2 dm^{-6};

$[Ag^+(aq)]$ in pure water $= 1.4 \times 10^{-5}$ mol dm^{-3}

For AgCl in 1.0 mol dm^{-3} NaCl(aq),

$2.0 \times 10^{-10} = [Ag^+(aq)] \times 1.0$

so now $[Ag^+(aq)] = 2.0 \times 10^{-10}$ mol dm^{-3}, which is very much less than in pure water!

21.07 The partition coefficient

A solute shaken with two immiscible solvents divides between the two solvents. At equilibrium, the solute is divided in a definite ratio called the **partition coefficient**, which is the ratio of the solubilities of the solute in the two solvents.

Worked example 21.03

What is the solubility of $Co(OH)_2$?

$K_{sp}[Co(OH)_2] = 6.3 \times 10^{-16}$ mol^3 dm^{-9}

How to get the answer:

Step 1: Write the equilibrium:

$Co(OH)_2(s) \rightleftharpoons Co^{2+}(aq) + 2OH^-(aq)$

Step 2: Write the K_{sp} expression:

$K_{sp} = [Co^{2+}(aq)][OH^-(aq)]^2 =$
6.3×10^{-16} mol^3 dm^{-9}

Step 3: Work out the relative concentrations:

$[OH^-(aq)] = 2[Co^{2+}(aq)]$

Step 4: Insert into K_{sp}:

$4[Co^{2+}(aq)]^3 = 6.3 \times 10^{-16}$ mol^3 dm^{-9}

Step 5: Evaluate the concentration.

$[Co^{2+}(aq)] = \sqrt[3]{\frac{6.3 \times 10^{-16}}{4}} = 5.4 \times 10^{-6}$ mol dm^{-3}

Step 6: Decide which ion concentration gives the solubility. $[Co(OH)_2] = [Co^{2+}]$ so solubility of $Co(OH)_2 =$
5.4×10^{-6} mol dm^{-3}

Partition coefficient = $\dfrac{\text{[equilibrium concentration in mol dm}^{-3}\text{ in first solvent]}}{\text{[equilibrium concentration in mol dm}^{-3}\text{ in second solvent]}}$

For example, if 5 g of solute dissolved in 100 cm³ of solvent A were shaken with 100 cm³ of solvent B until equilibrium is reached, the solute might divide so that just 1 g of solute is left in A. Solvent B therefore dissolves 4 g of solute.

$[\text{solute}]_A = (1/M_r) / 0.1$; $[\text{solute}]_B = (4/M_r) / 0.1$

partition coefficient = $\dfrac{(1/M_r) / 0.1}{(4/M_r) / 0.1} = 0.25$

Sample answer

Question:

5.0 g of solute is dissolved in 100 cm³ of solution. 100 cm³ of an immiscible solvent is available to extract the solute. The partition coefficient = 1.35. How much solute can be extracted if the immiscible solvent is added in two 50 cm³ portions? [6]

Answer:

For first extraction, let x = mass of solute remaining in solution and so $(5 - x)$ = mass extracted

partition coefficient = 1.35
= $(x/M_r) / 0.1 / ((5 - x)/M_r) / 0.05$ [1 mark]

$x = 3.65$ g [1 mark]

For second extraction, 1.35 g remains in solution [1 mark].

y = mass of solute remaining in solution and so $(1.35 - y)$ = mass extracted

$1.35 = y/0.1 / (1.35 - y)/0.05$ [1 mark], $y = 0.99$ g [1 mark]

total extracted = 4.64 g [1 mark]

TIP

More solute is extracted if the solvent is added in small portions rather than one amount but it takes more effort in separating small portions.

An aqueous mixture, including the desired product, may result after a reaction. An immiscible solvent, chosen to dissolve just the desired product, is shaken with the aqueous mixture and then allowed to settle in a separating funnel. The aqueous layer is discarded. The solvent containing the desired product is dried and fractionally distilled to give the pure product.

Progress check 21.04

1. Write K_{sp} expressions for the following substances:

   $NiCO_3$ $Cr(OH)_3$ PbI_2 $PbCrO_4$ $Mg(OH)_2$ $AgBr$

2. What is the solubility of $PbBr_2$ if $K_{sp} = 3.9 \times 10^{-5}$ mol³ dm⁻⁹?

3. What is the solubility of $PbBr_2$ in 0.1 mol dm⁻³ NaBr solution?

4. a 1,3-dinitrobenzene is 31 times more soluble in octan-1-ol than in water. What is the partition coefficient of 1,3-dinitrobenzene between octan-1-ol and water?

   b The solubility in water of 1,3-dinitrobenzene is 500 mg dm⁻³. 100 cm³ of a saturated aqueous solution of 1,3-dinitrobenzene is shaken with 100 cm³ of octan-1-ol until an equilibrium is established. How much 1,3-dinitrobenzene is left in the water?

   c How much more 1,3-dinitrobenzene can be extracted from 100 cm³ of a saturated aqueous solution if the octan-1-ol is added in two amounts of 50 cm³ instead of one amount of 100 cm³?

Revision checklist

Check that you know the following:

- explain and use pH, K_a, pK_a and K_w in calculations

- calculate pH for strong acids and bases and weak acids and buffer solutions

- describe and explain the changes in pH during acid base titrations and how to choose a suitable indicator

- explain how a buffer solution works including the buffering action in blood

- calculate the pH of a buffer solution

- define and use the solubility product in calculations

- show an understanding of the common ion effect

- state and carry out calculations using the partition coefficient

Exam-style questions

1. a Explain what a *buffer solution* is. [3]

   b What is the pH of a 0.10 mol dm^{-3} solution of ethanoic acid (CH_3CO_2H)?

   $K_a = 1.7 \times 10^{-5}$ mol dm^{-3} [2]

   c What are the pH values of the following mixtures?

   i   100 cm^3 of 0.10 mol dm^{-3} ethanoic acid mixed with 100 cm^3 of 0.1 mol dm^{-3} HCl.

   ii 100 cm^3 of 0.10 mol dm^{-3} ethanoic acid mixed with 150 cm^3 of 0.10 mol dm^{-3} NaOH.

   iii 100 cm^3 of 0.10 mol dm^{-3} ethanoic acid mixed with 50 cm^3 of 0.10 mol dm^{-3} NaOH. [7]

   Total: 12

2. a i   Define *partition coefficient*. [2]

   ii Describe how you can extract a solute from an aqueous solution using a solvent. [5]

   b 100 cm^3 of an aqueous solution contains 4.0 g of solute. 100 cm^3 of solvent with a partition coefficient of 3.8, is available to extract the solute.

   Calculate the mass of solute extracted by the following amounts of solvent:

   i   All 100 cm^3 of solvent is shaken with the aqueous solution. [2]

   ii The solvent is added as two separate 50 cm^3 volumes and these two are then combined. [3]

   iii Suggest how you could get the most solute from an aqueous solution. [2]

   Total: 14

Unit 22: Reaction kinetics

Learning outcomes

You should be able to:

- construct and use rate equations and explain *reaction order*, *rate constant* and *rate determining step*
- use concentration time graphs
- explain, calculate and use a half life
- calculate initial rates and deduce orders of reaction
- interpret graphs and calculate from other data to find orders of reaction and rate constants
- predict a reaction mechanism from a rate equation
- devise a method for studying a rate of reaction
- explain the effect of temperature on the rate constant
- outline the characteristics of homogeneous, heterogeneous and enzyme catalysis

22.01 Rate equations

We saw before that reaction rates depend on temperature, reactant concentrations and catalysts. For this discussion, we assume a constant temperature and look at the steps that the reactants take to make the products. Some steps may be slow and some fast; slow steps determine how fast the reaction proceeds and so we cannot decide how each reactant affects the reaction rate by looking at the overall equation.

Here is an example: iodine reacts with propanone in acid solution:

$$I_2(aq) + CH_3COCH_3(aq) \longrightarrow CH_2ICOCH_3(aq) + HI(aq)$$

The reaction rate depends on $[CH_3COCH_3(aq)]$ and [acid] but not on $[I_2(aq)]$.

In general, the reaction rate depends on reactant and catalyst concentrations raised to a power which is found experimentally:

Overall equation: $A + B \xrightarrow{\text{catalyst C}}$ products

Rate equation: rate = $k[A]^x[B]^y[C]^z$

- x, y and z are the **orders of the reaction** with respect to A, B and C respectively.
- $(x + y + z)$ = the overall order of the reaction.
- Square brackets mean concentration in mol dm^{-3}.

- k is the constant of proportionality in the experimentally determined rate equation and is called the **rate constant**.

The order of a reactant is the power to which the concentration is raised in the experimentally determined rate equation. It is found experimentally by altering one concentration whilst the others are kept constant.

If the above general reaction is investigated by altering [A], the other constant concentrations can become part of the rate constant:

$$\text{rate} = k[A]^x$$

If $x = 0$ this is called a zero-order reaction, if $x = 1$ it is a first-order reaction and if $x = 2$ it is a second-order reaction, with respect to A.

The rate of $(CH_3)_3CCl$ hydrolysis can be investigated using a large water concentration and a small measured amount of the halogenoalkane in ethanol as solvent. The $[H_2O(l)]$ hardly changes as it is large.

$$(CH_3)_3CCl + H_2O(l) \longrightarrow (CH_3)_3COH + HCl(aq)$$

The rate is found to be first order with respect to the halogenoalkane:

$$\text{rate} = k[(CH_3)_3CCl]$$

Concentration units are usually mol dm^{-3} and rate units are usually $\text{mol dm}^{-3}\text{s}^{-1}$ (although concentration and time units given in a particular question do need to be followed). The rate constant has dimensions depending on the order of the reaction.

For a first-order reaction:

$$(\text{mol dm}^{-3}\text{s}^{-1}) = (\text{units of } k)(\text{mol dm}^{-3})$$
$$\text{units of } k = \text{s}^{-1}$$

22.02 Concentration/time graphs

If the reactant concentration is found at measured time intervals and the values plotted graphically, the curve produced depends on the order of reaction (Figure 22.01).

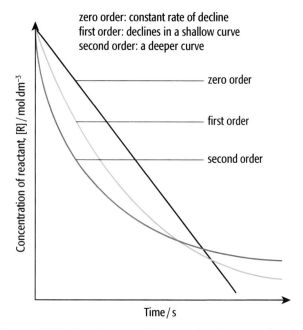

Figure 22.01 The shapes of the graphs for zero-, first- and second-order reactions

The zero-order reaction produces a straight line, but the other two produce curves which look similar. The gradient at any point is a measure of the reaction rate at that point. The gradient is found by drawing a tangent to the curve (Figure 22.02).

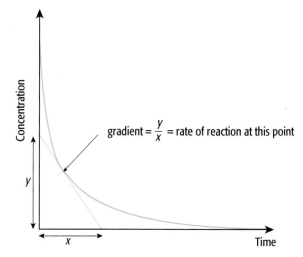

Figure 22.02 To find a gradient, draw a tangent to the curve

A zero-order reaction has a constant gradient:

$$\text{rate} = k$$

The gradient of a first-order reaction depends on the concentration:

$$\text{rate} = k[\text{reactant}]$$

The gradient also depends on concentration for a second-order reaction:

$$\text{rate} = k[\text{reactant}]^2$$

Plotting rates against concentrations shows the difference between the three orders (Figure 22.03).

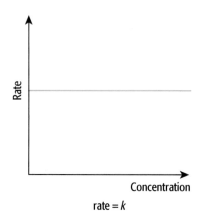 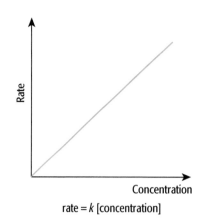 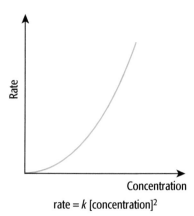

Figure 22.03 Rate/concentration graphs

Progress check 22.01

1. Find the units of k for zero- and second-order reactions.

2. For the following reaction and data in Table 22.01, plot a concentration/time graph. By taking gradients at five suitable points, plot a rate/concentration graph. Deduce the order of the reaction with respect to N_2O_5.

$$2N_2O_5(g) \longrightarrow 2NO_2(g) + O_2(g)$$

$[N_2O_5(g)]$ / mol dm^{-3}	t / s
0.2	0
0.142	50
0.112	100
0.088	150
0.065	200
0.050	250
0.037	300
0.029	350
0.023	400

Table 22.01 Data for the decomposition of $N_2O_5(g)$

3. The rate of the gaseous reaction between hydrogen and nitrogen monoxide is followed by monitoring the reduction in pressure. The $[H_2(g)]$ is constant at 0.02 mol dm^{-3} and the $[NO(g)]$ is much smaller. The data in Table 22.02 show the rates of reaction with NO(g) concentration. By plotting a rate/concentration graph, decide whether the value of m in the rate equation is 0, 1 or 2.

$$2H_2(g) + 2NO(g) \longrightarrow N_2(g) + 2H_2O(g)$$
$$\text{rate} = k[NO(g)]^m$$

$[NO(g)]$ / mol dm^{-3}	Rate / mol dm^{-3} s^{-1}
0.0040	4.0×10^{-7}
0.0035	3.06×10^{-7}
0.0030	2.25×10^{-7}
0.0025	1.56×10^{-7}
0.0020	1.0×10^{-7}
0.0015	5.63×10^{-8}
0.0010	2.50×10^{-8}
0.0005	6.25×10^{-9}

Table 22.02 NO(g) concentration and rate of reaction

TIP
When plotting graphs, label the axes clearly with the units and ensure you have linear scales. When drawing the line of best fit, use a sharp pencil and draw only one line and not several sketches.

22.03 Using half-life to find the order of reaction

The **half-life**, $t_{1/2}$, is the time taken for half a reactant concentration to be used up.

It is useful to distinguish between first- and second-order reactions:

- for a first-order reaction, the half-life is constant
- for a second-order reaction the half-life gets larger as the reaction proceeds.

Plotting a concentration/time graph and finding two half-lives enables you to distinguish between first- and second-order reactions (Figure 22.04).

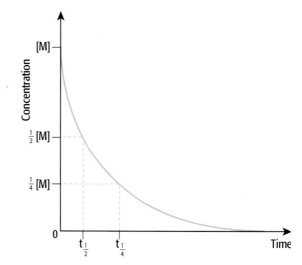

Figure 22.04 A concentration/time graph showing half-lives

For a first-order reaction, $t_{1/2} = (t_{1/4} - t_{1/2})$. For a second-order reaction, $t_{1/2} < (t_{1/4} - t_{1/2})$.

The half-life can also be used to evaluate the rate constant for a first-order reaction:

$$k = 2.3 \log_{10} 2 / t_{1/2} = 0.693 / t_{1/2}$$

> ## Progress check 22.02
>
> 1. For Progress check 22.01, question 2, find two consecutive half-lives and state whether they show the reaction is first or second order.
>
> 2. The reaction between hydrogen and nitrogen monoxide is first order with respect to hydrogen. Use the value of m you calculated in Progress check 22.01, question 3.
>
>   If $[NO(g)] = 0.0030 \, mol \, dm^{-3}$; $[H_2] = 0.02 \, mol \, dm^{-3}$; the rate $= 2.25 \times 10^{-7} \, mol \, dm^{-3} \, s^{-1}$.
>
>   Find a value for the rate constant. Include the units in your answer.

22.04 The initial rate method for finding an order of reaction

The initial rate is the reaction rate at the start of the reaction. This method depends on finding the initial rate for a range of experiments, each of which differs in the concentration of just one reactant. All other concentrations (and the temperature) are kept the same.

There are two ways of proceeding:

Draw concentration/time graphs for each experiment and draw the tangent at $t = 0$. However, drawing accurate tangents at this point can prove difficult. This is suitable for an experiment such as calcium carbonate reacting with aqueous hydrochloric acid. Carbon dioxide is given off and the volume of gas collected can be measured at known time intervals (Figure 22.05).

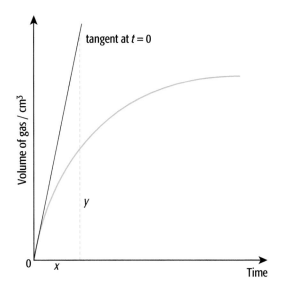

Figure 22.05 Volume of CO_2 against time

The initial rate of reaction is the gradient of the tangent drawn at $t = 0$.

For some reactions, we can use a 'clock' method which depends on letting each reaction mixture get to the same point when there is a very obvious visual change. For this to work, the reaction must use only small amounts of reactants before the visual change happens so that we can assume the reactant concentrations are unchanged from the initial concentrations. Each experiment is timed to the visual change. The reciprocal of this time is a measure of the rate of reaction.

This method is suitable for reactions in which iodine is produced. Iodine in low concentrations is pale yellow but with a small amount of starch produces an intense blue colour.

A range of experiments with known reactant concentrations including a measured amount of starch is set up. To each mixture, the same small measured amount of sodium thiosulfate solution is added; this reacts with any iodine produced. When the thiosulfate has been used up, the iodine and starch gives the strong blue coloration. The time taken to produce the colour is a measure of the initial rate of reaction – the longer the time, the slower the rate.

> **TIP**
>
> When dealing with data, take care to identify the question as either concentration/time or concentration/initial rate. Concentration/time data follow one concentration of an experiment with time; initial rate data show the times for a range of experiments to get to the same point and the rate of reaction is proportional to 1/time.

22.05 Calculating initial rates and rate constants

As we have seen, the initial rate from a concentration/time graph is the gradient of the tangent drawn at $t = 0$. Using the clock method, if $t =$ time for the colour to appear, $1/t$ is proportional to the rate of reaction.

The rate constant can be found once the orders have been established. The initial concentrations, orders and initial rate are inserted in the rate equation and the only unknown, k, can be evaluated.

Worked example 22.01

For the reaction, D + E ⟶ products, the initial rates for three experiments are shown in Table 22.03.

Experiment	[D] / mol dm^{-3}	[E] / mol dm^{-3}	Initial rate / mol dm^{-3} s^{-1}
1	0.050	0.30	1.80×10^{-4}
2	0.025	0.30	1.81×10^{-4}
3	0.050	0.15	8.92×10^{-5}

Table 22.03 Data of initial rates

Find the orders of reaction, write the rate equation and evaluate k.

How to get the answer:

Step 1: Choose two experiments using the same concentration for one reactant and find how the rate varies with the other reactant. From experiments 1 and 2, [D] is halved, the rate does not change and so the rate does not depend on [D]. The order with respect to D = 0.

Step 2: Repeat for the other reactants. From experiments 1 and 3, [D] stays the same but [E] is halved:

$1.80 \times 10^{-4} / 8.92 \times 10^{-5} = k(0.30)^e / k(0.15)^e$

$2.02 = 2^e$ so $e = 1$

The order with respect to E = 1.

Step 3: Use the orders to write the rate equation:

rate = $k[E]$

Step 4: Use the rate equation and data from one experiment in the table:

$1.80 \times 10^{-4} = k \times 0.30$

Step 5: Evaluate $k = 6.0 \times 10^{-4}$

Step 6: Put the units in the rate equation to find the units of k:

$(mol\,dm^{-3}\,s^{-1}) = (units\,of\,k)(mol\,dm^{-3})$

units of $k = s^{-1}$

Progress check 22.03

1 For the reaction between nitrogen monoxide and hydrogen at a higher temperature than we used before,

$2NO(g) + 2H_2(g) \longrightarrow N_2(g) + 2H_2O(l)$

use the initial concentration data in Table 22.04 and show that the rate equation is

rate = $k[NO(g)]^2[H_2(g)]$

Evaluate the rate constant.

Experiment	[NO(g)] / mol dm^{-3}	[H$_2$(g)] / mol dm^{-3}	Initial rate / mol dm^{-3} s^{-1}
1	0.53	0.195	0.483
2	0.53	0.0488	0.121
3	0.27	0.390	0.249

Table 22.04 Initial concentrations and rate for the reaction between NO and H$_2$

2. Bromine and nitrogen monoxide react together to form nitrosyl bromide.

$$2NO(g) + Br_2(g) \longrightarrow 2NOBr(g)$$

Use the initial concentration data in Table 22.05 to find the orders of the reaction and write the rate equation.

Experiment	[NO(g)] / mol dm^{-3}	[Br$_2$(g)] / mol dm^{-3}	Initial rate / mol dm^{-3} s^{-1}
1	0.0150	0.0050	1.35×10^{-4}
2	0.0150	0.0100	2.70×10^{-4}
3	0.0300	0.0100	1.09×10^{-3}

Table 22.05 Initial concentrations and rates for the reaction of NO(g) and Br$_2$(g)

22.06 Reaction mechanisms

The reaction **mechanism** is the series of steps that the reactants take to form the products. Much mechanistic information is found from the rate equation as one step is often much slower than the others and determines the reaction rate. The slow step is called the rate-determining step.

Sample answer

Question:

Suggest a mechanism for the reaction between nitrogen dioxide and carbon monoxide, explaining your working:

$$CO(g) + NO_2(g) \longrightarrow CO_2(g) + NO(g)$$

rate = $k[NO_2(g)]^2$ [8]

Answer:

The slow step determines the rate of reaction [1 mark].

The order with respect to NO$_2$ is 2 and so there must be a slow step involving two molecules of NO$_2$ [1 mark] and the other steps in the mechanism must be fast [1 mark].

The overall equation only has one NO$_2$ so a slow step involving two NO$_2$ molecules must be followed by a fast step producing one NO$_2$ [1 mark]. The fast step must include CO(g) and produce CO$_2$(g) [1 mark].

$$NO_2(g) + NO_2(g) \longrightarrow X \quad \text{slow}$$
$$X + CO(g) \longrightarrow NO_2(g) + CO_2(g) \quad \text{fast}$$

suggest X = NO$_3$(g) [1 mark]

suggested mechanism: $NO_2(g) + NO_2(g) \longrightarrow NO_3(g) + NO(g)$ slow

$$NO_3(g) + CO(g) \longrightarrow NO_2(g) + CO_2(g) \quad \text{fast [1 mark]}$$

Adding the two steps together:

$$NO_2(g) + NO_2(g) + NO_3(g) + CO(g) \longrightarrow NO_3(g) + NO(g) + NO_2(g) + CO_2(g)$$

which simplifies to

$$NO_2(g) + CO(g) \longrightarrow NO(g) + CO_2(g)$$

[1 mark] (for working)

An important principle of a mechanism is that the steps added together must give the balanced stoichiometric equation.

Progress check 22.04

The following mechanism is proposed for the reaction of nitrogen dioxide with ozone.

$$O_3 + NO_2 \longrightarrow NO_3 + O_2 \quad \text{slow}$$
$$NO_3 + NO_2 \longrightarrow N_2O_5 \quad \text{fast}$$

1. Write the stoichiometric equation.
2. Write the rate equation.

22.07 Techniques for studying rates of reaction

As we have seen, the two main ways of studying reaction rate are plotting a concentration/time graph and finding initial rates.

a To follow the progress of a reaction, at known time intervals

This can be done by:

- measuring the volume of a gas product
- mixing a large volume of reactant mixture and removing small portions to titrate

- keeping the pressure of a gas reaction constant and measuring the volume change
- following a colour change with a colorimeter
- measuring the conductivity if an ionic substance is produced or used up

b **To find the initial rate**

This can be done by:

- using all the above ways to draw a range of concentration/time graphs
- using the clock reaction.

22.08 The effect of temperature on the rate constant

In Unit 9, we saw that the reaction rate increases with temperature; the main reason is that more molecules have energy greater than the activation energy and so can react on collision.

The rate equation shows the concentration dependency of reaction rate; as the concentrations do not change with temperature, it must be the rate constant that has the temperature effect.

> **TIP** Make sure you can draw suitable Boltzmann distributions and, by using them, explain the effect of temperature and catalysts on a reaction.

22.09 Homogeneous and heterogeneous catalysis

a **Homogeneous catalysis**

A homogeneous catalyst is one which is in the same physical state as the reactants.

The peroxodisulfate electrode potential is more positive than that of iodine but the reaction of iodide with peroxodisulfate is slow as two negative ions have to collide:

$$2I^-(aq) + S_2O_8^{2-}(aq) \longrightarrow 2SO_4^{2-}(aq) + I_2(aq)$$

The electrode potential for Fe^{3+} lies in between those for iodine and peroxodisulfate so Fe^{3+} can oxidise iodide and the Fe^{2+} produced can be oxidised by peroxodisulfate. $FeCl_3(aq)$ acts as a catalyst.

Nitrogen dioxide catalyses the oxidation of sulfur dioxide to sulfur trioxide:

$$NO_2(g) + SO_2(g) \longrightarrow NO(g) + SO_3(g)$$

$$2NO(g) + O_2(g) \longrightarrow 2NO_2(g)$$

This fast production of SO_3 and its subsequent solution in water form acid rain.

The layer of high-level ozone is broken down by chlorine atoms:

$$Cl + O_3 \longrightarrow OCl + O_2$$

$$OCl + O \longrightarrow O_2 + Cl$$

Enzymes are water-soluble proteins in which the protein chain is folded and held in an almost spherical shape. They have a cleft, called the **active site**, into which the reactant molecule (called the **substrate**) fits exactly and where it reacts and releases the product. This is called the **lock-and-key mechanism** (Figure 22.06).

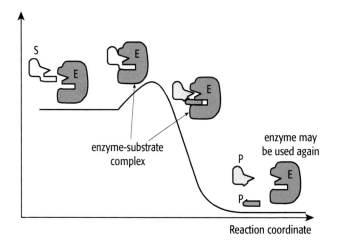

Figure 22.06 The lock-and-key model of enzyme catalysis

Only one substrate fits the active site, making enzymes highly specific catalysts, unlike transition metal catalysts.

b **Heterogeneous catalysis**

A heterogeneous catalyst is in a different phase, usually a solid, to that of the reactants. The reaction takes place on the catalyst surface so heterogeneous catalysts are used as a powder or fine mesh.

- The reactant molecules diffuse to the catalyst surface where they are adsorbed, making weak bonds.
- The reactant bonds are weakened so the activation energy is lowered.
- Neighbouring atoms and molecules on the catalyst surface form new bonds to produce the product, which is desorbed.

Iron in the Haber process and catalytic converters are heterogeneous catalysts.

Revision checklist

Check that you know the following:

- [ ] explain and use *rate equation, order of reaction, rate constant, half life of a reaction and rate-determining step.*
- [ ] draw and interpret concentration time graphs to find an order of reaction.
- [ ] find a half life and use it to find the order of reaction.
- [ ] calculate and use initial rates to construct a rate equation.
- [ ] devise a method to calculate a rate constant.
- [ ] suggest the relationship between a reaction mechanism, the rate equation and the overall equation and how to predict an order from a mechanism.
- [ ] devise techniques for studying rates of reaction.
- [ ] explain the effect of temperature on the rate of reaction.
- [ ] outline homogeneous and heterogeneous catalysis.
- [ ] knowledge of the catalysis involved in a range of reactions.

Exam-style questions

1. a How could the rate of the acid hydrolysis of ethyl ethanoate be followed? [6]

   $$CH_3COOCH_2CH_3(l) + H_2O(l) \longrightarrow CH_3COOH(l) + CH_3CH_2OH(l)$$

   b i Use the data in the table to find the orders with respect to HCl and $CH_3COOCH_2CH_3$. [2]

   ii Write the rate equation. [2]

[HCl] / mol dm^{-3}	[CH$_3$COOCH$_2$CH$_3$(aq)] / mol dm^{-3}	Relative rate / mol dm^{-3} min^{-1}
0.2	0.2	5.3×10^{-3}
0.1	0.2	2.6×10^{-3}
0.1	0.1	1.2×10^{-3}

Total: 10

2. a The decomposition of ammonia using an iron catalyst is a *zero-order reaction*.

   i Explain what *zero-order reaction* means.

   ii Suggest why this decomposition is a zero-order reaction. [4]

b The reaction A + B ⟶ products. The temperature and [A] are kept constant. Plot a graph of the following data and use it to find the order of the reaction with respect to B. Any construction lines should be clear. [7]

[B] / mol dm^{-3}	Time / min
1.0	0
0.850	4.0
0.715	9.5
0.595	14.8
0.500	19.8
0.424	25.0
0.330	34.0
0.248	40.0
0.210	45.0
0.155	53.0
0.120	60.0

Total: 11

Entropy and Gibbs free energy

Learning outcomes

You should be able to:

- explain entropy as the amount of disorder and predict whether a physical or chemical change has a positive or negative entropy change
- calculate entropy changes
- define and calculate the Gibbs free energy change for a reaction and state whether the reaction is spontaneous

23.01 Entropy

a **What is entropy?**

As we have already seen, substances have a particular energy associated with them, their enthalpy. Energy can neither be created nor destroyed although it can change in form; thus potential energy may be changed into kinetic energy which in turn may be changed into electrical energy. Hess's law depends on energy conservation.

Another quantity associated with substances indicates how their constituent particles and their associated energy are distributed. This is their **entropy**. The more disordered the particles and the greater the distribution of energy, the higher is the entropy. Particles in gases are more disordered than particles in liquids or solids and so gases have higher entropy values. The energy from burning fuel spreads out and warms a whole room and the entropy of the room increases.

During any **spontaneous change**, the entropy increases.

A spontaneous change is one which proceeds by itself without needing an outside source of energy to drive it. The entropy is the amount of disorder in a system.

During a chemical reaction, the entropy changes undergone by the reactants might increase or decrease, but if the entropy of the surroundings is also taken into account, it is found that the total entropy *always* increases.

For example, a pool of water evaporates even when it is well below the boiling point of water as long as more molecules can enter the air than will condense from the air – the two processes are random. The evaporation is driven by the movement of molecules from water to the less orderly vapour and the entropy change is positive.

Every substance has a positive entropy value. Solids have the lowest entropy values and gases the highest. A change of state from solid to liquid or from liquid to gas has a positive entropy change, because the particles become more disordered.

> **TIP**
> Entropy values are quoted in terms of joules per mole per degree, and not kilojoules per mole as for ΔH values.

At $-273\,°C$, the absolute zero temperature, elements and compounds are in their most ordered state and the entropy of all of them $= 0\,\text{J}\,\text{mol}^{-1}\,\text{K}^{-1}$. As the temperature is increased, the entropy increases.

b **Some entropy changes**

i If separate gas jars of nitrogen and oxygen are put into contact, the two gases spontaneously mix and spread through the whole of the available volume. The distribution of the molecules becomes more random and so the entropy has increased.

ii When a solid dissolves in a liquid, the particles spread out and the entropy increases.

iii Photosynthesis, in which carbohydrates are produced from CO_2 and H_2O, has a negative entropy change as there are far fewer product molecules than reactant molecules:

$$6H_2O(l) + 6CO_2(g) \longrightarrow C_6H_{12}O_6(s) + 6O_2(g)$$

Energy from sunlight is needed to drive the reaction.

c Calculation of entropy changes

Many entropy values are listed in data tables.

entropy change in a reaction = ∑(entropy of products) − ∑(entropy of reactants)

> ## Worked example 23.01
>
> What is the entropy change when calcium carbonate decomposes?
>
> **How to get the answer:**
>
> Step 1: Write the balanced equation:
> $$CaCO_3(s) \longrightarrow CaO(s) + CO_2(g)$$
>
> Step 2: Look up all the entropy values, $S^\ominus$:
> $$CaCO_3(s) = 92.9; CaO(s) = 39.7;$$
> $$CO_2(g) = 213.6$$
>
> Step 3: Note the units: $J\,mol^{-1}\,K^{-1}$.
>
> Step 4: Insert the values into the relationship:
> $$\Delta S^\ominus = 39.7 + 213.6 - 92.9$$
>
> Step 5: Calculate the answer:
> $$= +160.4\,J\,mol^{-1}\,K^{-1}$$
>
> Step 6: Check that the sign is correct. A solid produces a solid and a gas so the entropy change is expected to be positive.

> ## Progress check 23.01
>
> 1 Which of the following would have a positive entropy change?
>
>   a solid sodium chloride dissolving in water
>
>   b gaseous sulfur dioxide and oxygen reacting to produce liquid sulfur trioxide
>
>   c ethene molecules reacting to produce polyethene
>
>   d magnesium burning in oxygen
>
>   e water dissolving in ethanol
>
>   f crystallisation of copper sulfate from a solution.
>
> 2 Some standard entropy values in $J\,mol^{-1}\,K^{-1}$:
>
> $N_2 = 191.6 \quad H_2 = 131 \quad NH_3 = 192.3$
>
>   What is the entropy change of reaction for
> $$N_2(g) + 3H_2(g) \longrightarrow 2NH_3(g)?$$

23.02 Gibbs Free Energy

a Total entropy change

For a spontaneous change, the total entropy change must be positive as entropy always increases:

$$\Delta S > 0$$

If energy is lost from a reaction, the temperature of the surroundings increases and this means the entropy of the surroundings increases. Both the entropy change of the substances in the reaction and the entropy change in the surroundings need to be considered.

b Entropy change of the surroundings, ΔS_{surr}

The entropy change of the surroundings depends on both the energy given out from the reaction and the temperature of the surroundings; the same amount of energy has a far larger effect at a low temperature than at a high temperature.

This can be illustrated by the following: if you possess no money and then someone gives you a small amount, your wealth increases by an infinite percentage; if you already have a lot of money and then someone gives you a small amount, you would hardly notice! Giving you money has a far larger effect when you start off with none.

c Relating entropy and enthalpy changes

$$\Delta S_{total} = \Delta S_{surr} + \Delta S_{reaction}$$

As we saw above, ΔS_{surr} depends on both the amount of energy given to the surroundings (ΔH) and the temperature (T).

$$\Delta S_{surr} = -\frac{\Delta H}{T}$$

There is a negative sign as ΔH for an exothermic reaction is negative but the entropy change in the surroundings is positive.

Multiplying through the equation by $-T$ gives

$$-T\Delta S_{total} = -T\Delta S_{surr} - T\Delta S_{reaction}$$

$$= \Delta H - T\Delta S_{reaction}$$

d Gibbs free energy

'$T\Delta S$' is a quantity of energy. The American physicist, Willard Gibbs, realised that account needed to be taken for both the entropy change in the reaction and the entropy change in the surroundings. A new quantity, the Gibbs free energy change (ΔG) is defined to cover this:

$$\Delta G = \Delta H - T\Delta S_{reaction}$$

As the total entropy change for a spontaneous change is positive and $\Delta G = -T\Delta S_{total}$, the free energy change must be negative for a spontaneous change:

$$\Delta G < 0$$

TIP

The energy change for every reaction is in two parts – one due to enthalpy changes and the other due to entropy changes. The two together determine whether or not a reaction is spontaneous.

Worked example 23.02

Under what conditions is the Haber process for the production of ammonia a spontaneous reaction?

How to get the answer:

Step 1: Write the equation:

$$N_2(g) + 3H_2(g) \longrightarrow 2NH_3(g)$$

Step 2: Write down what you know about enthalpy changes. ΔH is negative as this is an exothermic reaction.

Step 3: Work out the sign of the entropy change of reaction. $\Delta S_{reaction}$ is negative as 4 moles of gas produce only 2 moles of gas.

Step 4: Use the relationship $\Delta G = \Delta H - T\Delta S_{reaction}$ for low temperatures. $T\Delta S < \Delta H$; the negative ΔH is dominant, so ΔG is negative and the reaction is spontaneous.

Step 5: Use the relationship $\Delta G = \Delta H - T\Delta S_{reaction}$ for high temperatures. $T\Delta S > \Delta H$; the negative ΔS is dominant, making the $-T\Delta S$ term large and positive, so ΔG is positive and the reaction is not spontaneous.

Progress check 23.02

$$H_2O(g) + C(s) \longrightarrow H_2(g) + CO(g)$$
$$\Delta H = +135 \text{ kJ mol}^{-1}$$

Under what conditions would the reaction be spontaneous?

Explain your answer.

Sample answer

Question:

Using the given values, work out $\Delta G^\ominus$ at 300 K and decide whether the following reaction is spontaneous: [4]

$$CH_3CH_2CH_2OH \longrightarrow CH_3CH(OH)CH_3$$
$$S^\ominus[CH_3CH_2CH_2OH] = 196.6 \text{ J mol}^{-1}\text{K}^{-1}$$
$$S^\ominus[CH_3CH(OH)CH_3] = 180.5 \text{ J mol}^{-1}\text{K}^{-1}$$
$$\Delta H^\ominus_f[CH_3CH_2CH_2OH] = -302.7 \text{ kJ mol}^{-1}$$
$$\Delta H^\ominus_f[CH_3CH(OH)CH_3] = -317.9 \text{ kJ mol}^{-1}$$

Answer:

$\Delta S^\ominus_{reaction} = 180.5 - 196.6 = -16.1 \text{ J mol}^{-1}\text{K}^{-1}$
[1 mark]

$\Delta H^\ominus_{reaction} = 302.7 - 317.9 = -15.2 \text{ kJ mol}^{-1}$
[1 mark]

$\Delta G^\ominus = \Delta H^\ominus - T\Delta S^\ominus_{reaction}$
$= -15.2 - 300 \times -16.1 / 1000$
$= -10.4 \text{ kJ mol}^{-1}$ [1 mark]

As $\Delta G^\ominus$ is negative, the reaction is spontaneous [1 mark].

Revision checklist

Check that you know the following:

- [ ] explain entropy and situations where entropy changes occur
- [ ] predict and calculate an entropy change
- [ ] define and calculate the standard Gibbs free energy change
- [ ] predict whether a reaction is spontaneous or not and the effect of temperature

Exam-style questions

1. a Which of these reactions has a positive entropy change?

   i $PCl_5(s) \longrightarrow PCl_3(l) + Cl_2(g)$

   ii $2Na(s) + 2H_2O(l) \longrightarrow 2NaOH(aq) + H_2(g)$

   iii $Ba^{2+}(aq) + SO_4^{2-}(aq) \longrightarrow BaSO_4(s)$

   iv $2Ca(s) + O_2(g) \longrightarrow 2CaO(s)$ [2]

   b Work out $\Delta G^{\ominus}$ at 298 K for the following two reactions:

   $PCl_5(s) \longrightarrow PCl_3(l) + Cl_2(g)$ [3]

   $2Mg(s) + O_2(g) \longrightarrow 2MgO(s)$ [3]

	$PCl_5(s)$	$PCl_3(l)$	$Cl_2(g)$	$Mg(s)$	$O_2(g)$	$MgO(s)$
$S^{\ominus}$/ J mol^{-1} K^{-1}	166.5	217.1	223.0	32.7	205.0	26.9
$\Delta H^{\ominus}_f$/ kJ mol^{-1}	−443.5	−319.7				−601.7

Total: 8

2. a Under what temperature condition is the following reaction spontaneous?

   $N_2O_4(g) \longrightarrow 2NO_2(g)$ is an endothermic reaction. [4]

   b i Under what temperature condition is the following reaction spontaneous?

   $CH_2=CH_2(g) + H_2(g) \longrightarrow CH_3CH_3(g)$ [3]

   ii At what temperature does it change from being spontaneous? [2]

	$CH_2=CH_2(g)$	$H_2(g)$	$CH_3CH_3(g)$
$S^{\ominus}$/ J mol^{-1} K^{-1}	219.5	130.6	229.5
$\Delta H^{\ominus}_f$/ kJ mol^{-1}	+52.2		−84.7

Total: 9

Unit 24: Transition elements

Learning outcomes

You should be able to:

- sketch d orbitals
- explain *transition element* and state the electron configurations of the elements and their ions
- describe the physical properties of the elements and contrast with those of calcium
- describe the tendency for variable oxidation states and predict some redox reactions of the elements using electrode potentials
- describe and explain complex formation, including stereoisomerism, and ligand exchange in terms of stability constants
- explain why many transition element ions are coloured

24.01 Electron configurations of the transition elements

In this series of elements, electrons fill the 3d orbitals which are higher in energy than the 3p and 4s orbitals. The five 3d orbitals (Figure 24.01) can accommodate ten electrons, giving ten transition elements.

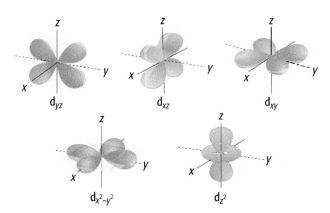

Fig. 24.01 The shapes of the five 3d orbitals.

Along the series there is little change in atomic radius and the ionisation energy rises gradually; the ions, formed by loss of the 4s electrons, are small. The electron configurations are shown in Table 24.01. There are two discrepancies, at Cr and Cu, in the regular filling of the 3d orbitals.

Cr has one electron in each of the 3d and 4s orbitals and Cu has one electron in the 4s but full 3d orbitals. This gives a symmetrical charge distribution, conferring stability, in each case.

Atom	Electron configuration
potassium	[Ar] $4s^1$
calcium	[Ar] $4s^2$
scandium	[Ar] $3d^1 4s^2$
titanium	[Ar] $3d^2 4s^2$
vanadium	[Ar] $3d^3 4s^2$
chromium	[Ar] $3d^5 4s^1$
manganese	[Ar] $3d^5 4s^2$
iron	[Ar] $3d^6 4s^2$
cobalt	[Ar] $3d^7 4s^2$
nickel	[Ar] $3d^8 4s^2$
copper	[Ar] $3d^{10} 4s^1$
zinc	[Ar] $3d^{10} 4s^2$
gallium	[Ar] $3d^{10} 4s^2 4p^1$

Table 24.01 The electron configurations of some of the elements of Period 4

TIP Although you can look at the Periodic Table, you will find it helpful to know the transition elements in the correct order and be able to write their electron configurations.

The 3d electrons enter 'inner orbitals' and it is the 4s electrons that are lost on ionisation. Once the 4s electrons have been lost, varying numbers of 3d electrons may be lost (Table 24.02).

ox. no.	Sc	Ti	V	Cr	Mn	Fe	Co	Ni	Cu	Zn
+7					MnO_4^-					
+6				CrO_4^{2-}						
+5			VO_2^+							
+4		TiO_2	VO^{2+}		MnO_2					
+3	Sc^{3+}		V^{3+}	Cr^{3+}		Fe^{3+}				
+2			V^{2+}		Mn^{2+}	Fe^{2+}	Co^{2+}	Ni^{2+}	Cu^{2+}	Zn^{2+}
+1									Cu^+	

Table 24.02 The ox. no. of the main ions of the transition elements

To be included as a transition element, an element needs at least one stable ion with a partially filled set of d orbitals. We can see in Table 24.02 that Sc produces Sc^{3+}; this has lost all its 3d electrons and thus scandium is not included. At the other end of the series, Zn produces Zn^{2+} with the electron configuration $[Ar]3d^{10}$. With no partially filled d orbital, zinc is not included.

The particular properties of transition elements shown by the eight remaining elements are:

- similar physical properties
- variable oxidation state and redox reactions
- coloured compounds
- catalytic activity
- complex formation

Progress check 24.01

Work out the oxidation numbers and give the corresponding electron configurations of the transition elements in the following compounds:

1. Cu_2I_2
2. $Fe_2(SO_4)_3$
3. $K_2Cr_2O_7$
4. $VOCl_2$
5. VCl_3
6. $Cu(NO_3)_2$.

TIP: You need to remember the five properties of transition elements and give suitable examples.

24.02 Catalysis

See Unit 22 for more information about catalysis.

Here are some important transition metal catalysts you have met:

- Fe in the Haber process
- V_2O_5 in the Contact process
- Ni in the hydrogenation of alkenes to produce margarine
- Fe^{3+} in the reaction between I^- and $S_2O_8^{2-}$

24.03 The physical properties of the elements

The melting points of the elements are higher than calcium, indicating they have giant structures. The electrical conductivity is good, indicating metallic bonding, so they are malleable and ductile, strong and shiny. They all have much higher densities than that of calcium.

24.04 The oxidation states and redox reactions of the elements

When forming ions, these elements lose the 4s electrons and varying numbers of 3d electrons. Highly charged and small, these ions are stabilised by species such as oxygen forming covalent or dative covalent bonds with them. Transition elements show a variety of oxidation states (see Table 24.02). From titanium to manganese, the highest oxidation state is the sum of the 4s and 3d electrons. After manganese, the +2 state becomes more important.

With variable oxidation states, redox reactions are important. Electrode potentials indicate the oxidising power in aqueous solution (see Unit 20):

$Fe^{3+}(aq) + e^- \rightleftharpoons Fe^{2+}(aq) \quad E^\ominus = +0.77\,V$

$Cr_2O_7^{2-}(aq) + 14H^+(aq) + 6e^- \rightleftharpoons 2Cr^{3+}(aq) + 7H_2O(l) \quad E^\ominus = +1.33\,V$

$MnO_4^-(aq) + 8H^+(aq) + 5e^- \rightleftharpoons Mn^{2+}(aq) + 4H_2O(l) \quad E^\ominus = +1.52\,V$

A more positive $E^\ominus$ value indicates a stronger oxidising agent and you can see that both acidified manganate(VII) and acidified chromate(VI) can oxidise $Fe^{2+}(aq)$ to $Fe^{3+}(aq)$.

a **Oxidation with manganate(VII)**

Reduction: $MnO_4^-(aq) + 8H^+(aq) + 5e^- \longrightarrow Mn^{2+}(aq) + 4H_2O(l)$

Oxidation: $Fe^{2+}(aq) \longrightarrow Fe^{3+}(aq) + e^-$

The oxidation equation needs to be multiplied by 5 to balance the electrons:

$MnO_4^-(aq) + 8H^+(aq) + 5Fe^{2+}(aq) \longrightarrow Mn^{2+}(aq) + 4H_2O(l) + 5Fe^{3+}(aq)$

As manganate(VII) is deep purple, $Mn^{2+}(aq)$ is very pale pink and $Fe^{3+}(aq)$ is yellow, this self-indicating reaction can be used to find Fe^{2+} concentrations by titration. The manganate(VII) solution is placed in the burette and run into a known volume of the Fe^{2+} solution. The mixture remains almost colourless until the end-point when the first permanent pink tinge appears.

b **Chromate(VI) ions**

In acidic solution these exist as orange $Cr_2O_7^{2-}$; in alkaline solution they exist as yellow CrO_4^{2-}. When acting as an oxidising agent, Cr(VI) ions are reduced to green Cr^{3+}. Acidified potassium dichromate(VI) is often used in organic chemistry to oxidise primary and secondary alcohols.

Worked example 24.01

A sample of $FeSO_4$ had partially oxidised to $Fe_2(SO_4)_3$. 0.56 g of the sample was dissolved in aqueous sulfuric acid to form 100.0 cm³ of solution. A 25.0 cm³ portion of this solution required 21.35 cm³ of 0.005 mol dm⁻³ $KMnO_4$ to reach the end-point. What is the mole percentage of Fe^{2+} in the amount of iron present in the sample?

How to get the answer:

Step 1: Work out the number of moles of $KMnO_4$ used:

$\dfrac{0.005 \times 21.35}{1000} = 1.0675 \times 10^{-4}$

Step 2: Work out the number of moles of $FeSO_4$ in 25.0 cm³ of solution:

$5 \times 1.0675 \times 10^{-4}$

Step 3: Work out the number of moles of $FeSO_4$ in 0.56 g of the sample:

$4 \times 5 \times 1.0675 \times 10^{-4}$

Step 4: Work out the mass of $FeSO_4$ in 0.56 g of the sample:

$2.135 \times 10^{-3} \times 151.9 = 0.324\,g$

Step 5: Work out the mass of $Fe_2(SO_4)_3$ in 0.56 g of the sample:

$0.56 - 0.324 = 0.236\,g$

Step 6: Work out the number of moles of Fe^{3+} in 0.56 g of the sample:

$2 \times \dfrac{0.236}{399.9} = 1.180 \times 10^{-3}$

Step 7: Work out the total moles of iron (Fe^{II} and Fe^{III}) in 0.56 g of the sample:

3.315×10^{-3}

Step 8: Mole percentage of Fe^{2+}:

$\dfrac{2.135 \times 10^{-3}}{3.315 \times 10^{-3}} = 64.4\%$

Progress check 24.02

1. a. Write the equation for the reaction between zinc and vanadium(III).

    $V^{3+}(aq) + e^- \rightleftharpoons V^{2+}(aq)$
    $E^\ominus = -0.26\,V$

    $Zn^{2+}(aq) + 2e^- \rightleftharpoons Zn(s)$
    $E^\ominus = -0.76\,V$

   b. Is the reaction represented by your equation able to proceed thermodynamically?

2. If acid is added to CrO_4^{2-}, the orange $Cr_2O_7^{2-}$ is formed. Write the balanced reversible equation for this conversion.

> **TIP**
> Learn the main oxidation states of chromium, manganese, iron and copper and the colours expected of the ions in solution.

24.05 Complex formation

Small, highly charged, transition metal ions strongly attract any species with a lone pair; the lone pair is used to form a dative covalent bond.

A **ligand** is a species with a lone pair of electrons which it donates to an ion and forms a dative covalent bond.

Ligands are species such as Cl^-, OH^-, H_2O and NH_3. Of these, NH_3 forms the strongest and H_2O the weakest bonds. Ligands are similar to bases and nucleophiles, which also have a lone pair of electrons used to form a covalent bond.

A **complex** is formed when a number of ligands form dative covalent bonds with a central metal ion.

Complex formation spreads out the ionic charge and stabilises the metal ion.

a **Charge and oxidation number**

The sum of the oxidation numbers of metal ion and ligands = charge on the complex.

> ### Worked example 24.02
>
> What is the oxidation number of Co in $[Co(NH_3)_5Br]^{2+}$?
>
> How to get the answer:
>
> Step 1: State the known ox.nos. and charges: charge on $NH_3 = 0$; $Br^- = -1$.
>
> Step 2: Equate the sum of charges and ox.nos. to the overall charge: ox.no.[Co] $+ 0 - 1 = +2$.
>
> Step 3: The answer: ox.no.[Co] $= +3$.

b **Shapes of complexes**

The **coordination number** is the number of attached ligands.

The coordination number is often six, but complexes with four or two ligands are also known. There are four common shapes (Figure 24.02).

Examples	Diagram	Shape	Bond angles
$[Cu(H_2O)_6]^{2+}$ $[Fe(CN)_6]^{3-}$		octahedral	90°

Examples	Diagram	Shape	Bond angles
$CuCl_4^{2-}$ $NiCl_4^{2-}$	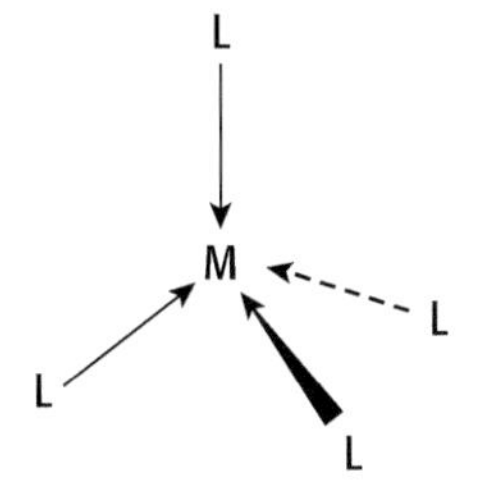	tetrahedral	109.5°
$[Pt(NH_3)_4]^{2+}$ $[PtCl_4]^{2-}$	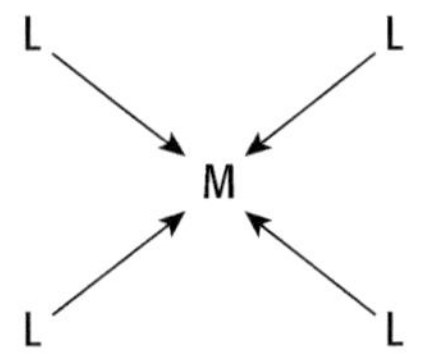	square planar	90°
$[Ag(NH_3)_2]^+$ $[CuCl_2]^-$	 L⟶M⟵L	linear	180°

Fig. 24.02 The most common complex shapes

The dative covalent bond is shown as an arrow drawn from the lone pair on the ligand pointing towards the transition metal ion.

The blue copper sulfate solution is due to the octahedral complex $[Cu(H_2O)_6]^{2+}$ (Figure 24.03).

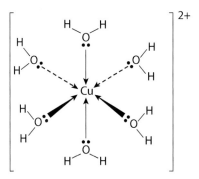

Fig. 24.03 The shape of $[Cu(H_2O)_6]^{2+}$

There is no rule as to which shape is taken up by a complex but most are octahedral. Even tetrahedral complexes usually change to the octahedral aquo complex when in water as more bonds are formed.

c **Isomerism**

If the surrounding ligands are different, the complex may show isomerism. The octahedral complex $[CoCl_2(NH_3)_4]^+$ shows *cis* and *trans* isomerism (Figure 24.04).

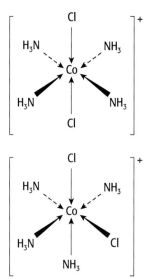

Fig. 24.04 The cis (lower) and trans (upper) isomers of a cobalt(III) complex

d **Polydentate ligands**

Some molecules act as ligands at two or more places in a chain.

i 1,2-Diaminoethane ($H_2NCH_2CH_2NH_2$) contains two nitrogen atoms, each with a lone pair of electrons; it is a **bidentate** ligand. Isomerism is again possible (Figure 24.05).

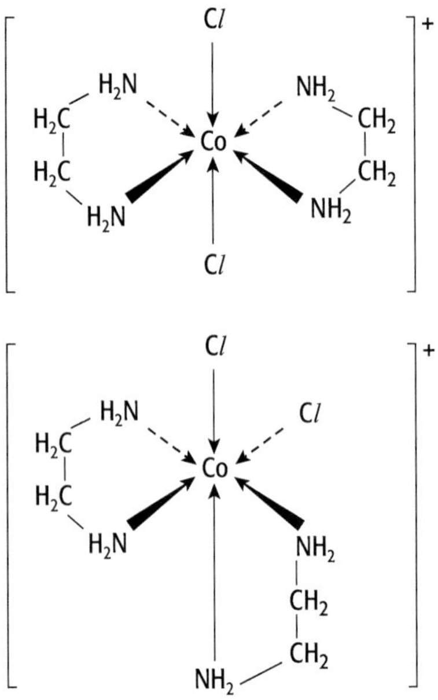

Fig. 24.05 The isomers of a cobalt(III) ion with a bidentate ligand

The second of these complexes, cis-[Co(en)$_2$Cl$_2$]$^+$ and also [Co(en)$_3$]$^{3+}$ are both chiral. They do not have a plane of symmetry.

ii The ligand usually known as EDTA (Figure 24.06) is a hexadentate ligand as it can donate six pairs of electrons.

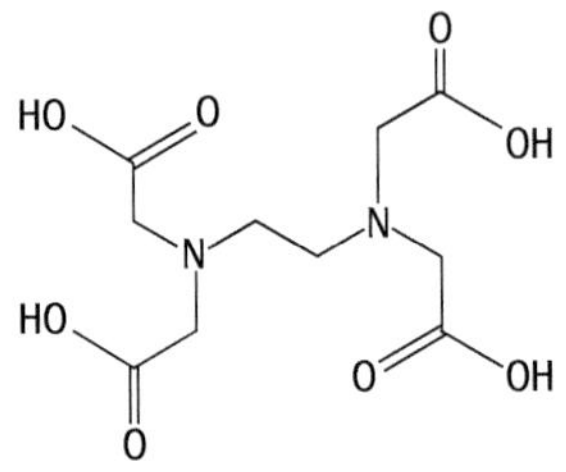

Fig. 24.06 EDTA

EDTA is usually used as the salt with a 4− charge. It wraps around a metal ion forming an octahedral complex as it has two N atoms and four carboxylate groups, all of which can act as ligands.

iii Some of the square planar complexes of platinum form anti-cancer drugs (Figure 24.07).

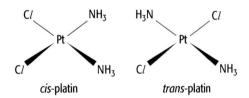

cis-platin   trans-platin

Fig. 24.07

When eaten, cisplatin slowly hydrolyses and the product forms bonds to the base guanine in separate DNA strands. Cross linking between DNA molecules destroys cancer cells.

Transplatin has no such effect and is toxic, so the anticancer medication is required to only contain the cis isomer.

iv The linear complex, [Ag(NH$_3$)$_2$]$^+$, is formed when silver chloride or silver bromide dissolves in aqueous ammonia.

e Ligand exchange reactions

i A stronger ligand, forming a stronger dative covalent bond, can replace a weaker one. When concentrated HCl is added to the blue [Cu(H$_2$O)$_6$]$^{2+}$(aq), it changes to yellow as the Cl$^-$ ligands replace the H$_2$O ligands:

[Cu(H$_2$O)$_6$]$^{2+}$ + 4Cl$^-$ ⇌ [CuCl$_4$]$^{2-}$ + 6H$_2$O(l)

Diluting with water moves the equilibrium to the left, returning the solution to the original blue colour. A stability constant, K_{stab}, can be written, with [H$_2$O] assumed constant:

$$K_{stab} = \frac{[CuCl_4^{2-}]}{[Cu(H_2O)_6^{2+}][Cl^-]^4}$$

The larger the value of K_{stab}, the further to the right is the equilibrium and the more stable is the complex.

ii NH$_3$ is a weak base. When it is added to blue [Cu(H$_2$O)$_6$]$^{2+}$(aq) a pale blue precipitate of [Cu(H$_2$O)$_4$(OH)$_2$] is produced. Excess ammonia dissolves the precipitate to form a deep blue solution as NH$_3$ ligands replace H$_2$O ligands to form a charged complex:

[Cu(H$_2$O)$_6$]$^{2+}$(aq) + 4NH$_3$(aq) ⇌ [Cu(NH$_3$)$_4$(H$_2$O)$_2$]$^{2+}$(aq) + 4H$_2$O(l)

The stability constant for the deep blue complex is much larger than the value for [CuCl$_4$]$^{2-}$.

iii Colourless ammonium thiocyanate solution and orange FeCl$_3$(aq) form a blood red complex which can be used as a sensitive test for iron(III).

[Fe(H$_2$O)$_6$]$^{3+}$(aq) + SCN$^-$(aq) ⇌ [Fe(H$_2$O)$_5$SCN]$^{2+}$(aq) + H$_2$O(l)

iv Ligands are exchanged one by one with a different stability constant for each equilibrium:

$[Cu(H_2O)_6]^{2+}(aq) + NH_3(aq) \rightleftharpoons$
$\qquad [Cu(H_2O)_5NH_3]^{2+}(aq) + H_2O(l)$
$\qquad K_1 = 1.8 \times 10^4 \, mol^{-1} \, dm^3$

$[Cu(H_2O)_5NH_3]^{2+}(aq) + NH_3(aq) \rightleftharpoons$
$\qquad [Cu(H_2O)_4(NH_3)_2]^{2+}(aq) + H_2O(l)$
$\qquad K_2 = 4.1 \times 10^3 \, mol^{-1} \, dm^3$

Progress check 24.03

1. What shape is $[Fe(H_2O)_5SCN]^{2+}$?
2. What is the oxidation state of the transition metal ions in the following complexes:

   $[Fe(CN)_6]^{4-}$ $Cu(NH_3)_2^+$ $[V(CN)_6]^{4-}$
   $[CrCl_6]^{3-}$ $[MnF_6]^{2-}$ $[Co(H_2O)_4Cl_2]$

3. Draw three-dimensional representations of the two isomers of $[Co(en)_3]^{3+}$ (en = 1,2-diaminoethane).

24.06 d orbitals and colour

a Octahedral complexes

The six ligands of octahedral complexes define three orthogonal axes. Three of the 3d orbitals have lobes pointing between the axes and two have lobes pointing along the axes (Figure 24.01). As the ligands donate electron pairs along the axes, they experience repulsion from electrons in the d orbitals with lobes pointing along the axes but there is less repulsion between the lone pairs and electrons in d orbitals with lobes pointing between the axes. The five degenerate 3d orbitals are split into two sets with different energies (Figure 24.08). (**Degenerate orbitals** are orbitals with the same energy.)

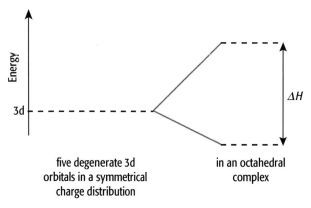

Fig. 24.08 The splitting of the 3d orbitals in an octahedral complex

For the ligand H_2O, the energy splitting, ΔE, is small and electrons are distributed between all the orbitals until they have to pair. For CN^-, the splitting is large and the lower energy orbitals are filled first. Fe^{III} has the electron configuration $[Ar]\, 3d^5$; in $[Fe(H_2O)_6]^{3+}$, there is one electron in each of the 3d orbitals but in $[Fe(CN)_6]^{3-}$, all five d electrons are in the lower orbitals and there is just one unpaired electron.

The frequency of radiation corresponding to $\Delta E = hf$ lies in the visible region.

(h = the Planck constant; f = frequency of light)

An electron in a lower energy 3d orbital can absorb light of just the right frequency and jump into a space in one of the higher energy orbitals. That absorbed frequency cannot be seen but all the other frequencies in white light can be seen; the complex is the complementary colour to the colour absorbed.

The electron loses its absorbed energy by molecular vibration, dropping down into the lower orbital without emitting any light so the whole process can happen again.

$[Cu(H_2O)_6]^{2+}$ absorbs in the red-orange region, the complementary colour being blue. In $[Cu(NH_3)_4(H_2O)_2]^{2+}$ the ΔE is larger and the frequency absorbed is larger. The absorption moves into the orange region and the complementary colour becomes dark blue (Figure 24.09).

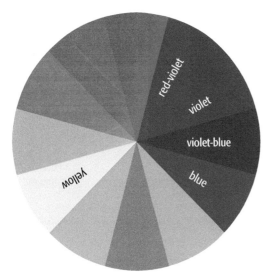

Fig. 24.09 The complementary colours are opposite to the ones absorbed

b Tetrahedral complexes

In a tetrahedral complex such as $[CuCl_4]^{2-}$, the 3d splitting is reversed and the three orbitals with lobes between the axes are at a higher energy than the two with lobes along the axes.

Progress check 24.04

1. Explain why TiO_2 is white.

2. In solution, potassium iodide reduces copper(II) sulfate to insoluble copper(I) iodide and some of the iodide is oxidised to iodine.
   a. Write a balanced equation for the reaction.
   b. What would be observed during the reaction?

3. $EDTA^{4-}$ is a strong ligand producing a large d–d splitting.
   a. What is the oxidation number of cobalt in $[CoEDTA]^-$?
   b. Write down the electron configuration of cobalt in $[CoEDTA]^-$.
   c. How many unpaired electrons does Co have in the complex $[CoEDTA]^-$?

Revision checklist

Check that you know the following:

- sketch the shapes of d orbitals
- state the electron configurations of transition elements and ions
- compare some physical properties with those of calcium
- predict the likely oxidation states of a transition element
- describe ligands, coordination number and the shapes of complexes
- describe how the isomerism of complexes arises
- describe and explain ligand exchange reactions and stability constants
- describe and explain Fe^{3+}/Fe^{2+}, MnO_4^-/Mn^{2+} and $Cr_2O_7^{2-}/Cr^{3+}$ as examples of redox systems
- describe the splitting of d orbitals into two energy levels in octahedral and tetrahedral complexes
- explain the origin of colour in transition element complexes and the effects of different ligands

Exam-style questions

1. a Write down the electron configurations of:

    i Fe

    ii Ni^{2+}. [2]

   b i What shape is $Ni(H_2O)_6^{2+}$? [1]

    ii Explain why $Ni(H_2O)_6^{2+}$ is green but $Ni(NH_3)_6^{2+}$ is blue. [5]

   c Suggest why $[Fe(H_2O)_6]^{3+}$ has more unpaired electrons in the d orbitals than $[Fe(CN)_6]^{3-}$. [5]

   Total: 13

2. a What would you see if separate portions of copper sulfate solution are reacted with NaOH(aq) and NH_3(aq). Write equations for any reactions which occur. [5]

   b $Fe^{3+}(aq) + e^- \rightleftharpoons Fe^{2+}(aq)$
   $E^\ominus = +0.77V$

   $Fe^{2+}(aq) + 2e^- \rightleftharpoons Fe(s)$ $E^\ominus = -0.44V$

    i Write the equation for the disproportionation (the same species is both oxidised and reduced in the same reaction) of Fe(aq)$^{2+}$. [2]

    ii Explain whether the disproportionation reaction is feasible. [2]

   Total: 9

Unit 25

Benzene and its compounds

Learning outcomes

You should be able to:

- [ ] describe and explain the shape of the benzene molecule in terms of delocalised π bonds
- [ ] understand how to name aromatic compounds
- [ ] describe the chemistry of arenes as exemplified by the following reactions of benzene and methylbenzene: substitution reactions with chlorine and bromine, nitration, alkylation, acylation, oxidation of side-chains, hydrogenation of the ring
- [ ] describe the electrophilic substitution mechanism forming nitrobenzene, bromobenzene and other compounds
- [ ] predict whether halogenation will occur in the side-chain or in the aromatic ring in arenes depending on reaction conditions
- [ ] apply knowledge relating to position of substitution in the electrophilic substitution of arenes
- [ ] describe the reactions of phenol with Na, NaOH(aq), Br$_2$(aq), diazonium salts
- [ ] describe how phenyl esters are made
- [ ] describe and explain the relative acidities of phenol, ethanol and water

25.01 The structure of benzene

Arenes are one of the homologous series of hydrocarbons that were mentioned in Unit 15. They are hydrocarbons that contain at least one benzene ring. The general term aromatic is used to describe non-hydrocarbon compounds (e.g. ketones, carboxylic acids) that contain a benzene ring (some of the earliest discovered aromatic compounds had pleasant odours – hence the name). This is to contrast them from aliphatic compounds, for example ethanoic acid, that do not contain a benzene ring.

The structure of benzene is a ring of six carbon atoms, each with a hydrogen atom attached. Each carbon atom uses three of its valence electrons to form σ bonds with adjacent atoms, leaving one electron in a p orbital. Rather than overlap in pairs to form normal C=C double bonds, these six p orbitals overlap to form one 6-centred 'delocalised' π orbital, which is represented in the skeletal formula by a circle inside the hexagon.

The alternative formula, showing three double bonds, was originally proposed by Kekulé in the 19th century. You may still find benzene structures drawn like this (especially on some Internet sites):

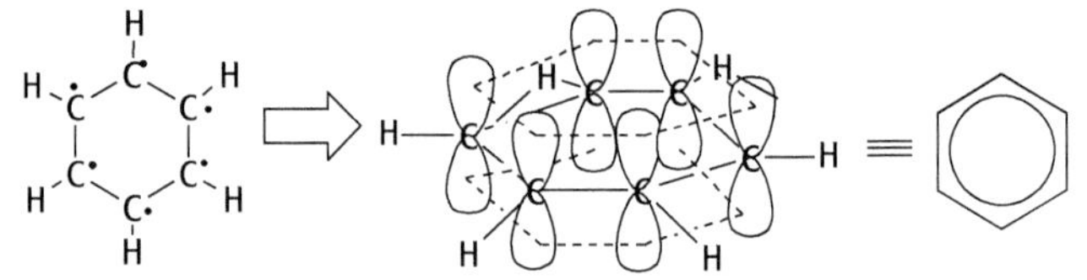

each carbon atom forms three σ bonds, leaving an electron in a p orbital

the six p-orbitals overlap to form a π bond whose electrons are delocalised over all six carbon atoms

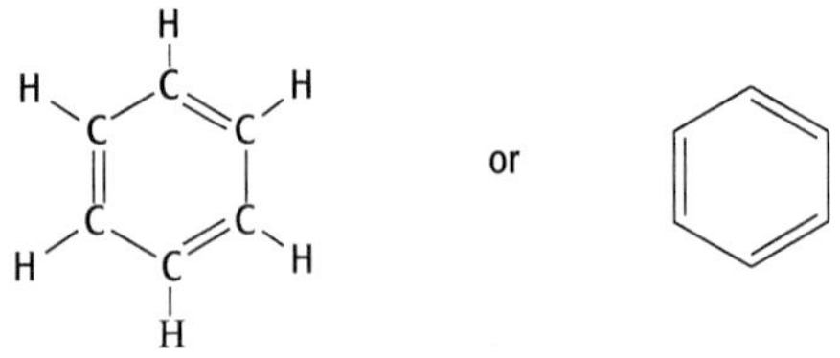

Kekulé's original formula for benzene

The delocalised π bond makes benzene more stable than expected.

25.02 Nomenclature and isomerism

Substituted benzenes show positional isomerism. There are three dichlorobenzenes, named as follows:

1,2-dichlorobenzene

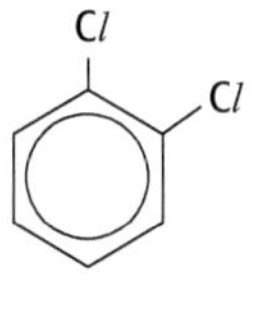

1,3-dichlorobenzene

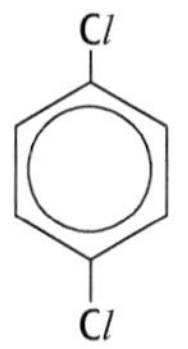

1,4-dichlorobenzene

If the substituent groups are different, the 'major' group is the last-named of the prefixes, and the ring numbering starts from that group, as the following examples illustrate (in each case the 'major' group is at the top of the ring):

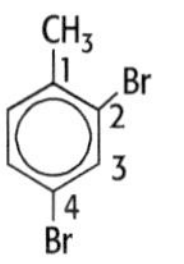

2,4-dibromomethylbenzene 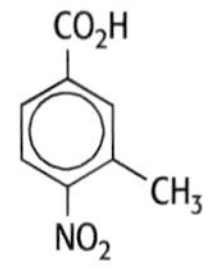   3-methyl-4-nitrobenzoic acid 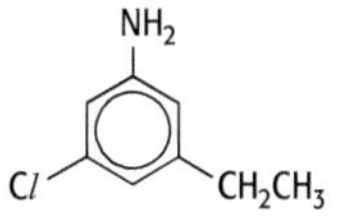   3-chloro-5-ethylphenylamine

25.03 The reactions of the benzene ring

a Hydrogenation

Benzene and other arenes can be **hydrogenated** by reaction with hydrogen at 200 °C and 60 atm pressure. These conditions are much more extreme than those sufficient for the hydrogenation of alkenes (H_2(g) at room temperature and pressure – see Unit 15):

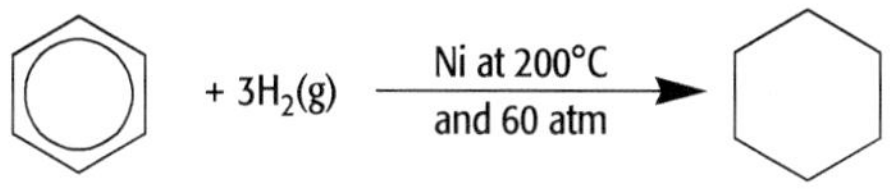

> **TIP**
> Think carefully whenever you draw a hexagonal ring: do you want to represent benzene or cyclohexane? In the rush of sitting an examination paper it is very easy to forget to draw the circle in the benzene ring. An examiner might think you are talking about the wrong compound.

Study the names given to the compounds mentioned in this chapter – the numbering of the carbon atoms round the ring starts with the major substituent as C-1.

b Halogenation

When warmed in the presence of anhydrous $AlCl_3$ or $FeCl_3$, chlorine or bromine undergo substitution reactions with arenes. Two examples are:

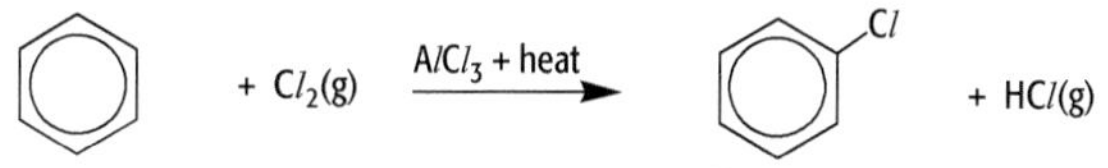

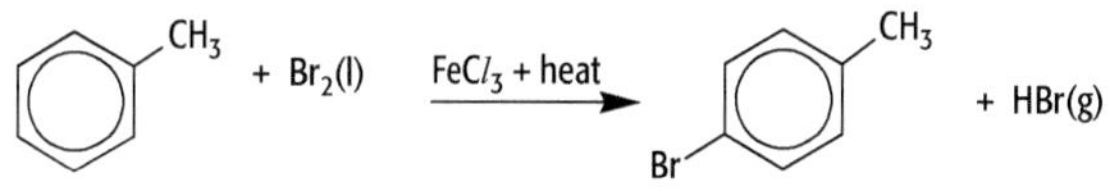

methylbenzene → 4-bromomethylbenzene

c Nitration

Arenes react with concentrated nitric acid in the presence of concentrated sulfuric acid to form nitro-arenes. Strict temperature control is necessary to avoid more than one nitro group substituting in the ring:

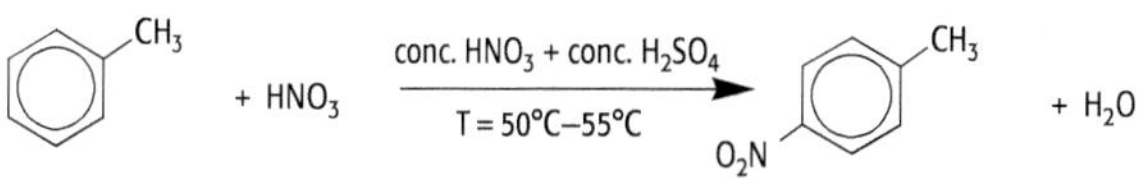

4-nitromethylbenzene

d Alkylation and acylation

Arenes react with chloroalkanes or acyl chloride when warmed with anhydrous $AlCl_3$, giving alkyl or acyl benzenes (these are known as **Friedel–Crafts reactions**):

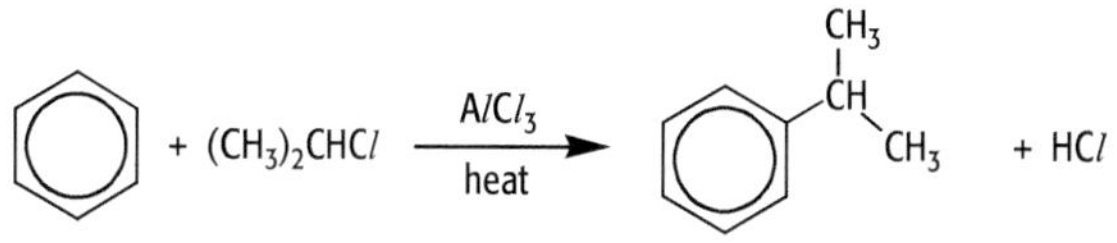

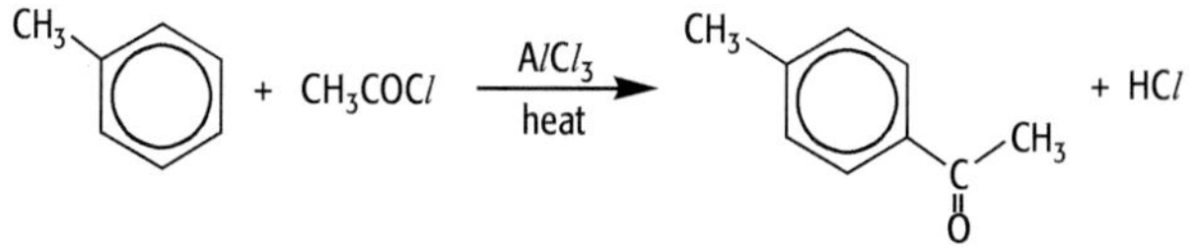

25.04 The mechanism of electrophilic substitution reactions

The halogenation and the nitration of arenes are examples of electrophilic substitutions. The benzene ring is electron-rich, with its π electron ring on its surface, so is attacked by electron-deficient cations. These electrophiles (electron-loving ions) are made when the relevant reagent reacts with the homogeneous catalyst.

a **Bromination**

- In the first step, Br_2 reacts with the catalyst $AlCl_3$ ($AlBr_3$, $FeCl_3$ or $FeBr_3$ can also be used) to give the electrophile Br^+. The initial interaction is a dative bond from a lone pair of electrons on bromine to a vacant orbital on the aluminium atom.

- This then reacts with benzene in a two-stage reaction, in which the intermediate **X** is a carbocation that has the four remaining π electrons delocalised over five carbon atoms of the ring. Intermediate **X** loses a proton to form the product, bromobenzene.

- In the last step, this proton reacts with the complex anion formed in the first step, to produce HBr(g) and to re-form the $AlCl_3$ catalyst, which can react with another Br_2 molecule.

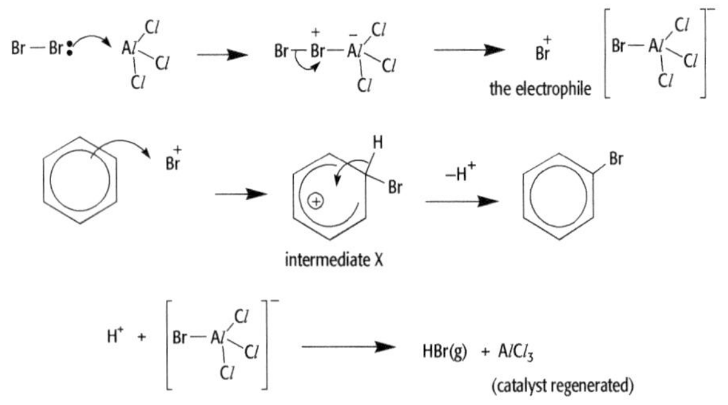

b **Nitration**

- In the first step, the nitric acid molecule is protonated by the concentrated sulfuric acid, and then loses water to form the nitronium cation. (The water is further protonated by another sulfuric acid molecule.)

- The NO_2^+ ion then reacts with benzene to form an intermediate cation **Y**, which, like intermediate **X** above, has the four remaining π electrons delocalised over five carbon atoms of the ring. Intermediate **Y** then loses a proton to form the product, nitrobenzene.

- The proton that is released can then protonate another nitric acid molecule and so on:

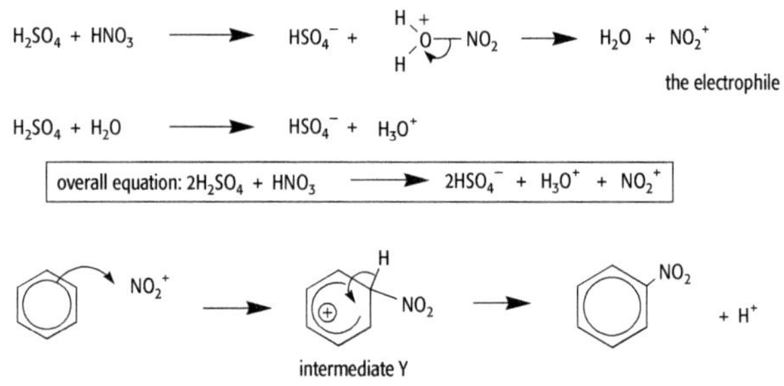

$$H_2SO_4 + HNO_3 \longrightarrow HSO_4^- + H_2O^+-NO_2 \longrightarrow H_2O + NO_2^+$$
the electrophile

$$H_2SO_4 + H_2O \longrightarrow HSO_4^- + H_3O^+$$

overall equation: $2H_2SO_4 + HNO_3 \longrightarrow 2HSO_4^- + H_3O^+ + NO_2^+$

intermediate Y

c **General mechanism**

There are many other electrophilic substitution reactions. In each one an electrophile E^+ is attacked by the π electrons to form an intermediate cation similar to **X** or **Y** above, and this cation then loses a proton to form the product:

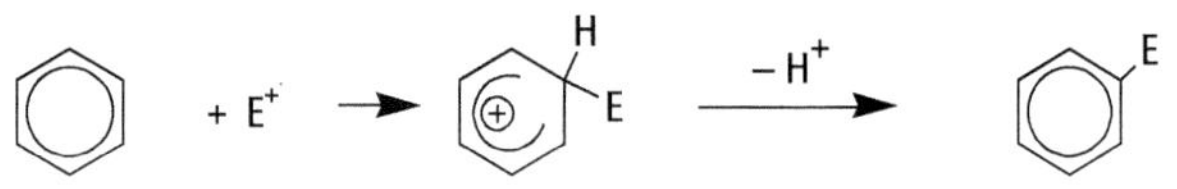

The electrophile in the Friedel–Crafts reaction is generated by the reaction between $AlCl_3$ and the alkyl or acyl chloride:

$$CH_3Cl + AlCl_3 \longrightarrow CH_3^+ + AlCl_4^-$$

> **Progress check 25.01**
>
> Suggest the mechanism for the formation of the electrophile and its subsequent reaction with benzene to form the product, for each of the following reactions.
>
> 1 The production of ethylbenzene from benzene, bromoethane and iron(III) bromide.
>
> 2 The production of ethanoylbenzene from benzene, ethanoyl chloride (CH_3COCl) and aluminium chloride. 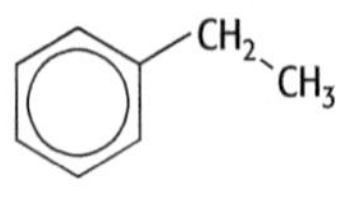
>
> ethylbenzene 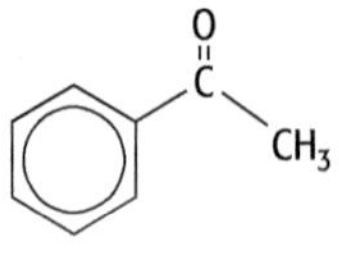   ethanoylbenzene

25.05 Orientation in electrophilic substitution

In benzene all six positions are equivalent: there is only one bromobenzene. In methylbenzene, however, there are three different positions round the ring that could be substituted by bromines, so there are three possible bromomethylbenzenes:

2-bromomethylbenzene   3-bromomethylbenzene   4-bromomethylbenzene

Which one of these isomers is formed when methylbenzene is brominated is determined by the methyl group already attached to the ring, rather than the nature of the incoming substituent. In general, groups that are electron-withdrawing cause the incoming substituent to be attached to the 3-position, whereas electron-donating groups produce a mixture of the 2- and 4-substituted compounds. Because methyl (like all alkyl groups) is electron-donating, the bromination of methylbenzene produces a mixture of 2-bromomethylbenzene and 4-bromomethyl benzene. The same is true when methylbenzene is nitrated:

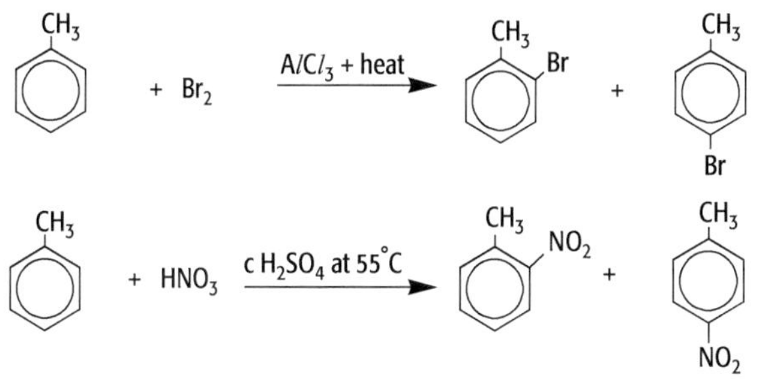

On the other hand, the electron-withdrawing CHO group causes the incoming substituent to be attached to the 3-position:

Table 25.01 summarises the effect of some substituents.

Electron-withdrawing groups		Electron-donating groups	
Name	Formula	Name	Formula
aldehyde	–CHO	methyl	–CH_3
ketone	–COR	alkyl	–R
carboxylic acid	–CO_2H	hydroxy (phenolic)	–OH
nitro	–NO_2	amino	–NH_2
ammonium	–NH_3^+		

Table 25.01 Some electron-withdrawing and electron-donating groups

Progress check 25.02

Suggest the structures of the products of the following reactions.

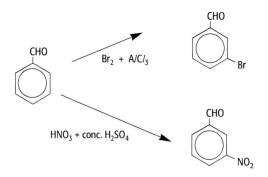

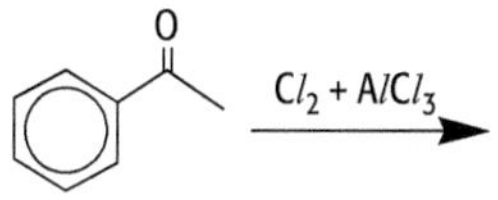

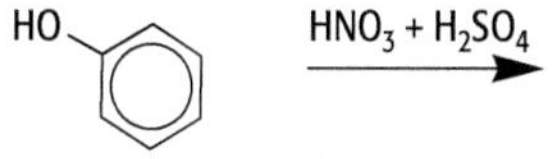

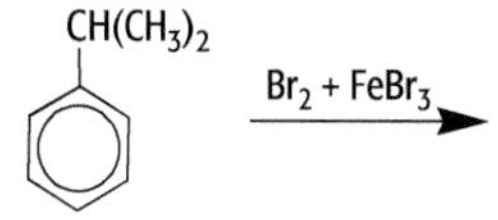

25.06 Two reactions of the side-chains in arenes

a Oxidation of the side-chain

The benzene ring is much more resistant to oxidation than alkyl side-chains. Heating an arene with a strong oxidant, such as alkaline potassium manganate(VII), causes all the side-chains to be completely oxidised to carboxylic acid groups, but leaves the benzene ring unchanged:

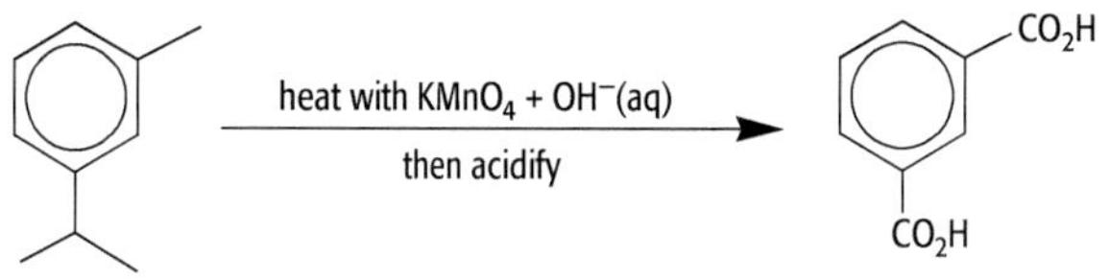

b Halogenation of the side-chain

Worked example 25.01

Suggest routes for synthesising the following two compounds from methylbenzene.

1. 3-bromobenzoic acid
2. 4-nitrobenzoic acid

How to get the answer:

Step 1: We need to use three different reactions here:

- to form a bromobenzene we need to heat with $Br_2 + AlCl_3$
- to form a nitrobenzene we need conc. $HNO_3 + H_2SO_4$ at 55 °C
- to form $-CO_2H$ from $-CH_3$ we need hot alkaline $KMnO_4$, followed by acidification.

Step 2: Consulting Table 25.02 (there is a similar one in the Data Booklet) we see that the $-CH_3$ group is 4-directing, whereas the $-CO_2H$ group is 3-directing. So the order in which the reactions are carried out is important: for 3-benzoic acid we need to oxidise *before* bromination; for reaction 2 we need to oxidise *after* nitration.

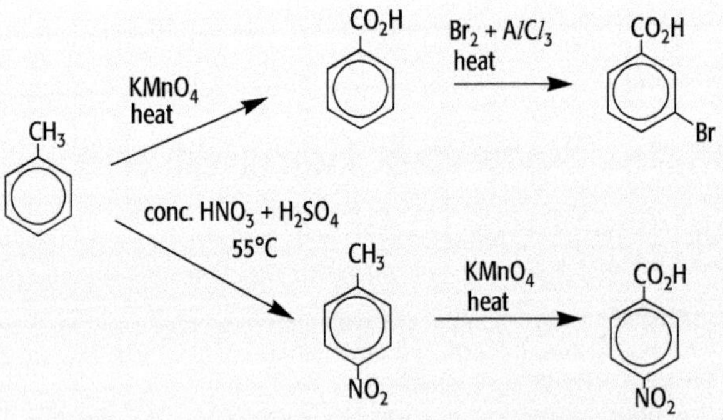

Under the influence of light or heat, and without any catalyst such as $AlCl_3$, chlorine or bromine will undergo a free-radical substitution in the side-chain of an arene. The hydrogen atoms on the carbon adjacent to the ring are replaced:

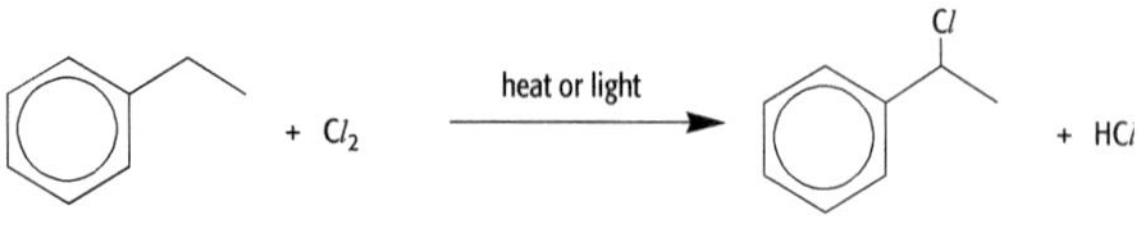

1-phenyl,1-chloroethane

Progress check 25.03

Suggest structures for compounds **A–C** in the following scheme:

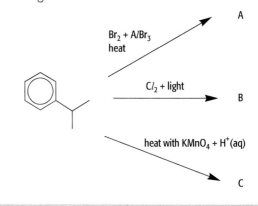

25.07 The properties and reactions of phenol

Aromatic compounds that contain a hydroxy group attached to the ring are called **phenols**. The presence of the benzene ring means that the reactions of the –OH group are different to the usual reactions of alcohols, and the presence of the –OH group on the ring has quite a dramatic effect on the ring's reactivity towards electrophiles.

a **Acidity**

Phenols are more acidic than alcohols, but less so than carboxylic acids:

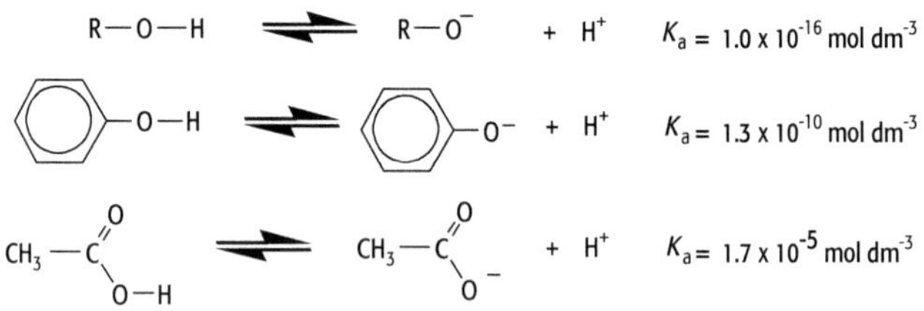

$R-O-H \rightleftharpoons R-O^- + H^+ \quad K_a = 1.0 \times 10^{-16}$ mol dm^{-3}

PhO-H $\rightleftharpoons$ PhO$^-$ + H$^+$ $\quad K_a = 1.3 \times 10^{-10}$ mol dm^{-3}

$CH_3COOH \rightleftharpoons CH_3COO^- + H^+ \quad K_a = 1.7 \times 10^{-5}$ mol dm^{-3}

The phenoxide anion is more stable than either a hydroxide or an alkoxide anion because its negative charge can be delocalised over the benzene ring:

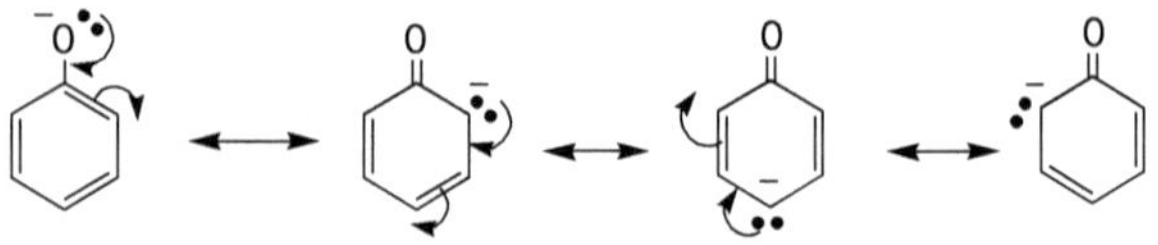

or

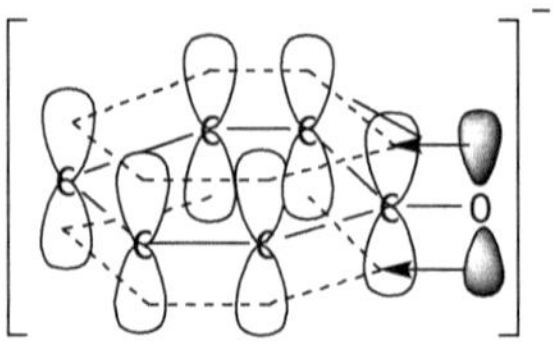

TIP

You may be asked to predict the effect of electron-withdrawing or electron-donating groups (see Table 25.02) on the acidity of phenol. As you might expect, electron-withdrawing groups on the ring enhance the acidity of phenol. For example, 4-nitrophenol is more than 500 times as acidic as phenol.

As a consequence of its acidity, phenol not only reacts like alcohols with sodium metal, giving off hydrogen gas:

C₆H₅OH + Na ⟶ C₆H₅O⁻Na⁺ + ½H₂(g)

sodium phenoxide
(a white solid)

but, unlike alcohols, it also dissolves in aqueous sodium hydroxide:

C₆H₅OH + NaOH ⟶ C₆H₅O⁻Na⁺ + H₂O

Phenol is visibly acidic: the pH of a 0.1 mol dm⁻³ solution in water is 5.4, so it will turn universal indicator solution yellow. An old name for phenol is carbolic acid.

b Esterification

Because of the delocalisation over the ring of the lone pair on the oxygen atom, phenol is not nucleophilic enough to undergo esterification in the usual way, that is, by heating with a carboxylic acid and a trace of concentrated sulfuric acid:

C₆H₅OH + CH₃CO₂H —heat with conc H₂SO₄→ [C₆H₅–O–C(=O)CH₃ + H₂O] ester not formed

Phenol can, however, be esterified by adding an acyl chloride, usually in the presence of a base:

C₆H₅OH + CH₃COCl —(NaOH)→ C₆H₅–O–C(=O)CH₃ + HCl

phenyl ethanoate

c Electrophilic substitution on the ring

Because of the delocalisation of the lone pair of electrons on oxygen, the ring of phenol is more electron-rich than that of the ring in benzene. Reaction conditions for electrophilic substitution can be much milder, and more substitution occurs:

C₆H₅OH —dilute HNO₃ (not conc. and no H₂SO₄ needed)→ 4-nitrophenol (+ 2-NO₂ isomer)

C₆H₅OH —Br₂(aq) (not Br₂(l), and no AlCl₃ needed)→ 2,4,6-tribromophenol (white precipitate)

The decolorisation of orange bromine water, with the formation of a white precipitate, is a good test for the presence of phenol (Figure 25.01) (or phenylamine – see Unit 27).

Fig. 25.01 A white precipitate is formed when orange bromine water is added to aqueous phenol

TIP

Remember that phenylamine (see Unit 27) also decolorises Br₂(aq) and gives a white precipitate. Alkenes also decolorise bromine water (but do not give a white precipitate). A unique test shown by alkenes but by no other functional group is the decolorisation of cold acidified KMnO₄(aq).

The bromine atoms only substitute at the 2, 4 and 6 positions, not the 3 or 5 position. If one of the three possible positions is blocked by another substituent, only two bromines are substituted:

[Structure: 2-methylphenol + 2Br₂(aq) → 2,4-dibromo-6-methylphenol + 2HBr]

A reaction of phenol that is not undergone by benzene is coupling with a diazonium salt:

[Structure: phenyldiazonium chloride + phenol → azo compound (4-hydroxyazobenzene) + NaCl, in NaOH(aq)]

This reaction is covered in more detail in Unit 27.

Sample answer

Question: Two compounds, **X** and **Y**, have the molecular formula C_7H_8O. Both react with sodium metal, but only **Y** reacts with NaOH(aq). When treated with ethanoyl chloride, compound **X** gives compound **V**, and **Y** gives compound **W**. **V** and **W** are isomers with the formula $C_9H_{10}O_2$. One of the isomers **X** or **Y** decolorises bromine water, and produces compound **Z** ($C_7H_6Br_2O$).

Suggest the structures of all lettered compounds **V–Z**. [5]

Answer:

X (benzyl alcohol) → V (benzyl ethanoate) [1] + [1]

Y (2-methylphenol) → W (2-methylphenyl ethanoate) [1] + [1]

Y + Br₂(aq) → Z (2,4-dibromo-6-methylphenol) [1]

Y could also be 4-methylphenol:

W ← Y (4-methylphenol) → Z (with Br₂(aq))

Revision checklist

Check that you know the following:

- The reagents and conditions needed to make the following compounds from benzene: bromobenzene, chlorobenzene, nitrobenzene, methylbenzene and ethanoylbenzene.

- The mechanism of these electrophilic substitution reactions.

- How methylbenzene can be converted into benzoic acid, phenylchloromethane and phenylbromomethane.

- How the position of substitution is determined by the group already bonded to the ring.

- How phenol reacts with Na, NaOH(aq), Br₂(aq), HNO_3(aq), diazonium salts and acyl chlorides.

- How to explain the relative acidities of phenol, ethanol and water.

Exam-style questions

1. a For each of the following reactions, draw the structure of the product and give its systematic name. [6]

   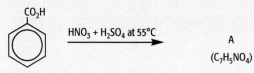

   b Compounds **D** and **E** can be obtained from methylbenzene by reacting it with chlorine under two different conditions.

   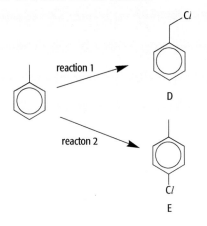

   i Suggest the reagents and conditions needed for reaction 1 and reaction 2. [2]

   ii Suggest a reaction you could use to distinguish between samples of **D** and **E**. You should describe the observations you would make with each compound. [2]

   Total: 10

2. a i Describe and explain how the acidities of phenol and phenylmethanol differ. [3]

   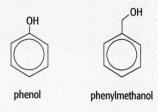

   ii Describe a test (reagent + observations) you could use to distinguish between phenol and phenylmethanol. [2]

   b How might the acidities of compounds **F** and **G** compare to that of phenol? Explain your answer. [2]

   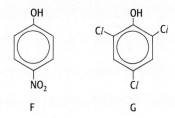

   c Phenyl ethanoate, **H**, cannot be prepared by the following reaction.

   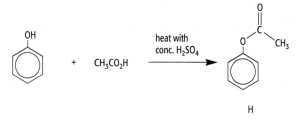

   i Explain why this reaction does not occur. [2]

   ii Suggest an alternative route to make **H** from phenol. [1]

   Total: 10

Unit 26: Carboxylic acids and their derivatives

Learning outcomes

You should be able to:

- [ ] explain the relative acidities of chlorine-substituted ethanoic acids
- [ ] recognise that methanoic and ethanedioic acids can be further oxidised
- [ ] describe the formation and hydrolysis of acyl chlorides
- [ ] describe the reactions of acyl chlorides with alcohols, phenol, ammonia and amines
- [ ] explain the relative ease of hydrolysis of acyl, alkyl and aryl chlorides
- [ ] describe the addition–elimination mechanism for acyl chlorides

26.01 The acidities of chlorine-substituted carboxylic acids

We explained in Unit 17 why carboxylic acids are more acidic than alcohols, because the negative charge on the anion can be delocalised over two electronegative oxygen atoms.

If the R group in $R-CO_2^-$ contains an electronegative atom, this will help to spread out even more the negative charge of the anion, and so cause the acid to be more dissociated. Table 26.01 shows some examples.

Name of acid	Formula of acid	pK_a
ethanoic acid	$CH_3 \rightarrow C(=O)OH$	4.76
chloroethanoic acid	$Cl \leftarrow CH_2 - C(=O)OH$	2.87
dichloroethanoic acid	$Cl_2CH - C(=O)OH$	1.26
trichloroethanoic acid	$Cl_3C - C(=O)OH$	0.66

Table 26.01 The pK_a values for some carboxylic acids

Progress check 26.01

Making use of the trends shown in Table 26.01, choose an appropriate pK_a for each of the following acids from the pK_a values given below. In each case explain your choice.

1. iodoethanoic acid
2. fluoroethanoic acid
3. trifluoroethanoic acid.

pK_a values: 0.59, 2.57, 2.87, 3.16, 4.87

26.02 The oxidation of methanoic and ethanedioic acids

Although most carboxylic acids do not undergo further oxidation, there are two that do.

a **Methanoic acid**

Methanoic acid contains a –CHO group, and has some of the reactions of aldehydes. When Fehling's solution or Tollens' reagent are warmed with methanoic acid, they are reduced:

Fehling's: $HCO_2H + 2Cu^{2+} + 6OH^- \longrightarrow CO_3^{2-} + Cu_2O(s) + 4H_2O$
    red ppt.

Tollens': $HCO_2H + 2Ag^+ + 4OH^- \longrightarrow CO_3^{2-} + 2Ag(s) + 3H_2O$
    silver mirror

b Ethanedoic acid

Ethanedoic acid is readily oxidised by warm acidified manganate(VII), turning the purple solution colourless:

$$5HO_2C\text{-}CO_2H + 2MnO_4^- + 6H^+ \longrightarrow 10CO_2 + 2Mn^{2+} + 8H_2O$$

26.03 Formation of acyl chlorides

Heating carboxylic acids with either phosphorus pentachloride or thionyl chloride produces acyl chlorides:

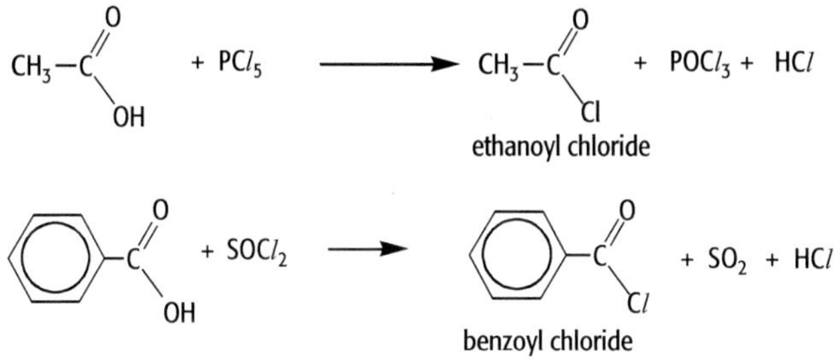

$CH_3\text{-}COOH + PCl_5 \longrightarrow CH_3\text{-}COCl$ (ethanoyl chloride) $+ POCl_3 + HCl$

$C_6H_5\text{-}COOH + SOCl_2 \longrightarrow C_6H_5\text{-}COCl$ (benzoyl chloride) $+ SO_2 + HCl$

26.04 The reactions of acyl chlorides with water, alcohols, phenols and amines

Acyl chlorides are readily hydrolysed by water – or even moist air – to give fumes of HCl(g) and the carboxylic acid:

$$RCOCl + H_2O \longrightarrow RCO_2H + HCl(g)$$

As the following scheme shows, they are useful intermediates in organic synthesis, being used to make esters (especially phenyl esters) and amides.

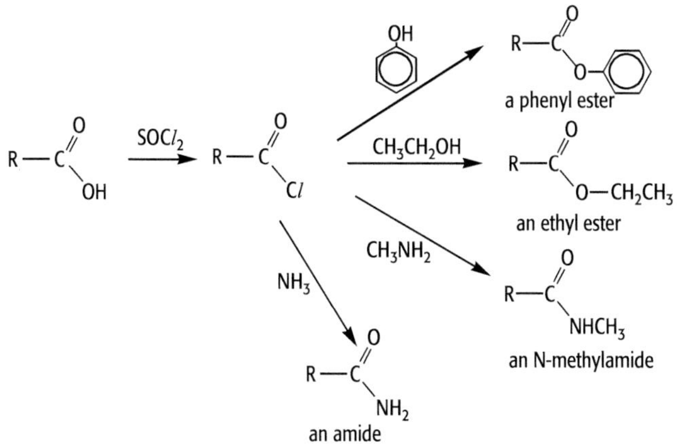

[Reaction scheme: RCO₂H →(SOCl₂) RCOCl, which reacts with:
- phenol (C₆H₅OH) → a phenyl ester
- CH₃CH₂OH → an ethyl ester (R-CO-O-CH₂CH₃)
- CH₃NH₂ → an N-methylamide (R-CO-NHCH₃)
- NH₃ → an amide (R-CO-NH₂)]

TIP: Learn to recognise reaction types: these are all nucleophilic substitution reactions, the nucleophile attacking the δ+ carbon atom of the C=O group, and displacing the chloride ion, Cl⁻.

Although alkyl esters can be made by the usual method (i.e. heating the carboxylic acid and the alcohol with concentrated H_2SO_4), this method does not work for phenyl esters, because phenol is not a strong enough nucleophile. Phenyl esters can be made in good yield by reacting acyl chlorides with phenols.

Progress check 26.02

Suggest reagents for two-step synthetic routes to the following compounds, starting with suitable carboxylic acids.

1 $(CH_3)_2CHCONHCH(CH_3)_2$

2

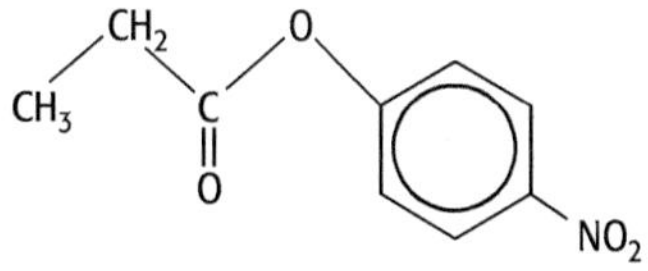

TIP: See Unit 30 on organic synthesis for further tips on how to devise synthetic routes.

26.05 The relative reactivities of various organochlorine compounds

As we saw in Unit 16, the reactivity of R–X increases from X = Cl to X = I. But even organochlorine compounds differ quite markedly in their reactivities.

For example, if we look at their reactions with water or NaOH(aq) (a stronger nucleophile), we find the results in Table 26.02.

Compound	Reaction with water	Reaction with NaOH(aq)
acyl chlorides, RCOCl	readily, at room temperature	very vigorous, at room temperature
tertiary chloroalkanes, R_3C-Cl	slowly, at room temperature, quicker on heating	speedily on warming
primary chloroalkanes, $R-CH_2Cl$	hardly at all at room temperature; slowly on heating	steadily on warming
chloroarenes, 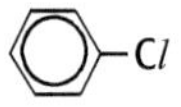 ⌬–Cl	no reaction	still no reaction

Table 26.02 The relative reactivities of organochlorine compounds

The reactivities are in the order RCOCl > R–Cl > Ar–Cl.

Why are acyl chlorides so much more reactive that chloroalkanes?

There are two reasons.

1. In acyl chlorides, the chlorine is bonded to a carbon atom that already has an electronegative oxygen atom attached. This makes the carbon very $\delta+$, so it is more prone to being attacked by the lone pair on the nucleophile (e.g. H_2O, NH_3).

2. Acyl chlorides have a double bond. This allows the nucleophile to bond with the carbon atom before the chloride ion has left. This extra bonding lowers the activation energy, and hence speeds up the reaction.

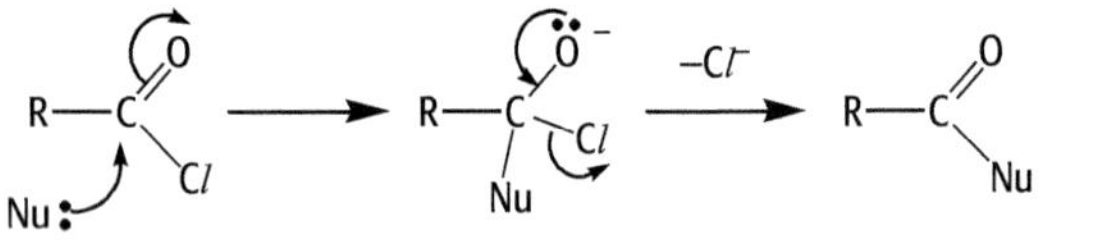

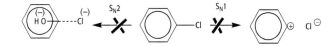

This is known as the addition–elimination mechanism.

Why are chloroarenes so much less reactive than chloroalkanes?

Here, again, there are two reasons.

1. One of the lone pairs on the chlorine atom can overlap with the π bond in the benzene ring. The overlap is less complete than, say, the overlap of the oxygen's lone pair in phenol (see Unit 25), because chlorine's lone pair is in a 3p orbital, which, being larger, overlaps with the 2π ring orbitals less effectively than the 2p lone pair on oxygen. But the effect is significant enough to make the C–Cl bond stronger, and less easy to break.

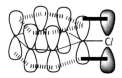

2. Because the ring hinders the attack of a nucleophile from the back of the C–Cl bond, the S_N2 route (see Unit 16) is unavailable for a nucleophilic substitution reaction. Since there is no means of stabilising the intermediate carbocation, the S_N1 route is also not favoured.

Worked example 26.01

Arrange the following halogeno compounds in order of reactivity, with the most reactive first.

A $(CH_3)_2CHCl$

B CH_3CH_2COCl

C $(CH_3)_3CBr$

D ⌬–Cl

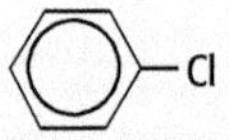

E ⌬–CH_2Cl

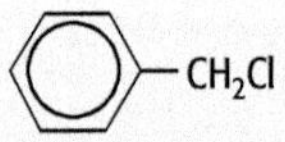

How to get the answer:

Step 1: We can see from Table 26.02 that acyl chlorides are the most reactive, and aryl chlorides are the least. So **B** will be first, and **D** last.

Step 2: From Unit 16, we know that tertiary halides are more reactive than primary halides, and primary halides are more reactive than secondary halides. We also know that bromides are more reactive than chlorides. So **C** is more reactive than **E**, and **E** more reactive than **A**.

Step 3: The order is therefore **B > C > E > A > D**.

Sample answer

Question:

Predict the products **A–D** of the following series of reactions. [4]

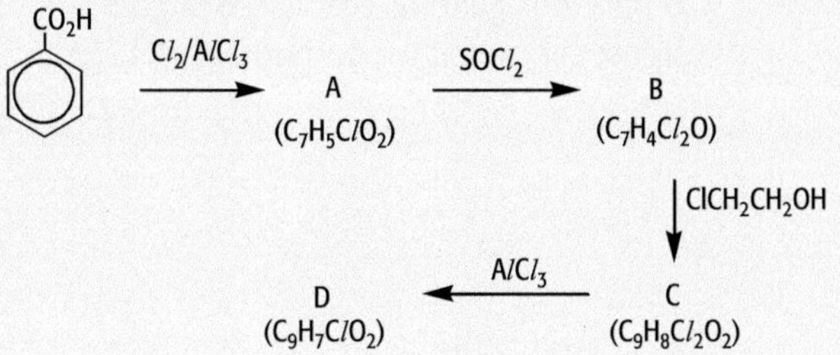

$$\text{C}_6\text{H}_5\text{CO}_2\text{H} \xrightarrow{\text{Cl}_2/\text{AlCl}_3} \mathbf{A}\ (\text{C}_7\text{H}_5\text{ClO}_2) \xrightarrow{\text{SOCl}_2} \mathbf{B}\ (\text{C}_7\text{H}_4\text{Cl}_2\text{O}) \xrightarrow{\text{ClCH}_2\text{CH}_2\text{OH}} \mathbf{C}\ (\text{C}_9\text{H}_8\text{Cl}_2\text{O}_2) \xrightarrow{\text{AlCl}_3} \mathbf{D}\ (\text{C}_9\text{H}_7\text{ClO}_2)$$

Answer:

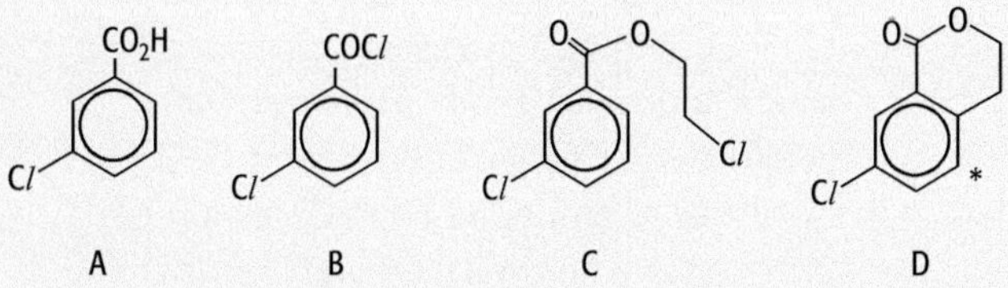

A: 3-chlorobenzoic acid (CO_2H, Cl meta)
B: 3-chlorobenzoyl chloride (COCl, Cl meta)
C: 2-chloroethyl 3-chlorobenzoate
D: chloro-isochroman-1-one (lactone)

[1 mark for each structure]

In **D** the Cl could be at the asterisked position instead of the one shown.

Revision checklist

Check that you know the following:

- [ ] How acyl chlorides are made.
- [ ] How methanoic and ethanedioic acids can be oxidised.
- [ ] Why chloro-carboxylic acids are more acidic than non-chorine-containing acids.
- [ ] The addition–elimination mechanism for the reactions of acyl chlorides.
- [ ] How acyl chlorides react with alcohols, phenols and amines.
- [ ] The explanation for the reactivity order acyl > alkyl > aryl chlorides.
- [ ] Explain why carboxylic acids are more acidic than phenol and alcohols

Exam-style questions

1. a i Describe and explain the relative acidities of ethanoic acid and chloroethanoic acid. [3]

   ii How would the acidities of the following acids compare to that of chloroethanoic acid? Explain your answers. [4]

   CH_2FCO_2H CH_2BrCO_2H $CHCl_2CO_2H$
   $ClCH_2CH_2CO_2H$

   b Suggest the structures of compounds **A–C** in the following scheme, explaining your reasoning.

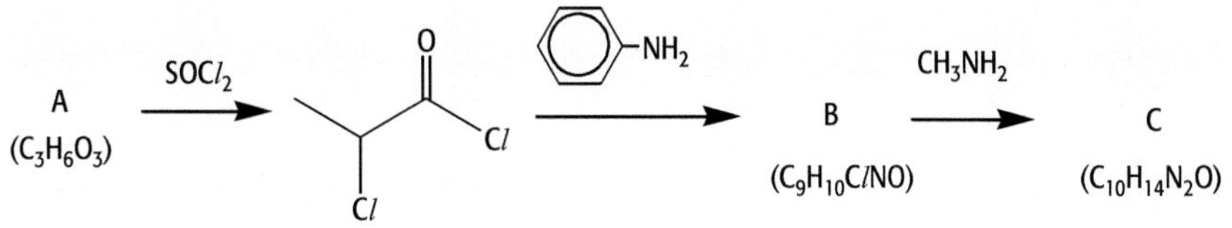

   $$\mathbf{A}\ (C_3H_6O_3) \xrightarrow{SOCl_2} CH_3CH(Cl)COCl \xrightarrow{C_6H_5NH_2} \mathbf{B}\ (C_9H_{10}ClNO) \xrightarrow{CH_3NH_2} \mathbf{C}\ (C_{10}H_{14}N_2O)$$

   [4]

   Total: 11

2 a When one mole of compound **D**, $C_4H_6O_4$, is hydrolysed, one mole of compound **E**, $C_2H_6O_2$, and two moles of compound **F** are produced. When E is warmed with acidified $K_2Cr_2O_7$(aq), compound **G**, $C_2O_4H_2$, is formed. Compound **F** produces a red precipitate when warmed with Fehling's solution. When one mole of compound **G** is heated with acidified $KMnO_4$, two moles of compound **H** are formed.

   i Suggest reagents and conditions for the conversion of **D** into **E** and **F**. [1]

   ii Suggest the structures of compounds **D**–**H**, explaining all the reactions involved. [7]

b The hydrolysis of compound **J**, $C_6H_8O_2$, produces compound **K**, $C_6H_{10}O_3$. When **K** is treated with hot acidified $KMnO_4$, the only product is compound **L**, $C_3H_4O_3$. Compounds **K** and **L** produce a yellow precipitate with alkaline aqueous iodine. Compounds **J** and **K** decolorise Br_2(aq). Suggest structures for compounds **J**–**L**. [3]

Total: 11

Organic nitrogen compounds

Learning outcomes

You should be able to:

- describe the various ways of making alkyl amines from halogenoalkanes, amides and nitriles, and how to make phenylamine from nitrobenzene
- describe and explain why amines are basic, and the relative basicities of ammonia, ethylamine and phenylamine
- describe the reactions of phenylamine with aqueous bromine and nitrous acid
- describe the formation of dyestuffs
- describe the formation of amides
- recognise that amides are neutral
- describe the hydrolysis and reduction of amides
- describe the formation of zwitterions from amino acids
- describe the formation of peptide bonds
- describe simply the process of electrophoresis

27.01 The synthesis of amines

a Aliphatic amines

There are three ways of making primary aliphatic amines. The following scheme also shows how the starting materials can be made from more readily available compounds:

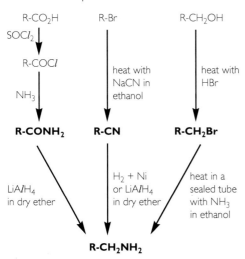

Worked example 27.01

Suggest synthetic routes to the following amines.

1. $CH_3CH_2CH_2CH_2NH_2$ from propan-1-ol
2. $CH_3CH_2CH_2NH_2$ from propanoic acid

How to get the answer:

Step 1: The first amine contains a chain of four carbon atoms, whereas propan-1-ol only contains three. So we need to use NaCN to add a carbon atom, and then reduce the –CN to –CH_2NH_2.

Step 2: NaCN only reacts with halogenoalkanes, so we need first to make 1-bromopropane from propan-1-ol.

Step 3: The second amine contains the same number of carbon atoms as propanoic acid, so we can first make the amide, and then reduce it.

Step 4: The complete routes are as follows:

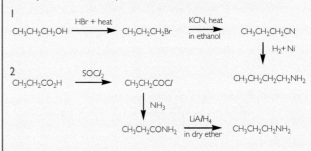

b Aromatic amines

Aromatic amines are usually made by reducing nitroarenes with tin and concentrated hydrochloric acid:

> **TIP** Sn + HCl is the only reducing agent that gives a good yield of phenylamine. Note that the tin is *not* a catalyst, but a reagent — it is used up during the reaction.

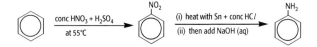

The NaOH(aq) is added after the reduction is complete, to liberate the free amine from its salt:

$$RNH_3^+Cl^- + NaOH \rightarrow RNH_2 + NaCl + H_2O$$

> **TIP** Note that it is not possible to make aromatic amines by reacting halogenobenzenes with ammonia, as the aryl–halogen bond is too strong to be broken by nucleophiles.
>
> C₆H₅Cl + 2NH₃ ⟶̸ C₆H₅NH₂ + NH₄Cl

27.02 Reactions of amines

a Reactions as bases

Like ammonia, all amines are weak bases, being protonated on the lone pair on their nitrogen atom. They react with acids to form salts:

$$CH_3CH_2\overset{..}{N}H_2 + HCl \longrightarrow CH_3CH_2-\overset{+}{N}H_3 + Cl^-$$

Those amines that are soluble in water form weakly alkaline solutions, just as ammonia does, due to partial reaction with the solvent, producing OH⁻ ions:

$$CH_3CH_2NH_2 + H_2O \rightleftharpoons CH_3CH_2NH_3^+ + OH^-$$

Aliphatic amines are slightly more basic than ammonia. However, phenylamine is considerably *less* basic than ammonia.

Basicity depends on the availability of the lone pair of electrons on nitrogen to form a dative bond with a proton. Electron donation from an alkyl group will encourage dative bond formation:

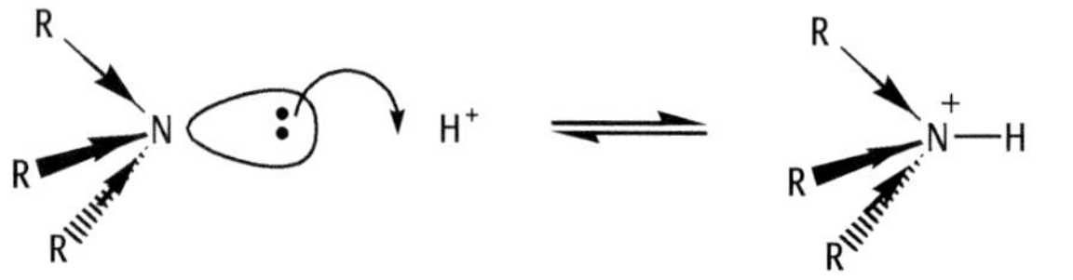

In phenylamine the lone pair of electrons on the nitrogen atom is delocalised over the benzene ring. The bonds around the nitrogen atom can take up a planar arrangement, with the nitrogen's lone pair in a p orbital, so that extra stability can be gained by overlapping this p orbital with the delocalised π bond of the benzene ring:

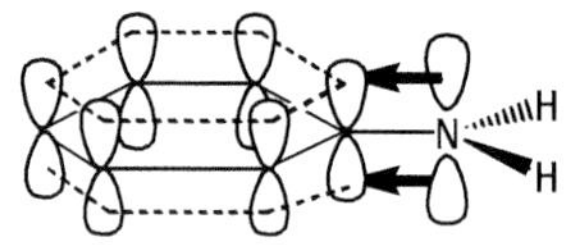

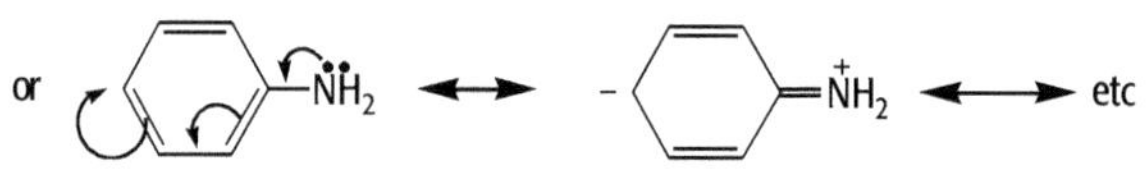

This overlap, which is similar to that described for phenol in Unit 25, has two effects on the reactivity of phenylamine:

- It causes the lone pair to be much less basic (see above).
- It causes the ring to be more electron rich, and so to undergo electrophilic substitution reactions much more readily than benzene.

This increased reactivity is similar to that of phenol, one example being the ease with which phenylamine decolorises bromine water:

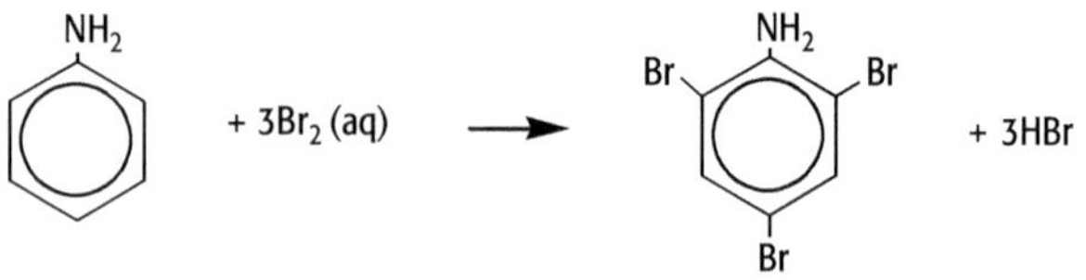

$$C_6H_5NH_2 + 3Br_2(aq) \rightarrow C_6H_2Br_3NH_2 + 3HBr$$

The decolorisation of orange bromine water, with the production of a white precipitate of the tribromo compound, is a distinctive test for the presence of phenylamine (or phenol – see Chapter 25).

b Reactions as nucleophiles

The lone pair on the nitrogen atom of amines makes them good nucleophiles. They react with alkyl halides and acyl halides.

$CH_3CH_2NH_2 + CH_3CH_2Cl \longrightarrow CH_3CH_2\text{-}NHCH_2CH_3 + HCl$

$C_6H_5\text{-}NH_2 + CH_3COCl \longrightarrow C_6H_5\text{-}NHCOCH_3 + HCl$

c The formation and reactions of diazonium salts

When phenylamine is reacted with a mixture of sodium nitrite and hydrochloric acid, at a temperature lower than 5 °C, a solution of a diazonium chloride is produced.

The sodium nitrite and hydrochloric acid react to give the unstable nitrous acid:

$$NaNO_2 + HCl \rightarrow NaCl + HNO_2$$

The nitrous acid then reacts with the phenylamine in acidic solution:

$C_6H_5NH_2 + HNO_2 + HCl \longrightarrow C_6H_5N_2^+ Cl^- + 2H_2O$

phenyldiazonium chloride

> **TIP**
> If you draw out the full bonding in the diazonium ion, make sure you put the + charge on the correct atom, with the right number of bonds. There are two mesomeric forms; you can choose either.
>
> Ph–N̈=N⁺ ↔ Ph–N⁺≡N (or just Ph–N₂⁺)

Phenyldiazonium chloride is unstable at temperatures above 10 °C. If warmed in water it produces phenol and nitrogen gas:

$C_6H_5N_2^+ + H_2O \xrightarrow{\text{warm to} >50°C} C_6H_5OH + N_2(g) + H^+$

The most important reaction of diazonium salts is to make **azo** dyes, by reacting them with phenoxide ions or aromatic amines. These are known as coupling reactions:

$C_6H_5OH + OH^- \xrightarrow{NaOH (aq)} C_6H_5O^- + H_2O$

$C_6H_5N_2^+ + C_6H_5O^- \xrightarrow{T < 5°C} C_6H_5\text{-}N=N\text{-}C_6H_4\text{-}OH$

an azo compound (this one is bright orange)

Azo dyes are used to dye clothes, and also for colouring foodstuffs and paints.

Fig. 27.01 The azo dye formed from phenyldiazonium and phenol

> ### Progress check 27.01
> Draw the structures of the amines that need to be diazotised to form the following dyes.
>
> 'aniline yellow'
>
> 'fast red A'
>
> 'sunset yellow'
>
> 'para red'

> **Sample answer**
>
> Question: 4-Nitrophenol can be made from benzene by the following steps. 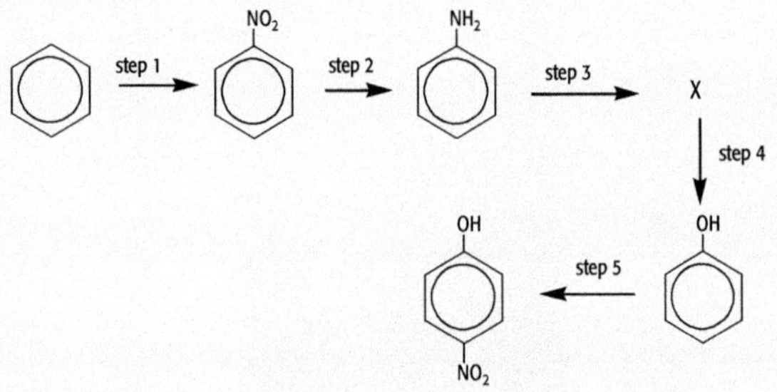
>
> benzene →(step 1)→ nitrobenzene →(step 2)→ aniline (NH₂) →(step 3)→ X →(step 4)→ phenol →(step 5)→ 4-nitrophenol
>
> a Suggest the structure of compound **X** in the above scheme. [1 mark]
>
> b Suggest reagents and conditions for the five steps in the above scheme. [7]
>
> **Answer:**
>
> a **X** is benzenediazonium chloride (C₆H₅N₂⁺ Cl⁻) [1 mark]
>
> b Step 1: $HNO_3 + H_2SO_4$ [1 mark], concentrated, at 55 °C [1 mark]
>
> Step 2: Sn + conc. HCl [1 mark]
>
> Step 3: $HCl + NaNO_2$(aq) [1 mark], at T < 5 °C [1 mark]
>
> Step 4: boil in water [1 mark]
>
> Step 5: dilute HNO_3(aq) [1 mark]

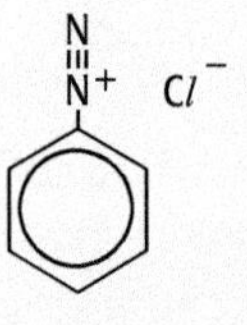

27.03 Properties and reactions of amides

Unlike amines, amides are neutral compounds, and do not form salts with dilute acids. This is because the lone pair on the nitrogen atom in amides is in a p orbital, and can overlap with the π orbital of the adjacent carbonyl group – the electrons are attracted to the electronegative oxygen atom:

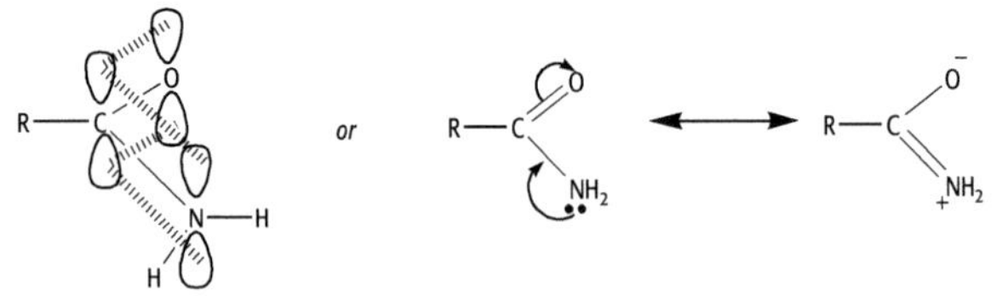

As we have seen above, amides can be prepared by reacting ammonia or an amine with an acyl chloride:

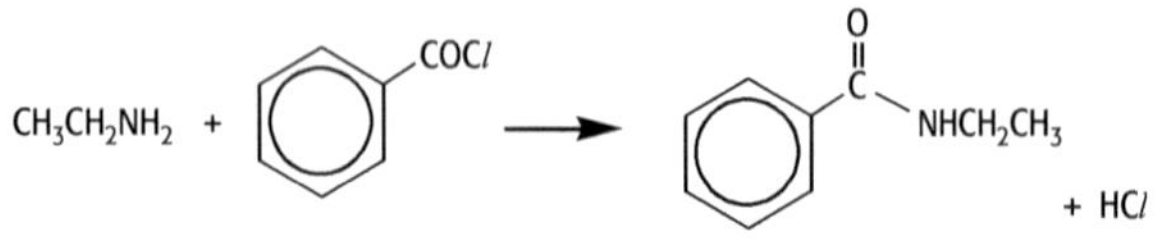

$CH_3CH_2NH_2$ + C₆H₅COCl → N-ethylbenzamide + HCl

There are two main reactions of amides.

a **Hydrolysis**

Hydrolysis can be carried out by heating the amide with dilute aqueous acid or alkali, and produces an amine (or its salt) and a carboxylic acid (or its salt).

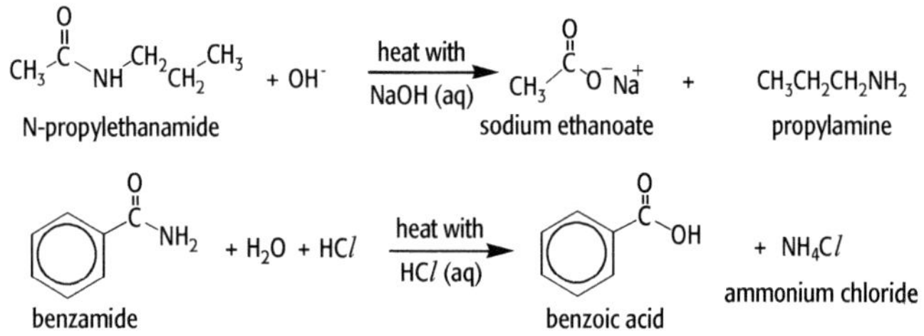

$CH_3CONHCH_2CH_2CH_3$ (N-propylethanamide) + OH⁻ →(heat with NaOH (aq))→ $CH_3CO_2^-Na^+$ (sodium ethanoate) + $CH_3CH_2CH_2NH_2$ (propylamine)

C₆H₅CONH₂ (benzamide) + H_2O + HCl →(heat with HCl (aq))→ C₆H₅CO₂H (benzoic acid) + NH_4Cl (ammonium chloride)

b **Reduction**

As mentioned above, amides can be reduced to amines with $LiAlH_4$ in dry ether.

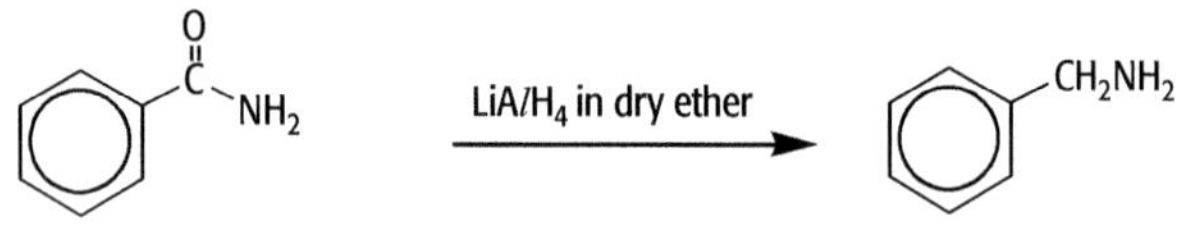

benzamide (C₆H₅CONH₂) →($LiAlH_4$ in dry ether)→ benzyl amine (C₆H₅CH₂NH₂)

27.04 Amino acids

Amino acids are important building blocks for peptides and proteins. They have the general formula

$$H_2N-CHR-CO_2H$$

where R can be H or an organic group.

There are about 20 different amino acids that make up naturally occurring proteins. The following are six of the more common amino acids:

glycine (Gly), alanine (Ala), serine (Ser), glutamic acid (Glu), lysine (Lys), cysteine (Cys)

Amino acids contain both the amino group and the carboxylic acid group, and so can react with either acids or bases:

$H_2NCH_2CO_2H + HCl \rightarrow Cl^-H_3N^+CH_2CO_2H$

$H_2NCH_2CO_2H + NaOH \rightarrow H_2NCH_2CO_2^-Na^+$

The two groups are of opposite chemical characteristics: the $-NH_2$ group is basic, whilst the $-CO_2H$ group is acidic. An internal acid–base reaction can take place, whereby a proton is transferred from the acidic $-CO_2H$ group to the basic $-NH_2$ group, to form a bipolar ion called a **zwitterion**.

A zwitterion is a molecule that contains both a cationic group and an anionic group.

An amino group on one amino acid can also react with the carboxylic acid group on another amino acid, to form an amide:

$H_2NCH_2CO_2H + H_2NCHRCO_2H \rightarrow$
$\quad H_2NCH_2CO-NHCHRCO_2H + H_2O$

An amide bond formed between two amino acids is called a **peptide bond**, and the 'dimer' is called a **dipeptide**.

> **TIP**
> Note that a dipeptide contains *two* amino acid residues, but only *one* peptide bond.

Further amino acids can form amide (peptide) bonds at each end of the dimer, and eventually a **polypeptide** chain can be formed.

part of a polypeptide chain

Like ordinary amides, peptides and polypeptides can be hydrolysed by heating them with HCl(aq).

Progress check 27.02

1. Draw the structure of the dipeptide made from the amino acids cysteine and glutamic acid.

2. Identify the amino acids, and the number of molecules of each, that will be formed by hydrolysing the following peptide.

27.05 Electrophoresis of amino acids and peptides

Electrophoresis is the movement of the electrically charged molecules (ions) of a compound in an electric field. When dissolved in a buffer at a particular pH, different amino acids are likely to be ionised to different extents. The overall charge on an amino acid depends on the pH of the solution it is dissolved in. For example, if glycine is dissolved in an acidic solution, its molecule will be protonated and it will exist as a cation:

$NH_2CH_2CO_2H + H^+ \rightarrow NH_3^+CH_2CO_2H$

If, however, glycine is dissolved in an alkaline solution, it loses a proton to become an anion:

$NH_2CH_2CO_2H + OH^- \rightarrow NH_2CH_2CO_2^- + H_2O$

In neutral solutions, it exists as its zwitterion, which although charged, has no *overall* charge:

$NH_2-CH_2-CO_2H \rightarrow NH_3^+CH_2CO_2^-$

Amino acids that contain a basic group in their side-chain, such as lysine, will exist as cations even in neutral solution:

$NH_2CH_2CH_2CH_2CH_2CH(NH_2)CO_2H + H_2O \rightleftharpoons$
$\quad ^+NH_3CH_2CH_2CH_2CH_2CH(NH_2)CO_2H + OH^-$

Amino acids that contain an acidic group in their side-chain, such as glutamic acid, will exist as anions even in neutral solution:

$HO_2CCH_2CH_2CH(NH_2)CO_2H + H_2O \rightleftharpoons$
$^-O_2CCH_2CH_2CH(NH_2)CO_2H + H_3O^+$

If a solution containing the above three amino acids were applied to the middle of a gel electrophoresis plate soaked in a buffer solution at pH 6 (where glycine has an overall charge of zero), and a potential difference is applied, we would see that the mixture would separate into three spots: the lysine would move towards the cathode (negative electrode); the glutamic acid would move towards the anode (positive electrode) and the glycine would not move at all (Figure 27.02).

Because each amino acid side-chain will affect the average charge on an amino acid molecule / ion to a different extent, and because the speed at which ions move through the gel depends on their size and shape as well as their charge (the larger a molecule is, the slower it moves), it is possible to separate and identify not only all of the amino acids but also small peptides derived from them.

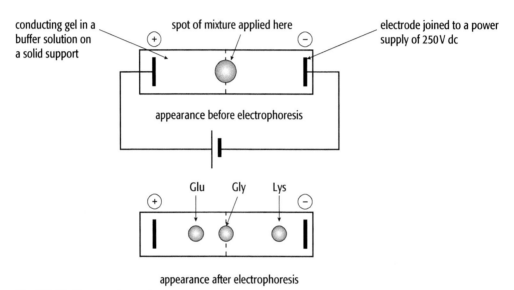

Fig. 27.02 Electrophoresis

Revision checklist

Check that you know the following:

- How alkyl amines are prepared from halogenoalkanes, amides or nitriles.
- How nitrobenzene is reduced to phenylamine.
- How their structures help to explain the relative basicities of ammonia, ethylamine and phenylamine.
- How phenylamine reacts with Br_2(aq) and HNO_2, and how phenol can be made from the resulting phenyldiazonium ion.
- recognise that amides are neutral
- How diazo dyestuffs are made.
- How amides are made from acyl chlorides, and how they are hydrolysed and reduced.
- How amino acids react as acids or bases, and how they form zwitterions.
- How peptides can be formed from amino acids.
- How electrophoresis works, to separate amino acids and peptides.

Exam-style questions

1 Study the following scheme and answer the questions below.

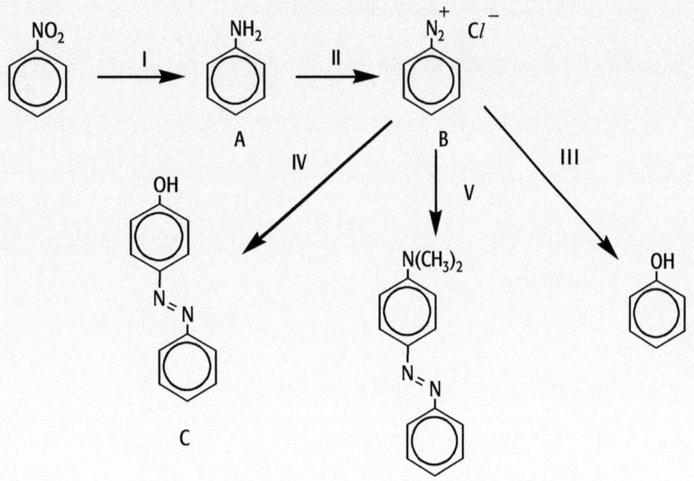

a State the reagents and conditions for reactions I–V. [5]

b To what *classes of compounds* do compounds **A**, **B** and **C** belong? [3]

c The following two compounds are made by coupling reactions. For each compound draw the structures of the two starting compounds required for their synthesis.

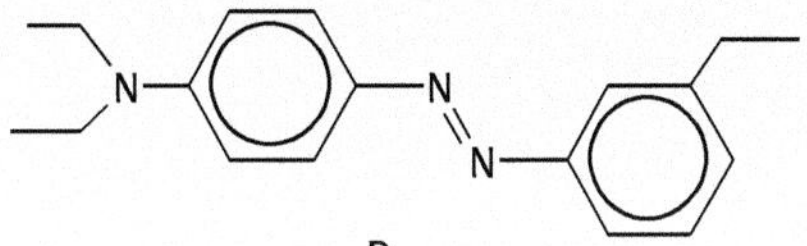

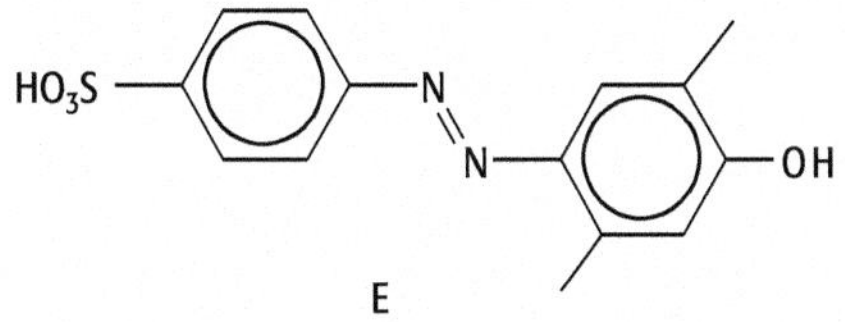

[4]

Total: 12

2 a 2-methylpropylamine, **G**, can be synthesised from 2-methylpropanoic acid by the following three-step reaction scheme.

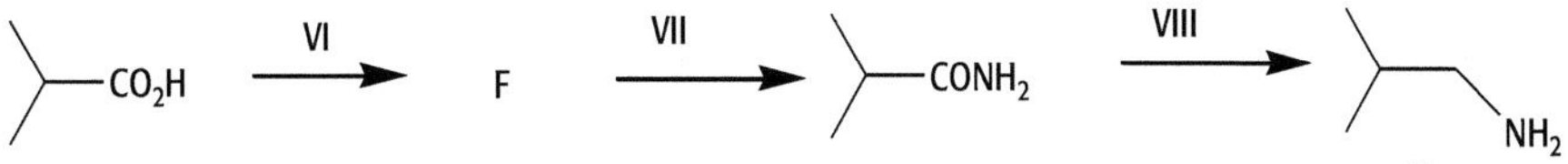

i Draw the structure of the intermediate **F**. [1]

ii State the reagents and conditions for each of the reactions VI–VIII. [3]

iii Draw the structure of the compound formed when **G** reacts with **F**. What *class of compound* is formed in this reaction? Write its molecular formula. [3]

b Below are the structures of two amino acids.

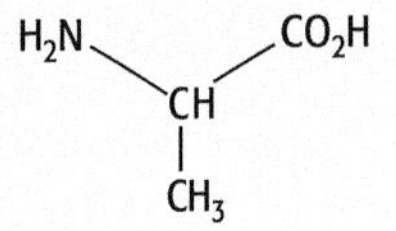

$$H_2N-CH(CH_3)-CO_2H$$
alanine (Ala)

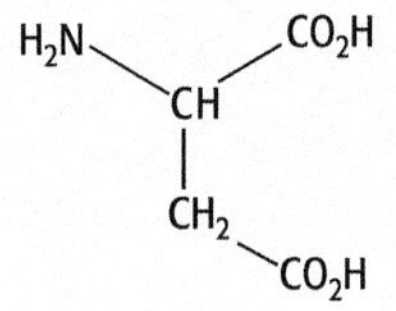

$$H_2N-CH(CH_2CO_2H)-CO_2H$$
aspartic acid (Asp)

i Draw the full displayed formula of the zwitterion of alanine. [2]

ii Draw the structure of the dipeptide Asp-Ala, showing the peptide bond in full. [2]

iii Draw the structure of aspartic acid that exists in an aqueous solution at pH 7. [1]

Total: 12

Polymerisation

Learning outcomes

You should be able to:

- describe the formation of polyesters and polyamides
- describe the characteristics of condensation polymerisation: *Terylene*, polypeptides, proteins, nylons and Kevlar
- deduce the repeat unit of a condensation polymer obtained from a given monomer or pair of monomers
- identify the monomers present in a given section of a condensation polymer molecule
- predict whether given monomers will form an addition or condensation polymer
- deduce whether a given polymer is an addition or condensation polymer
- relate the properties of a monomer to its type, and to the side chains and intermolecular forces it possesses
- explain base pairing in DNA
- distinguish between the primary, secondary and tertiary structures of proteins and explain how they are stabilised
- describe conducting polymers and adhesives such as epoxy resins and superglues
- recognise the different ways polymers can or cannot be biodegraded
- describe the hydrolysis of proteins

28.01 Condensation polymers

The polyalkenes mentioned in Unit 15 are called **addition polymers** because they are made by *adding* monomer units to a growing alkyl chain, and the molecular formula of an addition polymer is a multiple of the molecular formula of the monomer – both have the same empirical formula.

Another class of polymer is the **condensation polymers**. These are usually polyesters or polyamides, and are so called because when each ester or amide bond is formed, a molecule of H_2O (or sometimes HCl) is produced by a *condensation* reaction.

a **Polyesters**

If we react a diol (a compound having two –OH groups in each molecule) with a dicarboxylic acid (which has two –CO_2H groups in the same molecule), we can form an ester bond at each end of the diol, and similarly at each end of the diacid. A polymer is therefore produced. A common polymer of this type is the polyester known as Terylene™:

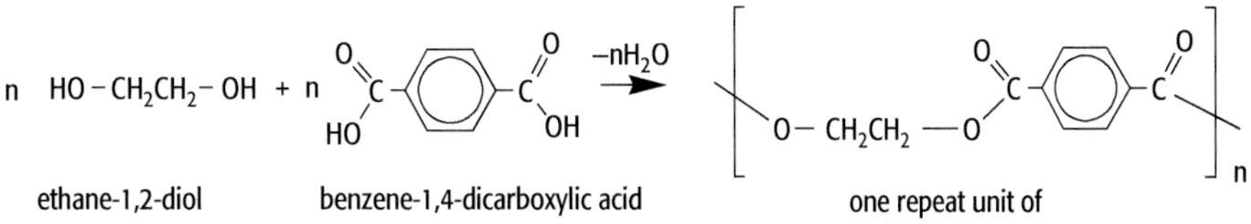

n HO–CH_2CH_2–OH + n HO–CO–C$_6$H$_4$–CO–OH $\xrightarrow{-nH_2O}$ [–O–CH_2CH_2–O–CO–C$_6$H$_4$–CO–]$_n$

ethane-1,2-diol   benzene-1,4-dicarboxylic acid   one repeat unit of *Terylene*, a polyester

187

The chemical structures of the diacid (or diacyl chloride) and the diol can be altered at will, to make polyesters with different properties. A general formation equation starting with acyl chlorides could be written as follows:

Terylene can be drawn into fine fibres and spun into yarn. Jointly woven with cotton or wool, it is a component of many everyday textiles. Cloth made from

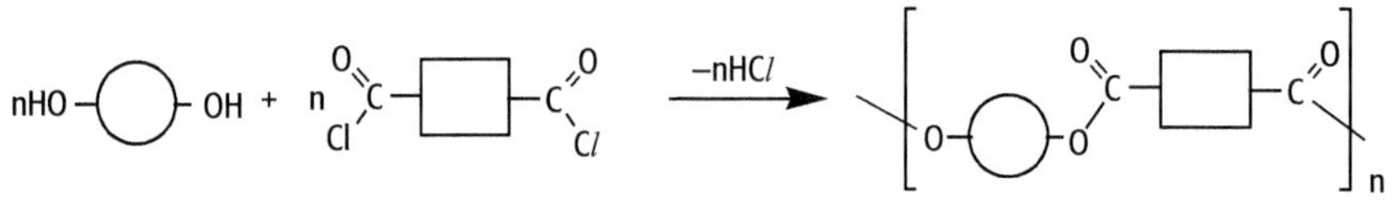

Notice that there are usually two different monomers that make up a polyester: a diacid (or diacyl chloride) and a diol. The repeat unit of a polyester contains one molecule of each – as is shown in the diagram above.

Terylene is hard wearing and strong, but under strongly alkaline or acidic conditions the ester bonds can be hydrolysed, causing the fabric to break up.

Progress check 28.01

1. Draw the repeat unit of the polymer formed when each of the following pairs of compounds is polymerised.

   a $CH_3CH(OH)CH_2CH(OH)CH_3$ and $CH_2(CO_2H)_2$

   b (catechol) and (CH(CH_3)(COCl)_2 type structure)

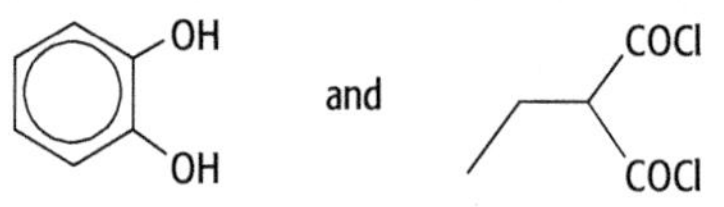

2. Draw the structures of the monomers of the following polymer.

   $[-OCH_2C(CH_3)_2CH_2OCO(CH_2)_2CO-]$

b Polyamides

Just as a diacid or diacyl chloride can react with a diol to produce a polyester, so they can react with a diamine to produce a polyamide.

Nylons are polyamides. The most common one is made by co-polymerising 1,6-diaminohexane and hexan-1,6-dioic acid:

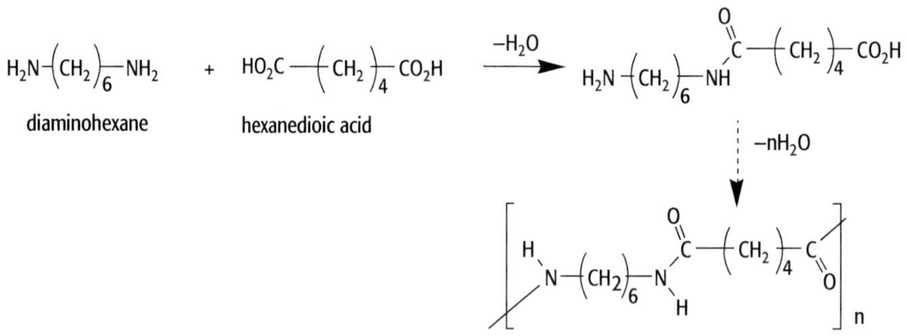

$H_2N-(CH_2)_6-NH_2$ + $HO_2C-(CH_2)_4-CO_2H$

diaminohexane   hexanedioic acid

$\xrightarrow{-H_2O}$ $H_2N-(CH_2)_6-NH-\overset{O}{\overset{\|}{C}}-(CH_2)_4-CO_2H$

$\downarrow -nH_2O$

$[-\overset{H}{\underset{}{N}}-(CH_2)_6-\overset{}{\underset{H}{N}}-\overset{O}{\overset{\|}{C}}-(CH_2)_4-\overset{O}{\overset{\|}{C}}-]_n$

nylon-6,6

Nylon 6 is unusual because it is made from only one monomer which contains both the amino and the carboxylic acid groups, 6-aminohexanoic acid:

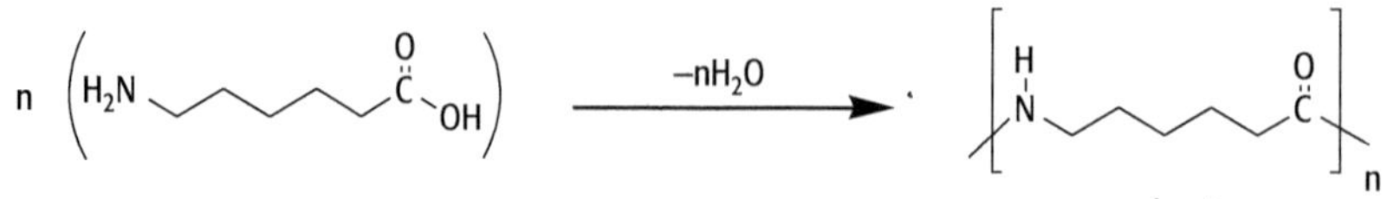

$n(H_2N-(CH_2)_5-COOH) \xrightarrow{-nH_2O} [-NH-(CH_2)_5-CO-]_n$

nylon 6

Figure 28.01 Climbers rely on nylon's elasticity and high tensile strength to minimise the effects of a fall

28.02 The relationship between a polymer, its repeat unit and its monomer(s)

We can recognise which type of polymer will be formed from given monomers by looking at their structures.

- If each of the monomers contains a C=C group, they are likely to produce an addition polymer.
- If the monomers contain two functional groups (carboxylic acid or acyl chloride, amine or alcohol) at the end of a chain of carbon atoms, they are likely to produce a condensation polymer (even if they also contain a C=C double bond).

If we are shown a section of a polymer chain, we can recognise what type of polymerisation has produced the polymer.

- If the chain contains only carbon atoms, the polymer is an addition polymer (side chains could contain ester or amide groups, however).
- If the chain contains oxygen or nitrogen, the polymer is a condensation polymer.

Worked example 28.01

Draw the structural formula of one repeat unit of the polymer formed from each of the following monomers or pair of monomers. State whether addition or condensation polymerisation takes place.

a $CH_2=CHCl$ and $CH_2=CH-CH_3$
b $HOCH_2CH_2OH$ and $HO_2CCH=CHCO_2H$
c $H_2NCH_2CH_2CH_2NH_2$ and $ClCOCH_2CH_2COCl$.

How to get the answer:

Step 1: To decide whether addition or condensation polymerisation has taken place, we look at the functional groups in the monomers.

In a there are C=C double bonds, but no alcohol or acid groups, so these will react by addition polymerisation.

In b and c there is one C=C bond, but in each case there is an acid or acyl chloride and an alcohol or an amine, so these will react by condensation polymerisation.

Step 2: To join the monomers in a, we break the two C=C bonds and join their ends:

Step 3: In b we form ester groups between the two monomers:

Step 4: In c we form amide groups between the two monomers:

> ## Progress check 28.02
>
> The following are *sections* (not necessarily repeat units) of addition or condensation polymers.
>
> Draw the monomers from which they are formed, and state whether addition or condensation polymerisation has taken place.
>
> 1. $-CH_2-CH(CN)-CH_2-CH(CN)-CH_2-$
> 2. $-CH_2-O-COCH_2CH_2CO-O-CH_2CH_2-O-$
> 3. $-CH_2-NH-CH_2CH_2CH_2CONH-CH_2CH_2CH_2CO-NH-$

28.03 Proteins

Proteins are naturally occurring polymers of amino acids. They are polyamides. An amide bond between two amino acids is called a **peptide bond**, and the polymers are known as **polypeptides**. Some proteins contain just one (very long) polypeptide chain, but most proteins are composed of several chains wrapped around each other.

In Unit 27 we looked at the structures of some amino acids, and saw how the polypeptide chains of proteins are made by joining together amino acid residues. Here we look in greater detail at the overall structure of proteins, and at some of their properties.

There are several levels of protein structure.

a **Primary structure**

The **primary structure** of proteins is the sequence of amino acids, covalently joined by peptide bonds, making up the polypeptide chain.

b **Secondary structure**

The **secondary structure** describes how some protein chains are held together by hydrogen bonding between peptide groups.

Because of the overlap between the lone pair of electrons on the nitrogen atom (in a p-orbital) and the π-bond of the carbonyl group next to it, the peptide group is rigid and planar. It exists in the *trans* arrangement (Figure 28.02).

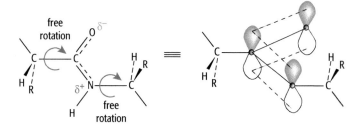

Figure 28.02 The peptide linkage is quite rigid and planar

This inflexibility of the peptide group allows parts of the polypeptide chain to take up an α-**helical** arrangement (see Figure 28.03). This causes some protein structures to be quite rigid. Strong hydrogen bonds can form between the N—H of one peptide group and the C=O of another one, four amino acids along the chain.

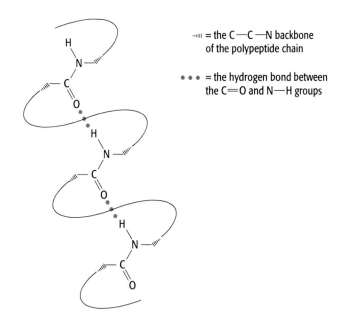

Figure 28.03 The α-helix

Another type of secondary structure is the **β-pleated sheet**. This is formed by the C=O group of one chain hydrogen bonding to the N–H group in another chain. The chains are described as *anti-parallel* because adjacent chains run in opposite directions (Figure 28.04).

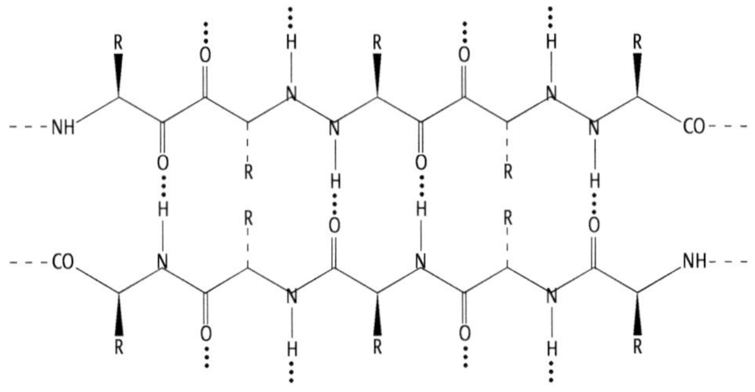

Figure 28.04 The β-pleated sheet

TIP

You should be able to draw from memory part of a β-pleated sheet, and also to explain the *principle* of hydrogen bonding forming the α-helix, though an accurate drawing of this will never be required.

c Tertiary structure

The **tertiary structure** involves interactions between the side-groups of amino acids.

There are four different interactions between side chains, depending on their different structures.

- van der Waals' attractions occur between non-polar side chains (e.g. glycine, alanine, valine and phenylalanine):

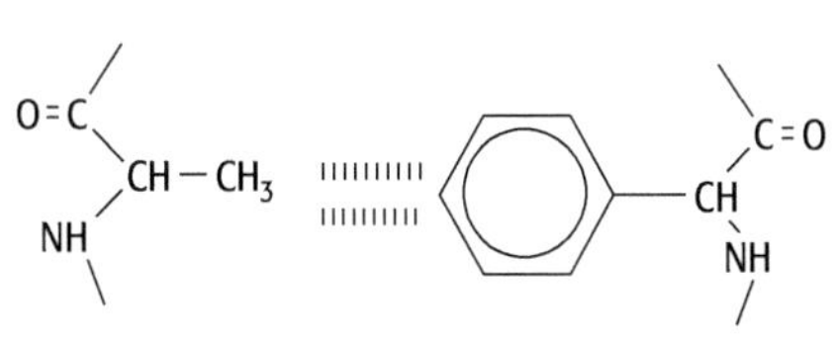

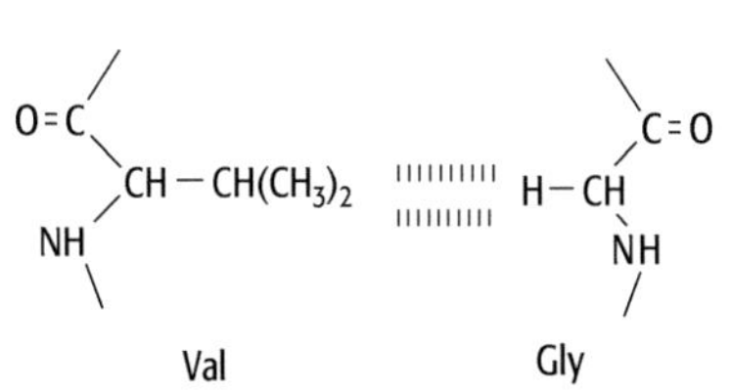

- Hydrogen bonding occurs between side chains that contain $-OH$ (e.g. serine and tyrosine), $-NH_2/-NH_3^+$ (e.g. lysine) and $-CO_2H/-CO_2^-$ (e.g. aspartic acid and glutamic acid):

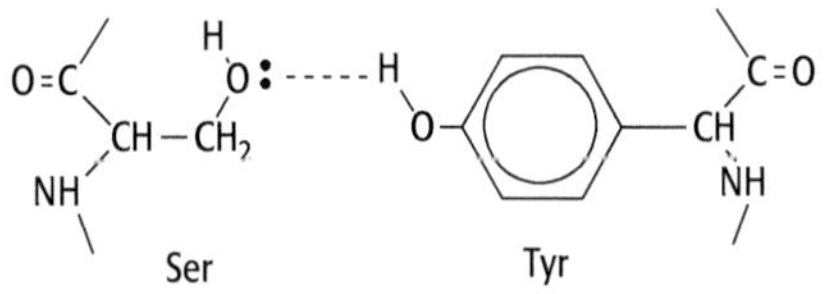

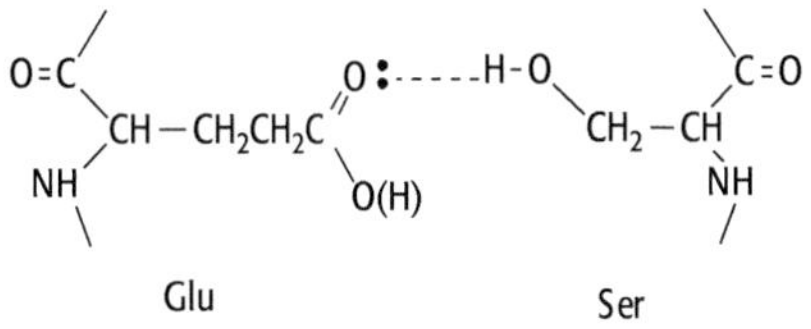

- Ionic attractions occur between side chains containing functional groups that exist in their anionic form at pH 7 (e.g. aspartic acid and glutamic acid) and those that exist in their cationic form (e.g. lysine and arginine):

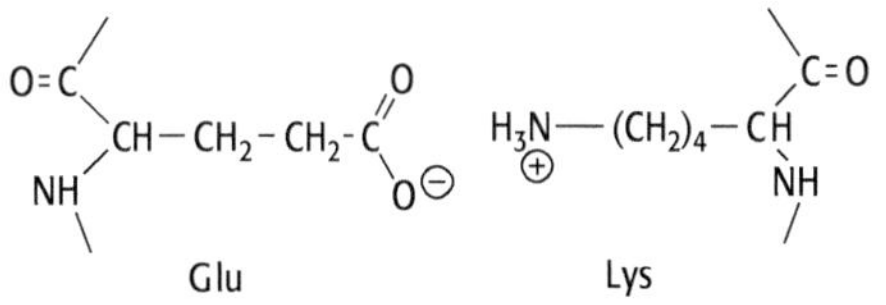

- Disulfide bridges can form between the side chains of two cysteine residues:

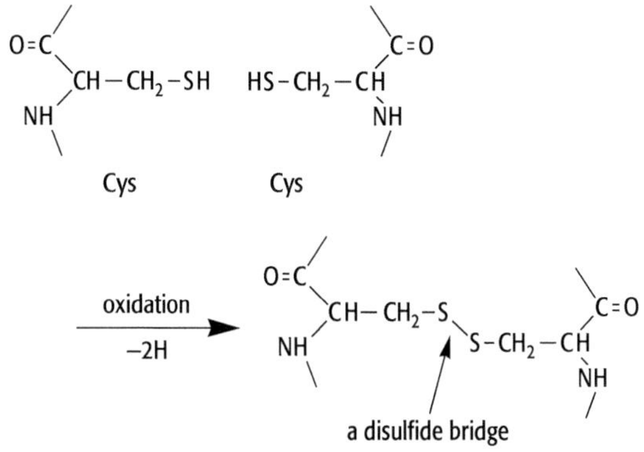

The tertiary structure is important in determining both the overall shape and the function of the protein. Many enzymes are globular (i.e. fairly spherical) proteins that are soluble in water. In these, the non-polar side chains are often pointing towards the central, hydrophobic, part of the molecule, whereas the hydrogen bonding and ionic groups are on the outside, where they can be solvated by water molecules.

The strongest of the tertiary interactions is the disulfide bridge. Once these are formed the various parts of the chain are drawn together, and this allows the other interactions to play their part in creating the overall three-dimensional structure.

Like all amides, proteins can be hydrolysed – the peptide bonds are broken and smaller fragments (e.g. dipeptides, tripeptides) and single amino acids are formed.

The hydrolysis can be carried out under much milder conditions if a proteolytic enzyme is used. For example the gut enzyme trypsin can hydrolyse peptide bonds at body temperature of 37°C and at nearly neutral pH.

28.04 DNA

DNA (deoxyribonucleic acid) consists of a primary structure, which is the covalent bonding that joins the monomers together, and a secondary structure, which is formed by hydrogen bonding between the chains.

The backbone chain of DNA consists of molecules of deoxyribose joined by phosphate groups. Bonded to each deoxyribose unit is a nitrogen-containing base (see Figure 28.05).

Figure 28.05 The monomer unit of DNA

Only four different bases occur in DNA: adenine (A), cytosine (C), guanine (G) and thymine (T). A single DNA molecule may contain thousands of monomer units. It is the order in which the bases are bonded to the deoxyribose–phosphate chain that determines the genetic information needed to construct specific proteins in the cell.

A DNA molecule contains two chains joined together by hydrogen bonding between the bases. Which base bonds to which is unique, and is determined by the number and orientation of the possible hydrogen bonds between them (Figure 28.06). Adenine bonds only to thymine, whereas cytosine bonds only to guanine. When DNA is duplicated during cell division, the chains unwind and each chain is used as a template for making two new chains. The unique pairing of the bases allows their order along the chains to be preserved.

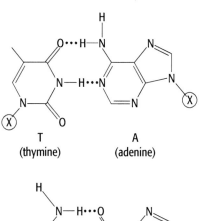

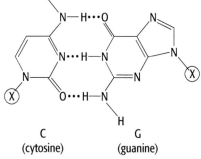

X = the deoxyribose in each of the chains

Figure 28.06 The pairing of the bases in DNA by hydrogen bonding

> **TIP**
> There is no need to learn the structures of the four bases of DNA, but you should remember that they hydrogen bond in the pairs A–T and C–G: the A–T pair is bonded by two hydrogen bonds, and the C–G pair by three hydrogen bonds.

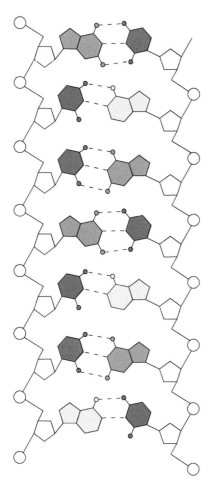

Figure 28.07 A portion of DNA showing the backbone of deoxyriboses (the pentagons) and phosphates (the circles), and the hydrogen bonding between the base pairs of adjacent chains

28.05 Properties of some polymers

a **Polyalkenes and PTFE**

The properties of polymers – density, melting point, strength, flexibility, hydrophilicity, conductivity, adhesiveness – all depend on their structures, the bonding within the polymer chain, and the side groups that are attached to it.

Simple examples are the two main forms of poly(ethene), LDPE and HDPE, which we saw in Unit 15.

The high electronegativity of fluorine means that its atoms keep a tight control on their bonding electrons, which are therefore much less polarisable than normal. Van der Waals' forces between fluorine atoms are small. If all the hydrogen atoms in poly(ethene) are replaced by fluorine atoms, the resulting polymer, polytetrafluoroethene (PTFE), makes an excellent non-stick surface for cooking pans and bridge bearings. (The 'stickiness' of materials is usually associated with the large van der Waals' forces between their molecules.)

$$3n\ \underset{\text{tetrafluoroethene}}{F_2C=CF_2} \longrightarrow \left[\underset{F\ F\ F\ F\ F\ F}{\overset{F\ F\ F\ F\ F\ F}{-C-C-C-C-C-C-}} \right]_n$$

3 monomer units of PTFE

b **Kevlar**

The synthetic polyamide Kevlar™ contains strong hydrogen bonding between its chains:

The tensile strength of Kevlar is five times that of steel, but it is quite flexible. It is used in tyres for cars and bicycles, bullet-proof body armour and the 'skins' on modern drumheads.

c **Non-solvent based adhesives**

Two common glues are the cyanoacrylates and the epoxyresins.

- The liquid inside a tube of cyanoacrylate glue ('superglue') is methyl 2-cyanopropenoate. In the presence of nucleophiles (H_2O molecules in moist air are often sufficient), addition polymerisation takes place to produce a solid which can use van der Waals' forces to bind with the articles that are being glued together:

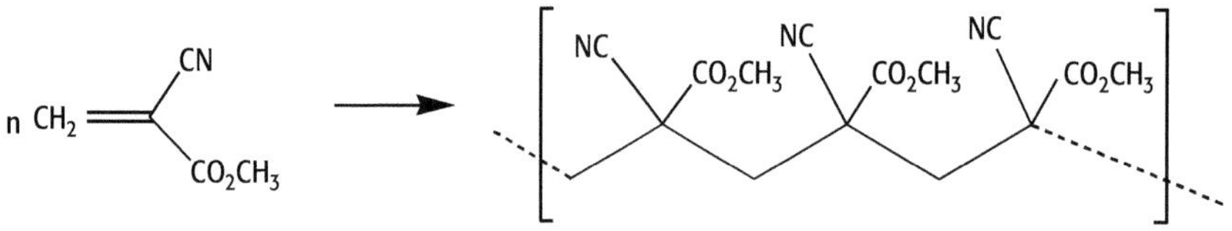

methyl 2-cyanopropenoate → 3 monomer units of polycyanopropenoate

- Epoxyresins are glues made from two liquid components. One component is a medium-length polymer containing epoxide groups. The other is a 'hardener' that contains a diamine. When the two components are mixed, the amino groups open up the epoxide rings and allow polymerisation to take place:

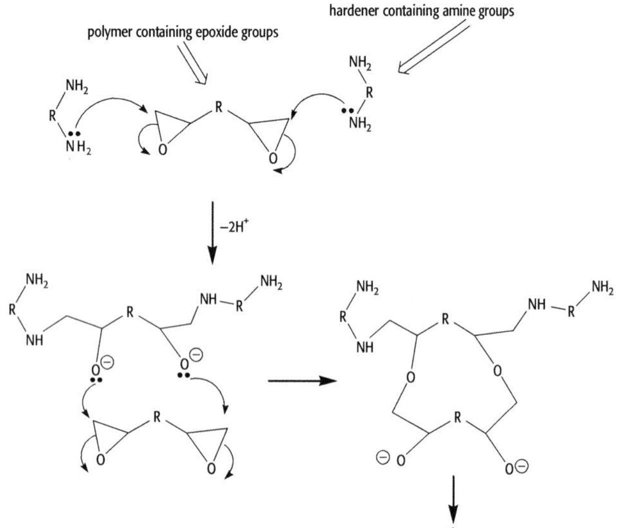

> **TIP**
> You can recognise a conducting polymer if its chain contains *all* sp^2 carbon atoms, and no sp^3 carbon atoms.

28.06 Degradable polymers

In Unit 15 we looked at the problems of the disposal of addition polymers. In general, condensation polymers are more easily biodegraded because they can undergo hydrolysis to re-form their monomers, which can often be metabolised by bacteria.

One way of overcoming the problems of polymer disposal is to make polymers that are easily biodegraded. An example is poly(2-hydroxypropanoic acid) (poly(lactic acid), PLA):

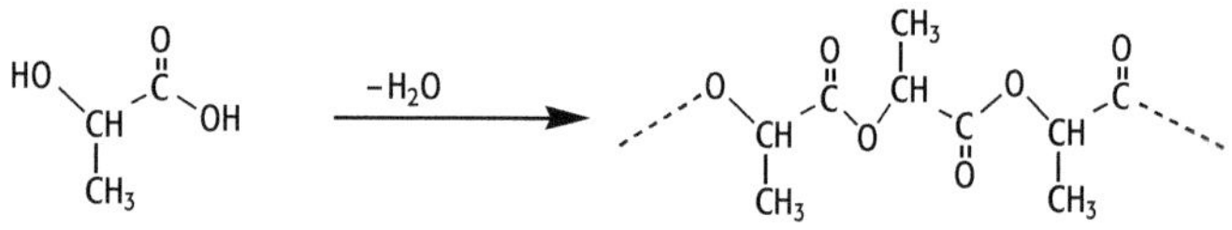

hydroxypropanoic acid → PLA ($-H_2O$)

d Conducting polymers

Conducting polymers are 'molecular wires', and are finding uses in light-emitting diodes (LEDs) and photovoltaic (PV) solar panels. They conduct an electric current through the delocalised π bond that encompasses the whole molecule. Two examples are polyacetylene (the simplest) and the polyphenylene vinylidenes:

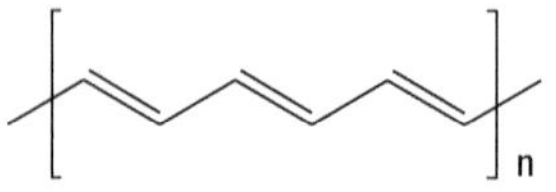

polyacetylene (polyethyne)

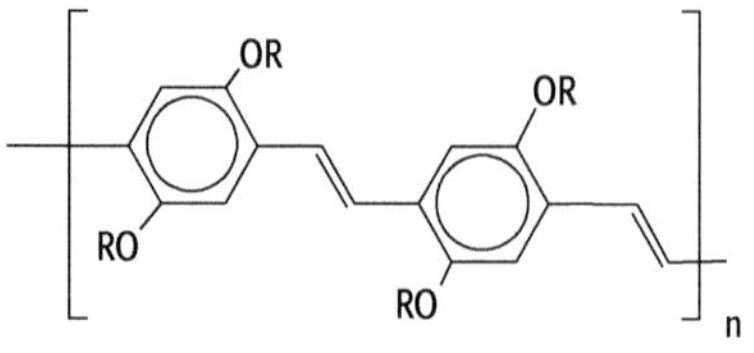

a disubstituted polyphenylene vinylene (PPV)

Hydroxypropanoic acid is a naturally occurring compound which is made industrially by the fermentation of corn starch or sugar. PLA is easily hydrolysed, either chemically or by esterase enzymes, to re-form hydroxypropanoic acid, which is readily metabolised by all living matter into CO_2 and water. The uses of PLA include replacing conventional plastics in the manufacture of bottles, disposable cups and textiles, but it is also used to make self-dissolving stitches for surgery.

Progress check 28.03

1. The repeat unit of another biodegradable polymer is as follows:

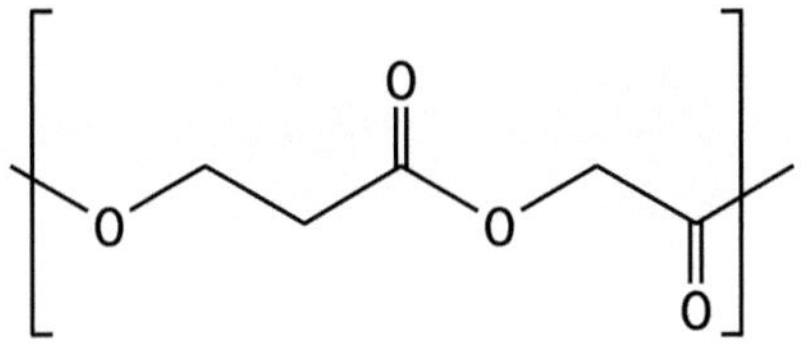

   [structure shown]

   Draw the structures of the products of the hydrolysis of this monomer.

2. Draw the monomers from which Kevlar can be formed.

3. Draw the monomers from which the following polymers are formed.

   a [structure with COCH$_3$ groups]

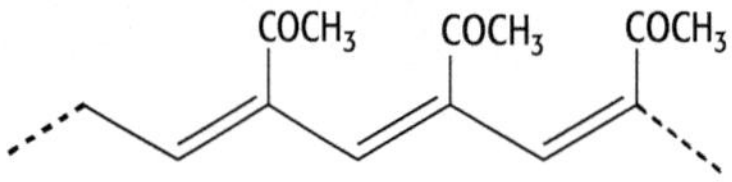

   b [structure with ester linkages]

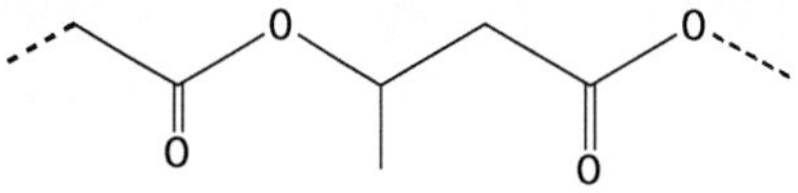

Sample answer

Question:

The following structure shows part of the chain in a water-absorbent polymer gel.

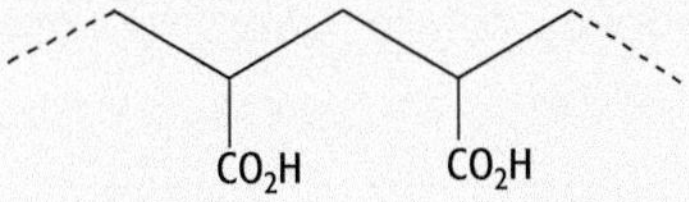

[structure showing chain with two CO$_2$H groups]

a i Name the type of polymerisation that has formed this polymer.

   ii Draw the structure of the monomer(s). **[2]**

b i What type of bonding will attract water molecules to this polymer gel?

   ii Draw a diagram to show at least *two* water molecules bonded to the gel. **[2]**

c Some polymers used in water-absorbent gels have their polymer chains cross-linked as indicated.

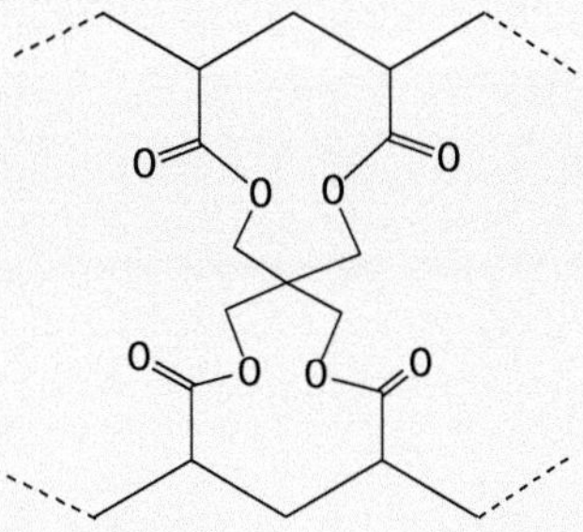

[cross-linked structure]

   i What functional group has been introduced during cross-linking?

   ii Draw the structure of the molecule used to cross-link the polymer.

   iii Suggest *one* effect on the properties of the polymer gel that the cross-linking would have. **[3]**

Answer:

a i addition **[1 mark]**

   ii CH$_2$=CH–CO$_2$H **[1 mark]**

b i hydrogen bonding **[1 mark]**

   ii [diagram showing hydrogen bonding with water molecules] **[1 mark]**

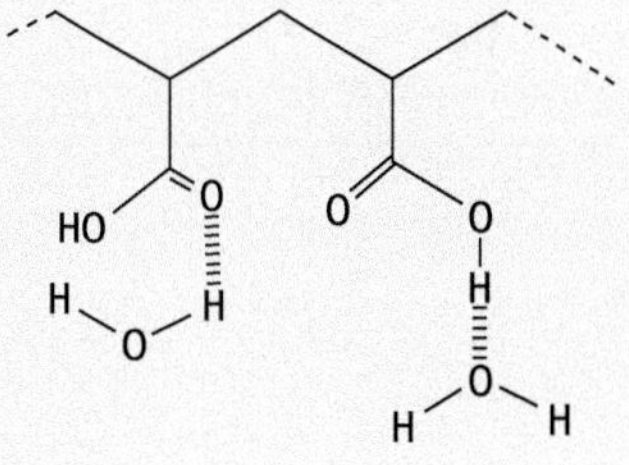

c i ester **[1 mark]**

   ii [structure of HOCH$_2$–C(CH$_2$OH)$_2$–CH$_2$OH] **[1 mark]**

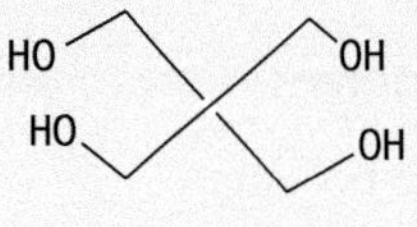

   iii increased rigidity **[1 mark]**

Revision checklist

Check that you know the following:

- How to recognise polyester and polyamides, and how they are formed, especially *Terylene*, proteins, nylon 6, nylon 6,6 and Kevlar.
- How to work out the repeat units of polymers, given their monomers.
- How to identify the monomers, given a section of the polymer chain.
- How to predict the type of polymerisation that given monomers will undergo, or that has produced a given section of polymer.
- How the properties of polyalkenes, PTFE and Kevlar are related to their structures.
- How the base pairing in DNA makes use of hydrogen bonding.
- The bonding involved in the primary, secondary and tertiary structures of proteins.
- The make-up of epoxyresins, superglues and conducting polymers.
- How various polymers can be degraded in the environment.
- Describe the hydrolysis of polyesters, polyamides and proteins.

Exam-style questions

1. a For each of the following polymers, state the type of polymerisation that forms it, and give the structures of its monomer(s).

    A, B, 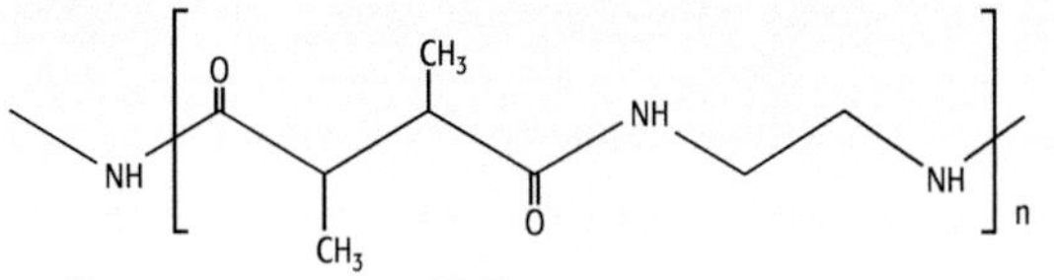 C [structures shown]

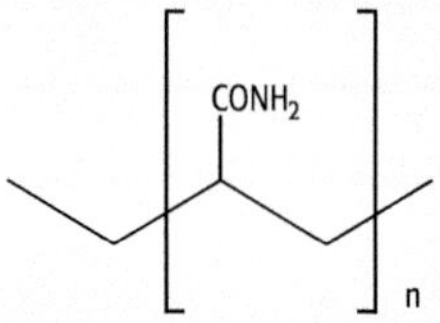

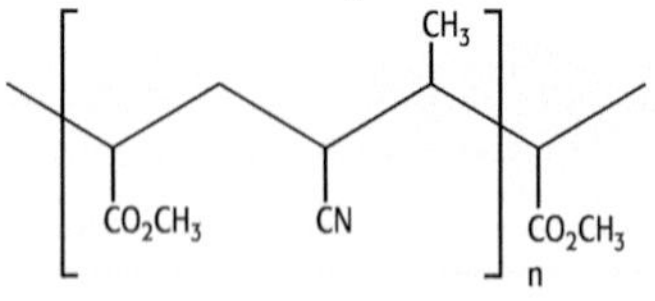

    [6]

   b Draw the repeat unit of an example of each of the following types of polymer:

      i   a conducting polymer

      ii a biodegradable polymer

      iii a 'superglue'. [3]

   Total: 9

2. a Explain the following terms, illustrating your answers by drawing part structures where appropriate.

      i   primary protein structure [2]

      ii the β-sheet [2]

      iii disulfide bridge. [2]

   b Explain in outline how the two strands of the DNA molecule are held together. [4]

   c State *one* similarity and *two* differences in structure between a typical protein and nylon 6,6. [3]

   d i   Name an example of a polyester. [1]

      ii Describe the reagents and conditions that convert a polyester into its monomers. [1]

   Total: 15

Analytical chemistry

Learning outcomes

You should be able to:

- explain and use the terms R_f value in thin-layer chromatography and retention time in gas–liquid chromatography
- interpret gas–liquid chromatograms in terms of the percentage composition of a mixture
- analyse an infra-red spectrum of a simple molecule to identify functional groups
- use a mass spectrum to deduce the Mr of a compound, the number of carbon atoms in a molecule and whether a compound contains chlorine or bromine
- analyse and predict a ^{13}C NMR spectrum in relation to molecular structure
- analyse and predict a ^1H NMR spectrum in relation to molecular structure, with regard to chemical shift values, relative peak areas and splitting patterns
- describe the use of tetramethylsilane as the chemical shift standard
- state the need for deuterated solvents, and describe proton exchange of O–H and N–H protons with D_2O

29.01 Chromatography

Chromatography relies on the distributing of compounds between a *mobile phase* and a *stationary phase* in order to separate the components of a mixture. The mobile phase is a liquid or a gas, and the stationary phase is either a solid onto which the solutes are adsorbed, or a thin layer of liquid on the surface of an inert solid.

The two types of chromatography we shall look at here are thin-layer chromatography (**TLC**) and gas–liquid chromatography (**GLC**).

a **Thin-layer chromatography (TLC)**

TLC relies on the principle of adsorption: the extent to which a compound is adsorbed onto a solid surface differs from one compound to another (Figure 29.01). The solid stationary phase is powdered silica or alumina, spread onto a glass or plastic sheet.

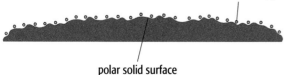

Figure 29.01 Separation by adsorption

The moving phase is usually an organic solvent such as ethanol or ether. The more polar a compound is, the more strongly it will be adsorbed onto the polar stationary phase so it will spend more time on the stationary phase than in the moving solvent, and so it does not move very far (Figure 29.02).

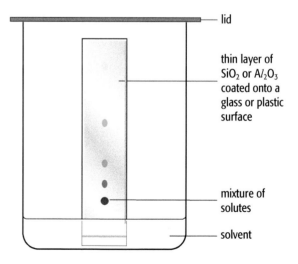

Figure 29.02 Thin-layer chromatography

Once the chromatogram has been run, the positions of the spots and their **retardation factors**, R_f values, can be measured, and compared to the R_f values of known reference compounds (Figure 29.03).

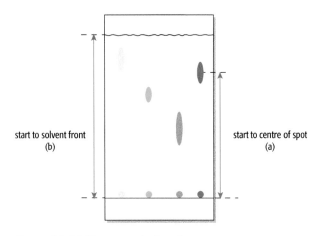

Figure 29.03 The retardation factor

The retardation factor, R_f, is defined by the equation:

$$R_f = \frac{a}{b} = \frac{\text{distance moved by solute}}{\text{distance moved by solvent}}$$

Each compound has a characteristic R_f value for a given solvent and a given solid phase.

> **TIP** If you are asked to calculate an R_f value from a drawing of a TLC plate, make sure you measure the distance from the starting line to the *centre* of the spot, to the nearest mm.

b Gas–liquid chromatography (GLC)

The apparatus for GLC is shown in Figure 29.04.

The column, made of glass or metal, is from 1 m to 3 m in length, and is packed with fine solid particles (brick dust is often used), coated with a high-boiling oil.

The components of mixtures that are separated by GLC need to have a reasonable vapour pressure at the temperature of the oven, which can be up to about 250 °C. The more volatile components travel through the column faster; the less volatile ones take longer to be carried through the column by the flowing gas, so have longer **retention times**.

The gas emerging from the column passes through a detector, which records the presence and amount of each component on a chart recorder or computer. On a chart print-out of the results, the area underneath each peak is often proportional to the amount of that component responsible for that peak.

Analytical GLC machines are routinely coupled to an in-line mass spectrometer. This allows the mass spectrum of each component in a sample to be taken, which enables further verification of the identity of the component.

29.02 Infrared spectroscopy

The frequency with which bonds in molecules vibrate is similar to the frequency of 'light' in the infrared region of the electromagnetic spectrum. If the frequency of incident radiation matches that of a particular bond vibration, the light will be absorbed as the bond vibrates more strongly. Bonds to electronegative atoms

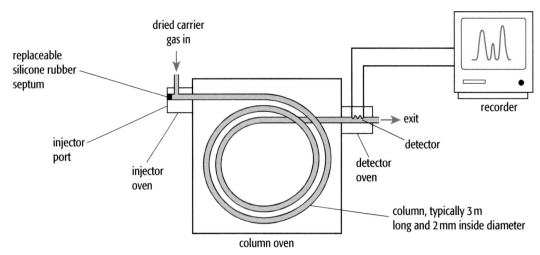

Figure 29.04 A diagram of GLC apparatus

like oxygen (e.g. in alcohols (O–H) and carbonyl compounds (C=O)) show very strong absorptions. Although an infrared spectrum shows a series of absorptions (see Figure 29.05), these are always referred to as *peaks* rather than troughs. Note also that the *x*-axis of infrared spectra runs from low frequencies on the right to higher frequencies on the left.

The infrared spectrum of a compound often enables us to identify the functional groups it contains, as each functional group has a characteristic absorption frequency. Some groups are listed in Table 29.01.

Bond	Frequency of absorption (wavenumber) / cm^{-1}
O–H	3200–3500
N–H	3400
O–H in RCO$_2$H	2500–3300
C–H	2800–3100
C=O	1700–1740
C=C	1600–1650
C–O	1100–1250

Table 29.01 Some IR absorption frequencies for organic groups

In molecules with more than half a dozen or so atoms there can be many different modes of vibration of the molecule as a whole. This leads to a complicated series of peaks in the region from 500 to 1500 cm^{-1}. Unless they are very strong (such as the C–O absorption), the peaks cannot be assigned to individual bonds, but their pattern is unique to the particular molecule. This is known as the 'fingerprint' region of the spectrum, and is very useful in detecting the presence or absence of a particular compound in a sample.

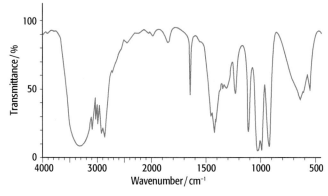

Figure 29.05 The infrared spectrum of propenol, CH$_2$=CHCH$_2$OH

The absorption of infrared radiation is important in another context: small molecules in the atmosphere (especially CO$_2$, CH$_4$, H$_2$O and CFCs) are responsible for the greenhouse effect. They absorb the infrared radiation being emitted from the surface of the Earth, and so prevent it from being lost to space. As result, the amount of heat lost is less than that gained from solar radiation, and the Earth warms up.

Worked example 29.01

Figure 29.06 shows the IR spectrum of compound **A**, with the formula C$_3$H$_6$O$_2$.

Determine which functional groups compound **A** contains and hence suggest its structure.

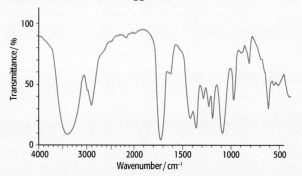

Figure 29.06 Infrared spectrum of compound **A**

How to get the answer:

Step 1: We first need to determine which peaks are significant: there are three to the left of the fingerprint region, and of those within the fingerprint region only the strongest should be considered. The four most significant ones are at 3400 cm^{-1}, 2900 cm^{-1}, 1720 cm^{-1} and 1100 cm^{-1}.

Step 2: Table 29.01 shows that the groups responsible for these peaks are as follows:

broad peak at 3400 cm^{-1} = O–H
peak at 2900 cm^{-1} = C–H
strong peak at 1720 cm^{-1} = C=O
strong peak at 1100 cm^{-1} = C–O

Step 3: Since **A** contains an –OH group (but *not* as part of a –CO$_2$H group), its one remaining oxygen must be in a ketone or aldehyde group. This fits with **A** being CH$_3$COCH$_2$OH, CH$_3$CH(OH)CHO or HOCH$_2$CH$_2$CHO.

Progress check 29.01

Figures 29.07 and 29.08 show the IR spectra of two other compounds **B** and **C**, with the formula $C_3H_6O_2$.

Use the information in Table 29.01 to suggest what functional groups the compounds contain and hence suggest the structure of each compound.

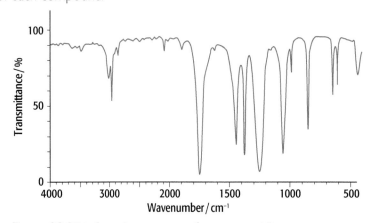

Figure 29.07 Infrared spectrum of compound B

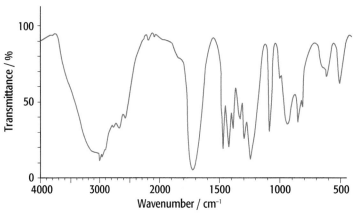

Figure 29.08 Infrared spectrum of compound C

29.03 Mass spectrometry

When high-energy electrons bombard a molecule in the gas phase, further electrons are knocked off to form positive ions. These ions have often been given enough energy by the impact to undergo fragmentation through bond breaking. A mass spectrometer measures the relative masses of molecules and their fragmented ions.

There are four main ways we can use the information a mass spectrum provides:

- The *m/e* ratio of the molecular ion gives us the M_r of the compound.

- Measuring the abundance ratio of its molecular ion M^+ peak and the $[M + 1]^+$ peak allows us to calculate the number of carbon atoms in a molecule.

- Measuring the abundance ratios of the $[M + 2]^+$, $[M + 4]^+$ and $[M + 6]^+$ peaks allows us to find out whether a compound contains chlorine or bromine atoms, and if so, how many of each.

- Looking at the fragments produced when an ion decomposes inside a mass spectrometer often allows us to work out the structure of a molecule.

a **Analysing the molecular ion**

If we vaporise an organic molecule and subject it to the ionising conditions inside a mass spectrometer, the mass/charge ratio (*m/e*) for the molecular ion (i.e. the ion with the highest *m/e* ratio) can be measured. This gives us a measure of the **relative molecular mass**.

b **The $[M + 1]^+$ peak**

The relative abundances of the two stable isotopes of carbon, ^{12}C and ^{13}C, are 98.9% ^{12}C and 1.1% ^{13}C. This means that out of every 100 methanol molecules (CH_3OH) about 99 molecules will be $^{12}CH_3OH$ and just one molecule will be $^{13}CH_3OH$. For ethanol, C_2H_5OH, the chances of a molecule containing one ^{13}C atom will have increased to about 2 in 100, because each C atom has a chance of 1 in 100 to be ^{13}C, and there are two of them. By measuring the abundance ratio of the $M+$ to $[M + 1]^+$ peaks, we can thus work out the number of carbon atoms the molecule contains. The formula we use is:

$$n = \frac{100}{1.1} \frac{A_{M+1}}{A_M}$$

where n = number of carbon atoms

A_{M+1} = the abundance of the $[M + 1]^+$ peak and

A_M = the abundance of the molecular ion peak.

> **TIP**
>
> You will need to remember this formula. You might be given the relative abundances of the M^+ and $[M + 1]^+$ peaks as numbers, or you might be expected to measure them on a given spectrum, using a ruler.

c **The $[M + 2]^+$ and $[M + 4]^+$ peaks**

Chlorine and bromine each have two stable isotopes. Table 29.02 shows their relative abundances.

Element	Isotope	Natural abundance	Approximate ratio
chlorine	^{35}Cl	75.5	3 : 1
	^{37}Cl	24.5	
bromine	^{79}Br	50.5	1 : 1
	^{81}Br	49.5	

Table 29.02 The abundances of the chlorine and bromine isotopes.

Any compound containing one chlorine atom, therefore, will have two 'molecular ion' peaks, one due to molecules containing ^{35}Cl and the other due to molecules containing ^{37}Cl. For example, the mass spectrum of chloroethane, CH_3CH_2Cl, will have peaks at *m/e* 64 (24 + 5 + 35 = 64) and at *m/e* 66 (24 + 5 + 37 = 66), corresponding to the species $CH_3CH_2{}^{35}Cl^+$ and $CH_3CH_2{}^{37}Cl^+$. The relative abundances of the two peaks will be in the ratio 3 : 1, which is the ratio of the abundances of the two Cl isotopes.

A similar situation occurs with bromine, although in this case the two molecular ion peaks will be of equal heights, since the isotopic abundance ratio of ^{79}Br : ^{81}Br = 1 : 1.

What happens when the molecule contains more than one halogen?

The simplest situation is for a molecule containing two bromine atoms. Take the molecule bromoethanoyl bromide, $BrCH_2COBr$. Each carbon can be attached to either a ^{79}Br or a ^{81}Br atom, and there is a (roughly) equal chance of either.

Worked example 29.02

The molecular ion peak of a compound has an *m/e* value of 106, with a relative abundance of 17%, and an $[M + 1]^+$ peak at *m/e* 107 whose relative abundance is 1.5%.

Calculate the number of carbon atoms in the molecule, and suggest its molecular formula.

How to get the answer:

Step 1: We apply the formula to find n:

$$n = \frac{100}{1.1} \times \frac{1.5}{17} = 8.02, \text{ so the compound has 8 carbon atoms.}$$

Step 2: The contribution of carbon atoms to $M_r = 8 \times 12 = 96$

This leaves $106 - 96 = 10$.

Step 3: The most likely other elements are hydrogen, oxygen and nitrogen, but since N = 14 and O = 16, the other element must be hydrogen only.

Step 4: Formula is therefore C_8H_{10}.

We therefore arrive at the four possibilities shown in Table 29.03, each of which is equally likely.

Formula	m/e value
$^{79}BrCH_2CO^{79}Br$	200
$^{79}BrCH_2CO^{81}Br$	202
$^{81}BrCH_2CO^{79}Br$	202
$^{81}BrCH_2CO^{81}Br$	204

Table 29.03 The molecular ions of bromoethanoyl bromide

There will therefore be three molecular ion peaks, with relative abundances of 1:2:1.

d Analysing molecular fragments

Given enough energy, the ionised molecules formed in a mass spectrometer can fragment. The m/e values of these fragments give us a clue as to the structure of the parent molecule.

For example, propanal (CH_3CH_2CHO) and propanone (CH_3COCH_3) are isomers with $M_r = 58$. Breaking the C–C bond next to the C=O group would give the following fragments, all of which can carry a positive charge by the loss of an electron, and hence show as peaks in the mass spectrum:

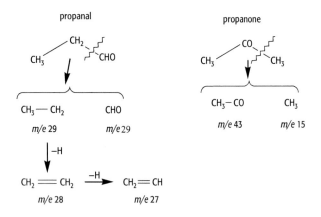

We therefore expect the mass spectrum of propanal to contain a strong peak at m/e 29, with others at m/e 28 and m/e 27, whereas we expect the mass spectrum of propanone to contain peaks at m/e = 15 and 43. This is in fact what is observed (Figure 29.09).

Progress check 29.02

1. Compound **D** contains the elements C, H and O. Its mass spectrum has a peak at m/e 132 with a relative abundance of 43.9 and a peak at m/e 133 with a relative abundance of 2.9. Calculate the number of carbon atoms in **D** and suggest its molecular formula.

2. Compound **E** has molecular formula $C_3H_6O_2$, and major peaks at m/e = 27, 28, 29, 45, 57, 73 and 74. Suggest formulae for these fragments, and a structure for **E**.

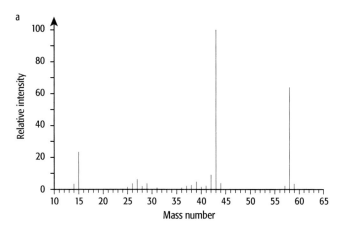

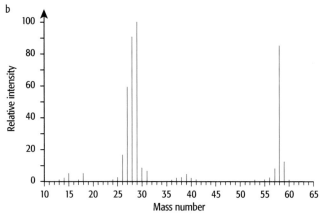

Figure 29.09 **a** Mass spectrum of propanone; **b** mass spectrum of propanal.

29.04 Nuclear magnetic resonance (NMR) spectroscopy

a The basis of NMR spectroscopy

Nuclei with an odd number of nucleons (e.g. ^{1}H and ^{13}C) have an overall magnetic moment, which can take up two possible orientations in an external magnetic field: aligned either with the field, or against it. If a nucleus is irradiated with radiation of the correct frequency, it can flip from one orientation to the other, and energy is absorbed.

The frequency of absorption depends on both the strength of the external field and the molecular environment of the nucleus.

b Analysing organic molecules

Protons in organic molecules absorb at different frequencies, depending on the electronegativity of groups nearby. This difference is measured by the **chemical shift**. For convenience, the hydrogen or carbon atoms in the compound tetramethylsilane, $(CH_3)_4Si$ (known as **TMS**), are used as a reference.

The chemical shift, symbol δ (delta), is the extent of the change in frequency of absorption, f. Chemical shift is defined as:

$$\delta = 10^6(f - f_{TMS}) / f_{TMS}$$

Values of δ are quoted in parts per million (ppm).

The values of some chemical shifts for hydrogen and carbon atoms within different organic groups are given in the CIE *Data Booklet*.

c ^{13}C NMR spectra

The ^{13}C NMR spectrum of a compound shows a single-peak absorbance at a different chemical shift value for every carbon atom in a different chemical environment. This can be illustrated in the spectrum of propyl methanoate, $HCO_2CH_2CH_2CH_3$, in Figure 29.10.

We can see that the spectrum consists of four peaks, corresponding to the four different carbon atom environments in the molecule. Consulting the CIE *Data Booklet*, we see that these can be assigned as follows:

peak at δ 161: carbonyl carbon, HCO–

peak at δ 66: CH_2 next to oxygen, $-O-CH_2-$

peak at δ 22: CH_2 in middle of chain, $-CH_2-$

peak at δ 10: CH_3 at end of chain, $-CH_3$

N.B. the heights of the peaks bear no relationship to the number of carbon atoms responsible for them, unlike ^{1}H NMR (see below).

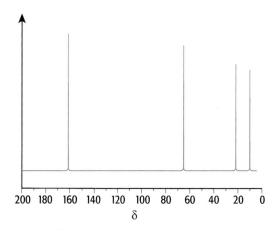

Figure 29.10 ^{12}C NMR spectrum of propyl methanoate

Progress check 29.03

1. Two isomers with the molecular formula $C_5H_{10}O$ are pentan-2-one and pentan-3-one. Predict the number of peaks in each of their ^{13}C NMR spectra.

2. The ^{13}C NMR spectrum shown in Figure 29.11 is of another ketone isomer of $C_5H_{10}O$. Use the spectrum to suggest its structure.

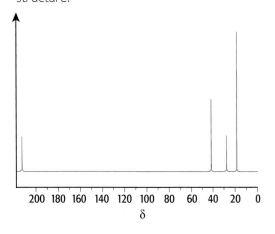

Figure 29.11 ^{13}C spectrum of $C_5H_{10}O$

d ^1H NMR

In general, the ^1H NMR spectrum of a compound is always more complicated that its ^{13}C NMR spectrum, but it can often tell us more about the structure of a compound. This is because of two features of a ^1H NMR spectrum not shown by a ^{13}C spectrum.

i The area underneath each peak is proportional to the number of hydrogen atoms responsible for that peak.

ii Hydrogen atoms on the adjacent carbon atom split the peak into a multiple peak. The splitting pattern tells us how many adjacent hydrogens there are (Table 29.04).

Number of adjacent hydrogens	Pattern of lines	Ratio of line intensities
0	singlet	1
1	doublet	1:1
2	triplet	1:2:1
3	quartet	1:3:3:1

Table 29.04 Some simple splitting patterns for ^1H NMR

The solvents used for ^1H NMR must not contain any protons, or else the signals due to protons in the sample would be swamped. Common solvents used are CCl_4, $CDCl_3$ and D_2O, where 'D' is deuterium (^2H).

Protons directly attached to oxygen or nitrogen atoms can be identified by **deuterium exchange**. If the compound containing them is dissolved in D_2O ('heavy water', $D = {}^2H$), the protons are exchanged with deuterium atoms in the water. The peaks due to the $-OH$ or $-NH_2$ protons disappear:

$$CH_3CH_2-OH + D_2O \rightleftharpoons CH_3CH_2-OD + HDO$$

> **TIP**
> The 'triplet + quartet' pattern seen in figure 29.12 is very typical of the ethyl group, CH_3CH_2-. The triplet due to the CH_3 group is usually very near to δ 1.1, but the δ value for the CH_2 quartet depends a lot on what the ethyl group is joined to (see table 6 in the CIE *Data Booklet*).

Worked example 29.03

Interpret the ¹H spectrum of ethyl ethanoate, $C_4H_8O_2$, shown in Figure 29.12.

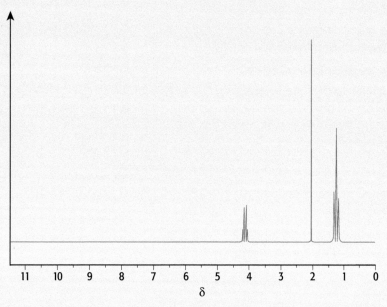

Figure 29.12 The ¹H NMR spectrum of ethyl ethanoate

How to get the answer:

Step 1: The spectrum shows peaks at three δ (ppm) values, suggesting there are three different types of proton. One type (at δ 2.1) has no H atoms adjacent to it; the other two do have adjacent H atoms.

Step 2: Using the CIE *Data Booklet*, we can assign the peaks as follows:

δ 4.1: 2H quartet, O–CH_2– split by the three adjacent H atoms on the –CH_3 group

δ 2.1: 3H singlet, CH_3–CO, with no adjacent H atoms

δ 1.3: 3H triplet, –CH_3 split by the two adjacent H atoms on the –OCH_2 group

Progress check 29.04

The spectrum shown in Figure 29.13 is that of compound **D**, $C_3H_6O_2$. Suggest the structure of **D** and identify the peak that would disappear if the sample were shaken with D_2O.

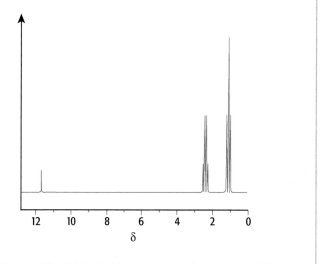

Figure 29.13 ¹H NMR spectrum of compound D

Sample answer

Question: The compounds CH_3COCH_2OH and $HCO_2CH_2CH_3$ are isomers.

a Describe *one* major difference and *one* similarity between their infrared spectra. [2]

b Describe the details of their ^1H NMR spectra, including approximate δ values, splitting patterns and how the spectra would change (if at all) when the samples were shaken with D_2O. [8]

Answer:

a Both will show C=O absorption at 1700–1740 cm^{-1} [1 mark]

CH_3COCH_2OH will also show an O–H absorption at 3200–3500 cm^{-1} [1 mark]

b CH_3COCH_2OH: 3 singlet peaks, at δ 2.2–3.0 (CH_3) [1 mark]

δ 3.2–4.0 (CH_2) [1 mark]

δ 0.5–6.0 (OH) [1 mark]

the OH peak would disappear in D_2O [1 mark]

$HCO_2CH_2CH_3$: one singlet peak at δ 9.3–10.5 (CHO) [1 mark]

quartet at δ 3.2–4.0 (CH_2, split by CH_3) [1 mark]

triplet at δ 0.9–1.7 (CH_3, split by CH_2) [1 mark]

no peaks would disappear in D_2O [1 mark]

Revision checklist

Check that you know the following:

- ☐ The principles of thin-layer and gas–liquid chromatography.
- ☐ The meaning of retardation factor (R_f) and retention time, and how to measure them.
- ☐ How to identify the functional groups in a molecule from its IR spectrum.
- ☐ How to determine the molecular mass and the number of carbon atoms in a molecule from the M^+ and $[M + 1]^+$ peaks in its mass spectrum.
- ☐ How the presence of an $[M + 2]^+$ peak indicates the presence of chlorine or bromine in a molecule.
- ☐ How to piece together the fragments in a mass spectrum to find the structure of a molecule.
- ☐ How a ^{13}C NMR spectrum indicates the number of different carbon atom environments in a molecule.
- ☐ Predict the number of peaks in a ^{13}C spectrum for a given molecule.
- ☐ How to interpret a ^1H NMR spectrum to work out the different types of proton present, their relative numbers and the number of protons on adjacent carbon atoms from the splitting pattern using the n+1 rule.
- ☐ The use of tetramethylsilane as the standard for chemical shifts.
- ☐ The use of deuterated solvents, and the identification of –OH and –NH protons using D_2O.

Exam-style questions

Data from the spectroscopy section of the *Data Booklet* will be needed to answer these questions.

1 The following figures show the mass, infrared, ¹H NMR and ¹³C NMR spectra of compound **A**.

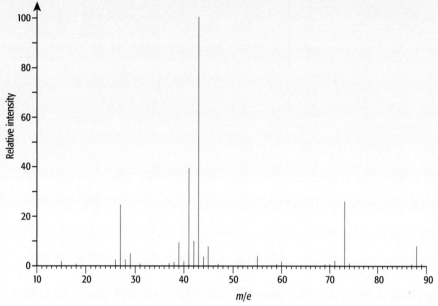

Figure 29.14 Mass spectrum of **A**

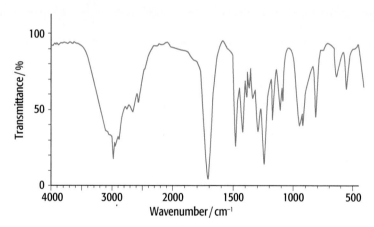

Figure 29.15 Infrared spectrum of **A**

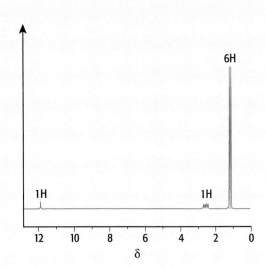

Figure 29.16 ¹H NMR spectrum of A

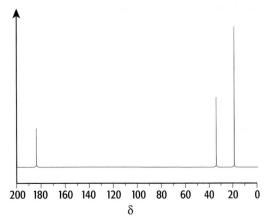

Figure 29.17 ¹³C NMR spectrum of A

Your answers to each of the following questions should explain how you arrive at your conclusion.

  a Use the mass spectrum to determine the M_r of compound **A**. [1]

  b Identify the peaks at 2500–3300 cm⁻¹, 1700 cm⁻¹ and 1230 cm⁻¹ in the infrared spectrum and use them to determine the functional group(s) in compound **A**. [4]

  c Use the ¹³C NMR spectrum to determine the number and type of the carbon atoms in a molecule of compound **A**. [3]

  d Use the ¹H NMR spectrum to suggest the structure of **A**. [4]

  e Suggest the structure of the ion at $m/e = 43$ in the mass spectrum. [1]

Total: 13

2 Two compounds, **B** and **C**, have the molecular formula $C_3H_6O_2$. Compound **B** boils at 140°C and readily dissolves in water to give a solution of pH 3. Compound **C** boils at 57°C and is much less soluble in water, giving a solution of pH 7.

  • The ¹³C spectra of both compounds show three absorptions.

  • The infrared spectra of both compounds show a peak in the region 1700–1750 cm⁻¹.

  • The ¹H NMR spectrum of **C** shows two singlet peaks.

  a Suggest structures for **B** and **C**. [2]

  b Suggest the chemical shift (δ) values for the three absorptions in each of the ¹³C NMR spectra. [6]

  c Suggest the chemical shift (δ) values in each of the ¹H spectra, and the splitting patterns for the peaks in the spectrum of **C**. [6]

  d Suggest a feature present in the infrared spectrum of one of **B** or **C** but absent in the spectrum of the other compound, that would allow you to distinguish between them. [1]

  e Which of the two compounds, **B** or **C**, would have the shorter retention time in a gas–liquid chromatograph? Explain your answer. [1]

Total: 16

Organic Synthesis

Learning outcomes

You should be able to:

- devise multi-stage synthetic routes for preparing organic molecules using the reactions in the syllabus
- recognise the different types of organic reaction
- describe the tests for different functional groups
- know the importance of making drug molecules as a single optical isomer

30.01 Devising synthetic routes

Most of the millions of organic compounds that chemists have made have used simple and readily available materials as their starting points. These could be naturally occurring compounds found in plants or the products from the refining of crude oil. In order to piece together a route from a common starting material to a target compound, we need to have an expert knowledge of the reactions of organic functional groups, so that we can decide which reactions are appropriate for our particular scheme.

It is useful to learn the reactions you have come across in this course in two ways:

- which products are formed during a particular reaction of a functional group
- which reactions could be useful in making a particular functional group

Worked example 30.01

Suggest how ethanoic acid could be made from ethene.

$$CH_2=CH_2 \Rightarrow CH_3CO_2H$$

How to get the answer:

This cannot be carried out directly; we need at least two steps in the transformation. We start cataloguing possible reactions from both ends.

Step 1: Reactions *of* ethene:

  a   reaction with HBr gives CH_3CH_2Br
  b   reaction with Br_2 gives $BrCH_2CH_2Br$
  c   reaction with H^+ and H_2O gives CH_3CH_2OH
  d   reaction with H_2 + Ni gives CH_3CH_3
  e   reaction with cold $KMnO_4$ gives $HOCH_2CH_2OH$

Step 2: Reactions that *produce* ethanoic acid:

  a   reaction of CH_3CN with $H^+ + H_2O$
  b   reaction of CH_3CO_2R with $H^+ + H_2O$
  c   reaction of CH_3CHO with $Cr_2O_7^{2-} + H^+$
  d   reaction of CH_3CH_2OH with $Cr_2O_7^{2-} + H^+$

Step 3: We can see that there is a reaction of ethene which produces the starting material for one of the reactions that produces ethanoic acid, so the synthetic route is as follows:

$$CH_2=CH_2 \longrightarrow CH_3CH_2OH \longrightarrow CH_3CO_2H$$

Worked example 30.02

Suggest a method for carrying out the following conversion.

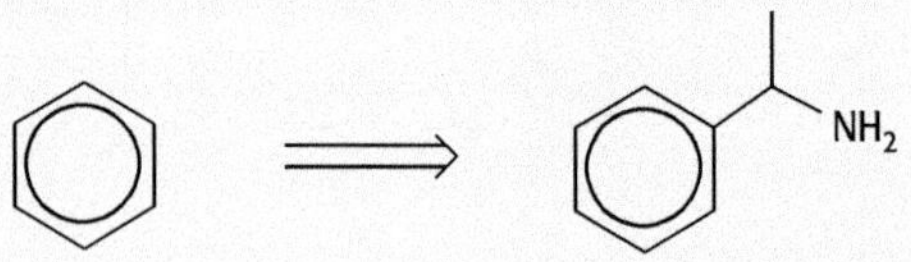

[benzene] ⟹ [1-phenylethylamine, PhCH(CH₃)NH₂]

How to get the answer:

Step 1: Consider the reactions of the starting compound, benzene. There is only one way to add a side chain to the benzene ring: the Friedel–Crafts reaction. There are two possible two-carbon reagents we could use: CH_3CH_2Cl or CH_3COCl.

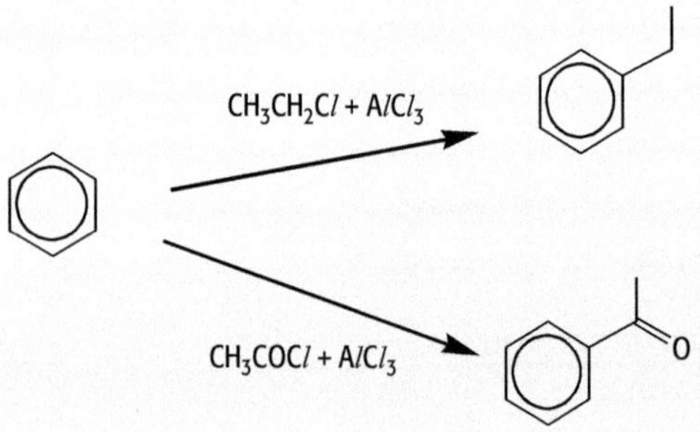

benzene + CH_3CH_2Cl + $AlCl_3$ → ethylbenzene

benzene + CH_3COCl + $AlCl_3$ → acetophenone (PhCOCH₃)

Step 2: Consider the reactions that produce the final amine. Aliphatic amines can be made by the following route:

$RBr + NH_3 \longrightarrow R\text{–}NH_2$

$RCN + H_2/Ni \longrightarrow RCH_2NH_2$

The second is no good in this case, as we do not have a –CH₂NH₂ group in the product.

Step 3: So an intermediate compound must be

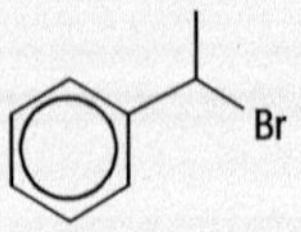

[PhCH(CH₃)Br]

Step 4: Monobromo compounds can be made by the following three methods:

- alkenes + HBr
- alcohols + HBr (or NaBr + H_2SO_4)
- alkanes + Br_2 in UV light

So there are three possible starting materials for the intermediate bromo compound:

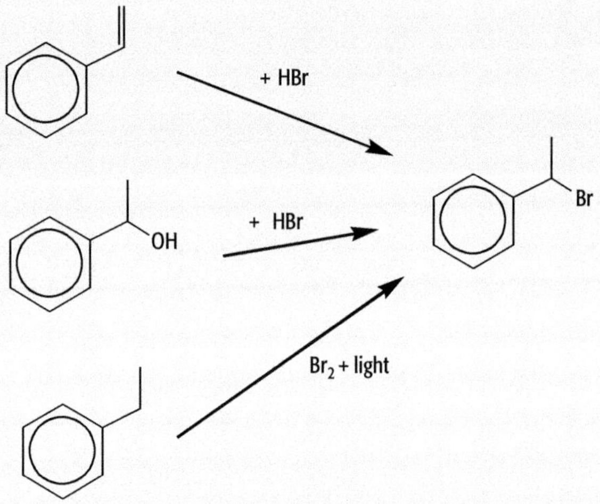

styrene (PhCH=CH₂) + HBr →
1-phenylethanol (PhCH(OH)CH₃) + HBr → PhCH(Br)CH₃
ethylbenzene + Br_2 + light →

Step 5: We have now arrived at a common point between the reactions of the starting material (benzene) and those that form the product amine: the common intermediate compound is ethylbenzene. The overall synthesis can be carried out in the following three steps.

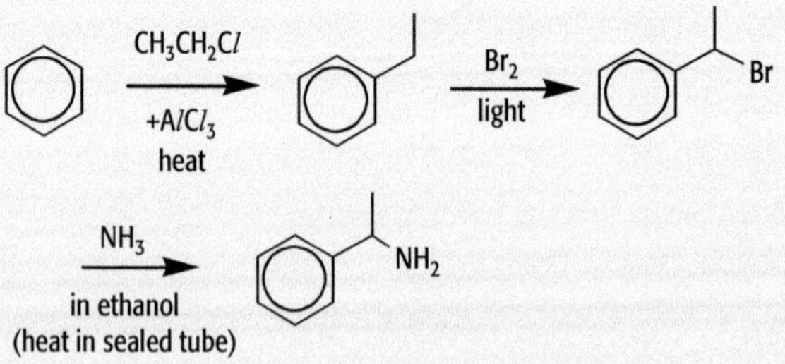

benzene $\xrightarrow[\text{heat}]{CH_3CH_2Cl, +AlCl_3}$ ethylbenzene $\xrightarrow[\text{light}]{Br_2}$ PhCH(Br)CH₃ $\xrightarrow[\substack{\text{in ethanol}\\ \text{(heat in sealed tube)}}]{NH_3}$ PhCH(NH₂)CH₃

> **TIP**
>
> You can see how important it is to learn the transformations of one organic functional group into another. Always learn these as *general* reactions of functional groups (e.g. those of *all* primary alcohols) rather than as reactions of a *specific* compound (e.g. those of just ethanol).

Progress check 30.01

Suggest routes to convert the following starting materials into the stated products:

1. $CH_3CH_2OH \Rightarrow CH_3CH(OH)CO_2H$
   (3 steps)

2. $CH_3CH=CH_2 \Rightarrow (CH_3)_2CHCH_2NH_2$
   (3 steps)

3. C₆H₅– ⇒ C₆H₅–NH–C(=O)–CH₃ (3 steps)

30.02 The design of drugs

Various methods are used to decide which molecules may be useful as drugs:

- The natural substrate can be used to suggest compounds that could mimic its effect.
- If the molecular structure of a traditional remedy, such as the active ingredient of a plant extract, is known then that structure can be used as a basis for further development.

The following formulae show the chemical similarities between the synthetic drug salbutamol (used in the treatment of asthma), the body's own drug adrenaline, and the plant extract ephedrine (known as *ma huang*, and which has been used for over 2000 years in China).

salbutamol

adrenaline

ephedrine

Since the enzymes and receptors with which the drugs interact are built up from chiral amino acids, their active sites are also chiral. Usually only one optical isomer of a drug is physiologically active. The other isomer is often either inactive, or worse, responsible for harmful side-effects. It is therefore important for a pharmaceutical company to design a synthesis that will produce just the desired isomer.

Progress check 30.02

1. How many chiral centres are there in **a** salbutamol; **b** ephedrine?

2. Draw the stereochemical formulae of the isomers of salbutamol, and decide which isomer would be the physiologically active one. Explain your reasoning.

30.03 Chemical tests for functional groups

Throughout the previous units we have listed the reactions of, and the tests for, each functional group as we have come across them. In Table 30.01 we collect together all the tests the other way round, based on the reagent used.

Reagent	Observation	Functional group present
universal indicator solution	turns red	$R-CO_2H$
universal indicator solution	turns blue	$R-NH_2$
Na(s)	fizzes/gas (H_2) evolved	ROH, ArOH (phenol) or RCO_2H
NaOH(aq)	substance dissolves	ArOH or RCO_2H
Na_2CO_3(aq)	substance dissolves and gas (CO_2) evolved	RCO_2H
PCl_5(s)	fizzes/gas (HCl) evolved	ROH or RCO_2H
warm with $Cr_2O_7^{2-}$ + H^+	turns from orange to green	RCH_2OH, $R_2CH(OH)$, RCHO
adding H_2O drops	fizzes/gas (HCl) evolved	RCOCl
MnO_4^- + H^+ (cold)	decolorises	alkene, C=C
Br_2(aq)	decolorises	alkene, C=C
Br_2(aq)	decolorises and white ppt.	ArOH or $ArNH_2$
2,4-DNPH	orange ppt.	RCHO or R_2CO
warm with Fehling's solution	red ppt.	RCHO
warm with Tollens' reagent	silver mirror	RCHO
I_2 + OH^-(aq)	pale yellow ppt.	$RCOCH_3$ or $RCH(OH)CH_3$

Table 30.01 Testing for functional groups

TIP

If in a question you are given the results of various tests, you should *always* explain what information they give you about which functional groups are present, even if you have already worked out what the 'unknown' compound is from other evidence.

Progress check 30.03

1. Suggest structures for the following compounds.

   a. Compound **A**, $C_4H_8O_3$, fizzes when Na_2CO_3(aq) is added, forms a pale yellow ppt. with I_2 + OH^-(aq), and on warming with acidified dichromate turns the colour of the reagent from orange to green.

   b. Compound **B**, C_4H_6O, gives an orange ppt. with 2,4-DNPH but does not react with Tollens' reagent. It decolorises $KMnO_4$(aq).

2. Suggest a reagent that could distinguish between the following pairs of isomers. State what you would observe with each compound.

   a. [butanal] and [butan-2-one]

   b. [3-methylaniline] and [benzylamine]

   c. [benzoyl chloride] and [3-chlorobenzaldehyde]

30.04 Recognising reaction types

We can recognise what type of reaction is taking place during a chemical transformation either by looking at the reagent involved, or by looking at how the functional group has changed.

Sample answer

Question:

State the *types of reaction* occurring in each of the steps I–VI. [6]

Answer:

I is an electrophilic aromatic substitution reaction. [1 mark]

II is a reduction. [1 mark]

III is a nucleophilic substitution. [1 mark]

IV is a free radical substitution. [1 mark]

V is a nucleophilic substitution. [1 mark]

VI is a hydrolysis (of two groups at once: an amide and a nitrile). [1 mark]

> **TIP**
> Some reactions can have more than one description. For example, most hydrolysis reactions are also nucleophilic substitutions; the reaction of an alkene with hydrogen + nickel is both an addition reaction and also a reduction.

Progress check 30.04

Identify the reaction types in the following scheme.

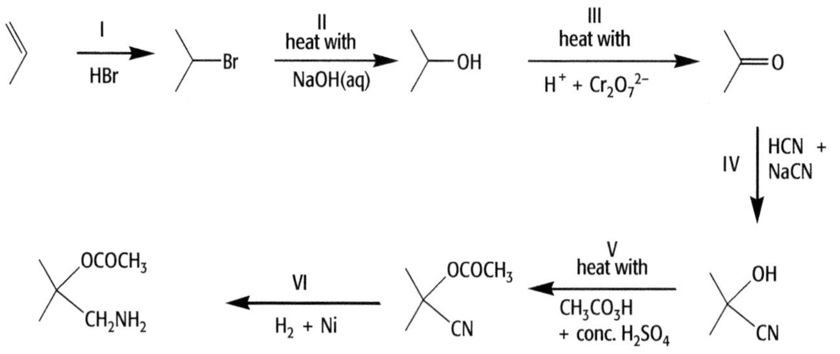

Revision checklist

Check that you know the following:

- [ ] The importance of chirality in the design of drugs.
- [ ] How to design a multi-stage organic synthesis.
- [ ] How to identify the various types of organic reaction.
- [ ] How to test for the different functional groups.

Exam-style questions

1. Compound **A** has the molecular formula $C_9H_{16}O$. When **A** is heated with concentrated acidified $KMnO_4(aq)$, two compounds are produced: **B**, $C_5H_{10}O$, and **C**, $C_4H_6O_3$. The results of five tests carried out on these three compounds are given in Table 30.02.

Test reagent	Results of test with ...		
	compound A	compound B	compound C
cold dilute $KMnO_4(aq)$	decolorises	no reaction	no reaction
Tollens' reagent	silver mirror formed	no reaction	no reaction
$I_2(aq)$ + $OH^-(aq)$	no reaction	no reaction	yellow precipitate
2,4-DNPH reagent	orange precipitate	orange precipitate	orange precipitate

Table 30.02

a. Explain the functional group that each of the above four reagents tests for. [4]

b. Suggest the structures of these three compounds. [3]

c. When compound **A** is treated with HBr, two different positional isomers with the molecular formula $C_9H_{17}BrO$ are formed.

Draw the structures of these two isomers, indicating with an asterisk any chiral carbon atoms they contain. [4]

Total: 11

2. The plant extract containing ephedrine has been used in China as an anti-asthmatic and stimulant since ancient times. Ephedrine can be synthesised in the laboratory by the following route.

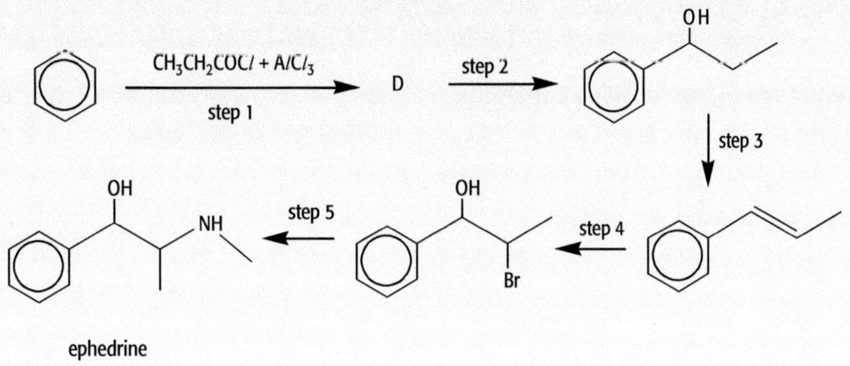

a Suggest the structure of intermediate compound **D**. [1]

b Suggest reagents and conditions for steps 2–5. [5]

c State the *types of reaction* that are occurring in steps 1–5. [5]

d Deduce the number of optical isomers that have the same structure as ephedrine. Explain your answer. [2]

e State, giving a reason in each case, whether or not ephedrine will be soluble in:

  i NaOH(aq)

  ii HCl(aq). [2]

Total: 15

Practical Skills 1

This course will test your skill in manipulating apparatus, how well you present your data, your analysis and evaluation. It also tests your powers of accurate observation, and your ability to make logical conclusions from those observations.

You should be able to:

- plan an experiment or investigation
- record measurements and present observations clearly
- reflect on your collected data to reach conclusions
- evaluate your experimental method and suggest improvements

1 Plan an experiment

You could be given equipment and asked to use it to find the answer to a question.

Here is an example.

Find the enthalpy of neutralisation of HCl(aq) with NaOH(aq) given the following chemicals and apparatus:

$1.0\,mol\,dm^{-3}$ HCl

$1.0\,mol\,dm^{-3}$ NaOH

thermometer measuring to $1.0\,°C$

$50\,cm^3$ graduated cylinder

$100\,cm^3$ beaker

Before you start any practical work, you need to be clear what you are going to do and what you need to measure. Put your thoughts down on paper.

i   You need to add the acid and alkali. How much of each? How will you measure the volume? What vessel will you add them to?

ii You need to measure a temperature change. The initial temperature needs to be taken. Are the two solutions at the same temperature? What will you do if they are not? When do you take the temperatures?

iii Write out what you are going to do and what you are going to measure and how you are going to measure it.

Progress check P.01

Plan an experiment to compare the enthalpy changes when iron and zinc are separately dissolved in aqueous hydrochloric acid.

You have available samples of iron and zinc; $1.0\,mol\,dm^{-3}$ HCl; $100\,cm^3$ beaker; thermometer graduated to $1.0\,°C$; balance; $50\,cm^3$ graduated cylinder.

2 Observation and recording

a **Specified experiments**

You may be asked to carry out specified experiments to investigate an unknown substance or substances. It is important to follow the instructions and record each observation carefully.

Here is an example.

You are provided with solutions A and B, AgNO$_3$(aq) and NH$_3$(aq).

The observations need to be recorded at the time they are made.

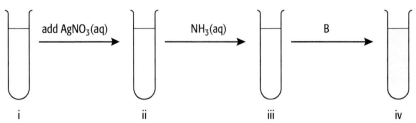

	Experiment	Observation
i	A	colourless solution
ii	add AgNO$_3$(aq)	white precipitate
iii	add NH$_3$(aq)	precipitate dissolves to form a colourless solution
iv	add B	yellow precipitate

b **Accuracy**

Another aspect of this skill is taking measurements to the right degree of accuracy:

- With a thermometer graduated in °C, the best you can do is to take a measurement to +/− 0.5 °C.

- A graduated cylinder with markings for every 1 cm^3 can be read to +/− 0.5 cm^3.

- A burette graduated in 0.1 cm^3 can be read to +/− 0.05 cm^3.

- The timing on a stopwatch is probably dependent on your own reaction time, so the larger the time measured, the lower is the proportion which is uncertain due to your reaction time and the measured value is more accurate.

- A beaker marked for 100 cm^3 would have an uncertainty in this value of +/− 5 cm^3.

> **TIP**
> When using any particular measuring instrument, the smaller the amount you measure, the larger will be the percentage uncertainty in the value.

Record your measurements clearly as you take them.

c **Titration procedure**

One question that you may encounter is about titrations, so it is important that you know how to carry these out accurately.

i The *conical flask* needs to be clean and rinsed with deionised water but it does not need to be dry.

ii The *pipette* needs to be rinsed out with the solution which is to be measured (and then discard that portion of the solution). Fill it so that there are no bubbles in the solution and the bottom of the meniscus is level with the gradation mark. Allow the solution to empty naturally into the conical flask (do not blow the solution out) and at the end touch the end of the jet onto the top of the solution.

iii The *burette* needs to be rinsed with the solution to be measured before filling it. Make sure the jet is filled below the tap and there are no air bubbles round the tap or in the cylinder.

iv During the titration, swirl the conical flask continuously to ensure the solutions are mixed. Take a 'rough' titre (to the nearest 0.5 cm^3) and several 'accurate' titres, each recorded to the nearest 0.05 cm^3. The average of the accurate titres can be recorded to 2 d.p. (e.g. the average of 20.10 cm^3 and 20.05 cm^3 is 20.08 cm^3).

v Read the burette with your eye horizontal to the meniscus.

d **Some titration tips**

i Acid–base titrations: remember to check the end-point colour change of the indicator provided.

ii KMnO$_4$: this is usually put into the burette and the volume is read from the *top* of the meniscus.

iii I$_2$/thiosulfate: the thiosulfate goes into the burette. When the I$_2$ colour in the conical flask has faded to a light straw colour, the starch indicator is added to give an intense blue colouration.

e **Rates experiments**

Swirl the conical flask vigorously at the start of the experiments to ensure mixing.

> **TIP**
> Stir a solution well before taking a temperature. Keep the thermometer in the solution whilst reading it and read the scale with your eye horizontal to the grading mark.

3 Reaching conclusions

When you have carried out all the experiments and collected all the data, you can come to some conclusions. This depends on your chemical knowledge. You need to write your conclusions clearly and methodically.

a What you should know
- The reactions of the following cations: NH_4^+, Mg^{2+}, Al^{3+}, Ca^{2+}, Cr^{3+}, Mn^{2+}, Fe^{2+}, Fe^{3+}, Cu^{2+}, Zn^{2+}, Ba^{2+}.
- The reactions of the following anions: CO_3^{2-}, NO_3^-, NO_2^-, SO_4^{2-}, SO_3^{2-}, Cl^-, Br^-, I^-.
- Tests for the following gases: NH_3, CO_2, Cl_2, H_2, O_2, SO_2.

Sample answer

Question:

For the experiment in section 2, above, make suitable observations and draw conclusions. [7]

Answer:

	Experiment	Observation	Conclusion
i	A	colourless solution [1 mark]	
ii	add $AgNO_3(aq)$	white precipitate [1 mark]	a white precipitate with $AgNO_3(aq)$ is a test for a chloride: A is a chloride [1 mark]
iii	add $NH_3(aq)$	precipitate dissolves to form a colourless solution [1 mark]	$AgCl$ dissolves in $NH_3(aq)$; this confirms the chloride [1 mark]
iv	add B	yellow precipitate [1 mark]	this is the very insoluble AgI: B is an iodide [1 mark]

b Examination

During examination you will be able to use the wet test observations / inferences table provided. However, you do need to familiarise yourself with it so that you can use it constructively to help you.

4 Evaluation

a A better way?

Some experiments allow you to make choices. Your evaluation allows you to put forward a better way of carrying out an experiment. Think of each piece of apparatus or procedure in turn.

For example, in the neutralisation experiment …

- Would different volumes of solutions have been better? Could you have used larger volumes, and what effect would this have?
- Would a more accurate thermometer have given a more accurate result? (It may be reasonable to decide that the thermometer was accurate enough but you do need to say why.)
- Is there a significant heat loss when using a beaker? Would a lid or lagging have prevented some heat loss?

Which of these effects gives the most inaccuracy to your value?

b When plotting a graph
  i Choose a suitable scale to fill as much of the graph paper as possible but which can be read with ease.
  ii Label the axes clearly with what is being plotted and the units.
  iii Use the origin as a 'point' if appropriate.
  iv Plot a smooth line/curve that best fits the points and point out any anomalous points.

Practical Skills 2

This course tests the following skills:
- planning
- analysis, conclusions and evaluation

All that has been written above for AS level applies also to A level but at A level you are assumed to be competent at carrying out practical work. At this level you are therefore expected to be able to reach a higher level of skill in planning experiments, analysing results, reaching a suitable conclusion and evaluating the work.

1 Planning

a You need to be able to:
- put forward a safe and efficient procedure which would lead to a reliable result
- state how you would carry it out and point out any risks involved
- explain what apparatus is required
- identify both the independent variable and the dependent variable
- predict what you expect to happen and explain why your procedure will be effective
- identify any variables which need to be controlled

b Now you need to think about the method in detail:
- What volumes of solutions will you use?
- How will you put the reaction mixtures together?

c Finally show how the data gathered might be displayed – perhaps in a table in this case.

Progress check P.02

Four unlabelled colourless liquids are known to be $CH_3CH_2OCH_2CH_3$, $CH_3CH_2CH_2CH_2OH$, $CH_3CH_2CH(OH)CH_3$ and $(CH_3)_3COH$.

Suggest a series of experiments which will allow you to identify them.

2 Analysis and conclusions

You need to be able to:
- recognise the significance of observations from qualitative tests
- complete tables of the data to enable conclusions to be drawn

For quantitative work, you also need to be able to:
- identify and carry out any calculations which will enable you to draw conclusions
- analyse data from experimental work and from spectra to draw conclusions
- draw graphs showing the independent variable on the x-axis and the dependent variable on the y-axis

3 Evaluation

You need to be able to:
- identify any unexpected, anomalous results or values and suggest why they are anomalous
- decide whether the controlled variables really have been controlled (e.g. was the temperature really constant?)
- decide whether the results are reliable enough to make any conclusion; has repetition of the experimental work led to the same result?
- identify any weaknesses in the experimental method and in the apparatus (e.g. 'the calorimeter should have been lagged' or 'the reactant should have been powdered so it reacted quickly') and explain what effect these weaknesses would have on the final result

Exam-style questions

When magnesium nitrate(V) and calcium nitrate(V) are heated, they decompose to form the metal oxide, nitrogen(IV) oxide and oxygen. Nitrogen(IV) oxide is an acidic gas that reacts completely with alkalis.

You are to plan an experiment to investigate the relative rate of decomposition of the nitrates by collecting the oxygen produced.

Nitrogen(IV) oxide can have severe effects on breathing, particularly if you suffer from asthma.

You are provided with the separate anhydrous nitrates and usual laboratory apparatus and reagents.

a i   Write the equation for the thermal decomposition of magnesium nitrate(V). [1]

   ii Calculate the volumes of nitrogen(IV) oxide and of oxygen produced when 0.1 mol of magnesium nitrate(V) decomposes. [1]

   (Assume 1.0 mol of any gas occupies 24.0 dm³ at room temperature and pressure.)

b Draw and label the apparatus you would need to carry out the decomposition, absorb the nitrogen(IV) oxide as it is produced and then collect the oxygen. [4]

c Decide what maximum volume of oxygen you will collect and then calculate the masses of magnesium nitrate(V) and calcium nitrate(V) which need to be heated to give this volume. (A_r[O] = 16.0; A_r[N] = 14.0; A_r[Mg] = 24.3) [2]

d Explain what measurements you would need to make to ensure the comparison is as fair as possible. [4]

e Describe how you would use your measurements to show the relative rates of oxygen production. [3]

f How would you ensure the experiment was carried out safely? [1]

Total: 16

Answers to Progress Check questions

Unit 1

1.01

1. a 2.01×10^{23}
   b 8.60×10^{22}
   c 4.30×10^{22}
2. a 5.09 g
   b 22.5 g
   c 5.6 g
   d 1.00 g
3. a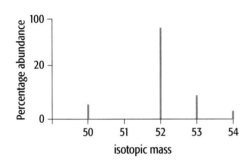
   b 52.1
4. a y-axis = percentage abundance; x-axis = isotopic mass
   b 6
   c 106.5

1.02

1. C_2H_5NO
2. Fe_3O_4

1.03

1. $P_4(s) + 5O_2(g) \rightarrow 2P_2O_5(s)$
2. $SiCl_4(l) + 2H_2O(l) \rightarrow SiO_2(s) + 4HCl(aq)$
3. $Li_2CO_3(s) \rightarrow Li_2O(s) + CO_2(g)$
4. $Cu(s) + 4HNO_3(conc.) \rightarrow Cu(NO_3)_2(aq) + 2NO_2(g) + 2H_2O(l)$
5. $3Cu(s) + 8HNO_3(aq) \rightarrow 3Cu(NO_3)_2(aq) + 2NO(g) + 4H_2O(l)$

1.04

1. 1.43 g
2. 1.20 g
3. 3.5 g
4. CH_2O
5. C_2H_5NO
6. 15.6 g dm^{-3}
7. 6.2×10^{-4} mol dm^{-3}

Unit 2

2.01

1. a 1.9×10^8
   b protons and neutrons
   c they are very strong
2.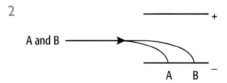

2.02

1.

Symbol	Species	Number of protons	Number of neutrons	Number of electrons
$^{138}_{56}Ba^{2+}$	barium ion	56	82	54
$^{37}_{17}Cl^-$	chloride ion	17	20	18
$^{132}_{54}Xe$	xenon atom	54	78	54
$^{86}_{38}Sr^{2+}$	strontium(II) ion	38	48	36

2. the chromium atom with a mass number 50

Unit 3

3.01

1. a spherical
   b two
2. a 9
   b 18

3.02

1. - The ionisation energy increases across the period as the nuclear charge increases.
   - There is a decrease from Mg to Al as the electron added enters the 3p orbital which is further from the nucleus than the 3s orbital.
   - There is an increase from Al to P as the nuclear charge increases and the electrons enter the three 3p orbitals singly.
   - There is a decrease from P to S as the electron has to pair in a 3p orbital and then there is an increase to Ar as the nuclear charge increases.

2.

Element	First ionisation energy / kJ mol^{-1}
Na	496
Li	520
Mg	738

3.03

1. F is

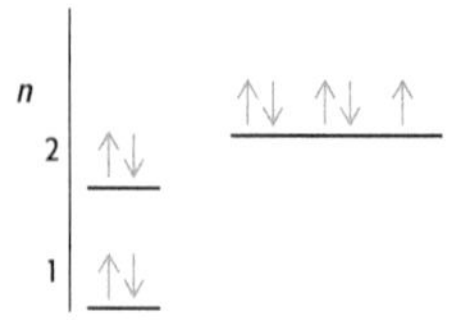

   [energy level diagram: n=1 (↑↓), n=2 (↑↓) (↑↓ ↑↓ ↑)]

   Sc is

   [energy level diagram: n=1 (↑↓), n=2 (↑↓) (↑↓ ↑↓ ↑↓), n=3 (↑↓) (↑↓ ↑↓ ↑↓), n=4 (↑)]

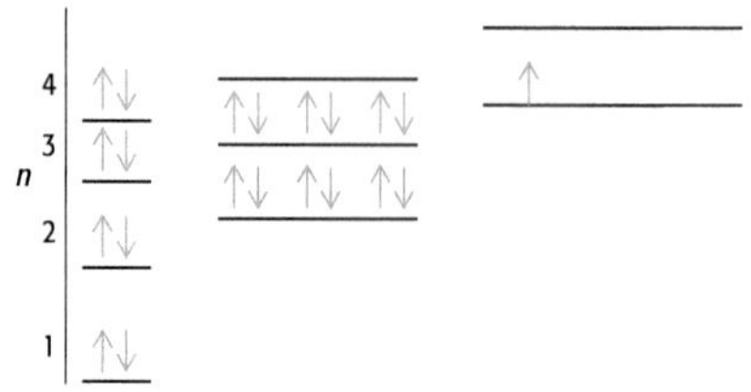

2. $Si^{2+}(g) \rightarrow Si^{3+}(g) + e^-$
3. two outer electrons

3.04

F	$1s^2 2s^2 2p^5$
F$^-$	$1s^2 2s^2 2p^6$
N	$1s^2 2s^2 2p^3$
Ca^{2+}	$1s^2 2s^2 2p^6 3s^2 3p^6$
S	$1s^2 2s^2 2p^6 3s^2 3p^4$
S^{2-}	$1s^2 2s^2 2p^6 3s^2 3p^6$
K	$1s^2 2s^2 2p^6 3s^2 3p^6 4s^1$

Unit 4

4.01

1. Na$^+$ $1s^2 2s^2 2p^6$
   Cl$^-$ $1s^2 2s^2 2p^6 3s^2 3p^6$
   Li$^+$ $1s^2$
   O^{2-} $1s^2 2s^2 2p^6$
   Mg^{2+} $1s^2 2s^2 2p^6$

2. [dot-and-cross diagrams: K$^+$ [Br]$^-$; Al^{3+} 3[F]$^-$; Ca^{2+} [Cl]$^-$; Ba^{2+} [O]$^{2-}$; 2 Cs$^+$ [O]$^{2-}$]

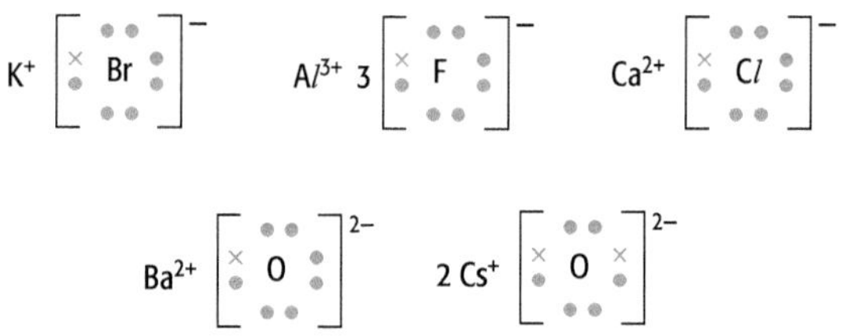

4.02

1. [dot-and-cross diagrams of H−C−C−O−H (ethanol-like), O=C=O (CO$_2$), and C$_2$H$_4$]

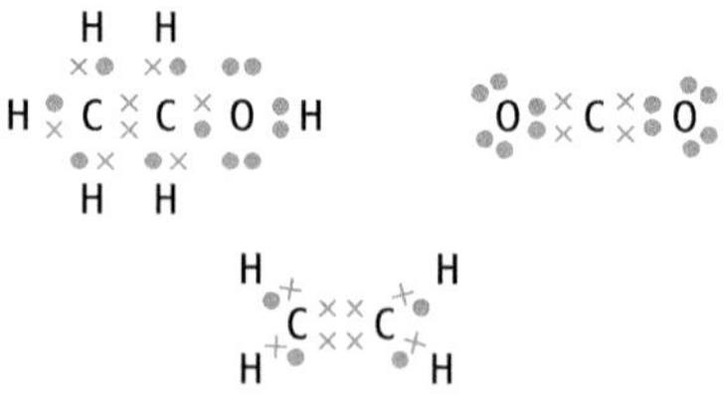

2. In CO$_2$, each O has two lone pairs and two shared pairs each C has four shared pairs of electrons.

   In C$_2$H$_4$, each C has four shared pairs of electrons each H has one shared pair of electrons.

3. [dot-and-cross diagrams of NH$_3$ and BF$_3$]

4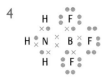

4.03

1 C is surrounded by four bond pairs which repel each other and so CH_4 is tetrahedral, with a bond angle of 109.5°.

N is surrounded by one lone pair and three bond pairs. Repulsion between a lone pair and a bond pair is larger than repulsion between two bond pairs and so the bond angle is smaller than 109.5°. The molecule is pyramidal with a bond angle of 107°.

O is surrounded by two lone pairs and two bond pairs. Repulsion between two lone pairs is greater than between a bond pair and a lone pair H_2O is bent with a bond angle of 104°.

2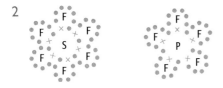

3 In C_2H_6 each C is sp^3 hybridised as four bonds are made. Bond angle = 109.5°.

4 $^1CH_3{-}^2CH{=}^3CH_2$; C-1 is sp^3 hybridised with bond angles = 109.5°; C-2 and C-3 are sp^2 hybridised with bond angles = 120°.

5 F_2O is bent with bond angle = 104.5°.

CCl_4 is tetrahedral with bond angles = 109.5°.

NH_4^+ is tetrahedral with bond angles = 109.5°.

ClF_5 is octahedral with one position occupied by a lone pair; bond angles = 90°.

XeO_3 is pyramidal, surrounded by three double bond pairs and one lone pair; bond angles greater than 110°.

4.04

1 CO_2 is linear; CCl_4 is tetrahedral; NF_3 is pyramidal; SF_6 is octahedral; SO_2 is bent; SO_3 is planar triangular.

2 NF_3 and SO_2

3 Both exist as small covalent molecules. The molecules are held together by weak van der Waals' forces. $SnBr_4$ has more electrons than $SnCl_4$ and so the van der Waals' forces are stronger and take more energy to overcome.

4 Mg has two outer electrons but Na only has one. The electrostatic attraction between Na^+ and the delocalised electrons is smaller than between Mg^{2+} and the delocalised electrons and so the energy needed to overcome the attractive force is less in Na than in Mg.

5 A = Ca (metals conduct at room temperature)

B = iodine (only van der Waals' forces between molecules and these are easily overcome)

C = calcium iodide (the high boiling point is indicative of having strong electrostatic attractions between ions; the lack of electrical conductivity at room temperature shows that it is not a metal)

Unit 5

5.01

1 Hydrogen; it has only weak van der Waals forces between the molecules which is the weakest intermolecular force.

2 Hydrogen fluoride as it has hydrogen bonds between the molecules and hydrogen bonds are stronger than van der Waals or permanent dipole forces.

5.02

1 3.81×10^{-3}

2 56.6

5.03

1 a D

b A, C

c C

d B is ionic and E is metallic.

2 Ca as it has more electrons which can be delocalised.

Unit 6

6.01

1. $C_2H_6(g) + 3½O_2(g) \rightarrow 2CO_2(g) + 3H_2O(l)$
2. $Mg(s) + ½O_2(g) \rightarrow MgO(s)$
3. $CH_3CO_2H(l) + 2O_2(g) \rightarrow 2CO_2(g) + 2H_2O(l)$
4. $CH_3COOC_2H_5(l) + 5O_2(g) \rightarrow 4CO_2(g) + 4H_2O(l)$

6.02

ΔH_c[butanone] = $-1770\,kJ\,mol^{-1}$

6.03

1. $-288\,kJ$
2. $-124\,kJ$
3. $-43\,kJ$

6.04

1. HNO_3 is a strong acid and fully ionised; reaction is $H^+(aq) + OH^-(aq) \rightarrow H_2O(l)$

   H_2S is a weak acid and has to ionise, which requires energy, before the neutralisation reaction can take place.
2. $-54.8\,kJ\,mol^{-1}$
3. Same temperature. Less energy is given out but heats a smaller volume of solution.
4. weighted average initial temperature = 20 °C

   temperature change = 4.4 K

   final temperature = 24.4 K

6.05

1. $-9\,kJ\,mol^{-1}$
2. $-818\,kJ\,mol^{-1}$
3. $+269\,kJ\,mol^{-1}$
4. $-30\,kJ\,mol^{-1}$

Unit 7

7.01

CO_2	$O = -2, C = +4$
CH_4	$H = +1, C = -4$
NH_3	$H = +1, N = -3$
HCl	$H = +1, Cl = -1$
P_2O_5	$O = -2, P = +5$
ClO_2	$O = -2, Cl = +4$
$KClO_3$	$O = -2, K = +1, Cl = +5$
HNO_3	$O = -2, H = +1, N = +5$
CrO_4^{2-}	$O = -2, Cr = +6$
$Ni(OH)_2$	$O = -2, H = +1, Ni = +2$
CO_3^{2-}	$O = -2, C = +4$
NH_4VO_3	$H = +1, O = -2, N = -3, V = +5$

7.02

1. a $Fe^{3+} + e^- \rightarrow Fe^{2+}$

   $I^- \rightarrow ½I_2 + e^-$

   b $Fe^{3+} + I^- \rightarrow Fe^{2+} + ½I_2$
2. a $VO^{2+}(aq) + e^- \rightarrow V^{3+}(aq)$

   $Sn(s) \rightarrow Sn^{2+}(aq) + 2e^-$

   b $Sn(s) + 2VO^{2+}(aq) + 4H^+(aq) \rightarrow 2V^{3+}(aq) + Sn^{2+}(aq) + 2H_2O(l)$

7.03

1. $8H^+(aq) + 2VO_2^+(aq) + 3Zn(s) \rightarrow 2V^{2+}(aq) + 3Zn^{2+}(aq) + 4H_2O(l)$
2. $2H_2S + SO_2 \rightarrow 2H_2O + 3S$
3. $2MnO_4^-(aq) + 5SO_2(aq) + 2H_2O(l) \rightarrow 2Mn^{2+}(aq) + 5SO_4^{2-}(aq) + 4H^+(aq)$
4. $S_2O_8^{2-} + 2Fe^{2+} \rightarrow 2SO_4^{2-} + 2Fe^{3+}$

Unit 8

8.01

1.
   a $PCl_5(g) \rightleftharpoons PCl_3(g) + Cl_2(g)$
   b $2HI(g) \rightleftharpoons H_2(g) + I_2(g)$
   c $N_2(g) + 3H_2(g) \rightleftharpoons 2NH_3(g)$
   d $H_2(g) + CO_2(g) \rightleftharpoons H_2O(g) + CO(g)$
   e $2SO_2(g) + O_2(g) \rightleftharpoons 2SO_3(g)$

2.
   a left to right
   b left to right
   c right to left
   d left to right
   e right to left

3.
   a right to left
   b not at all
   c left to right
   d not at all
   e left to right

4.
   c left to right
   d right to left
   e left to right

8.02

1 and 2

a $K_c = \dfrac{[HI(g)]^2}{[H_2(g)][I_2(g)]}$; no units

b $K_c = \dfrac{[H_2(g)]^{\frac{1}{2}}[I_2(g)]^{\frac{1}{2}}}{[HI(g)]}$; no units

c $K_c = \dfrac{[NH_3(g)]^2}{[N_2(g)][H_2(g)]^3}$; $mol^{-2}\,dm^6$

d $K_c = \dfrac{[N_2O_4(g)]}{[NO_2(g)]^2}$; $mol^{-1}\,dm^3$

3. $K_c = 9.48\,mol^{-2}\,dm^6$

4. number of moles H_2O = number of moles CO = 0.0174

   number of moles H_2 = number of moles CO_2 = 0.0325

8.03

1. 0.7 moles of B; 0.3 moles of C
2. 1.7
3. 2.06 atm for A and B; 0.88 atm for C
4. $K_p = \dfrac{p_C}{p_A p_B}$
5. $K_p = 0.21\,atm^{-1}$

8.04

Acids are HNO_3, H_2O, CH_3COOH, H_3O^+, H_2SO_4, $H_2NO_3^+$

Unit 9

9.01

1. a

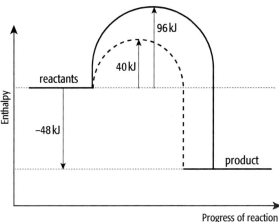

   b $\Delta H_r = +48\,kJ\,mol^{-1}$
   c $E_a = 144\,kJ\,mol^{-1}$

2. a Monitor the volume of hydrogen gas produced in known time intervals.

   b i Rate increases; as the concentration increases, the number of particles per unit volume increases and so the rate of collision increases.

   ii Rate decreases; the surface area of the zinc decreases and so the rate of collisions decreases.

   iii Rate increases; surface area of zinc increases so the rate of collisions increases.

3 a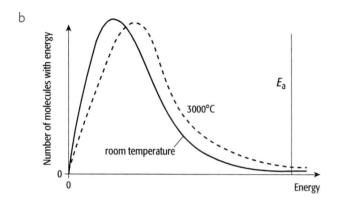

b

E_a is large and so at room temperature there are a negligible number of molecules with enough energy to react on collision. At the higher temperature, more particles have energy greater than the activation energy and so will react on collision.

9.02

1 a The Boltzmann distribution is similar to that in Fig 9.05. At a higher temperature, the particles have more energy and so more of them have energy greater than E_a and can react on collision. The rate of successful collisions increases.

b The Boltzmann distribution is similar to that in Figure 9.06. The activation energy is lowered by the Pt catalyst and so more molecules have energy greater than the activation energy and can react on collision. Again the rate of successful collisions increases.

2 a homogeneous

b heterogeneous

c heterogeneous

3 a The rate increases as the reaction mixture warms up from the heat produced; more molecules will have energy greater than the activation energy.

b Boiling will denature the enzyme and it will stop working so the rate will decrease and the reaction will stop.

c Catalase is specific to one reaction.

Unit 10

10.01

1 The ionisation energy increases across the period because the nuclear charge is increasing and each successive electron is added to the same shell. A*l* has a lower ionisation energy than expected as the electron is added to a new sub-shell, the 3p, which is slightly further from the nucleus than the 3s electrons. S has a lower ionisation energy than expected; the electrons added to form A*l*, Si, P enter the three 3p orbitals singly but, with S, the next electron has to pair up and there is a small amount of repulsion.

2 $2Na(s) + \frac{1}{2}O_2(g) \rightarrow Na_2O(s)$

$Mg(s) + \frac{1}{2}O_2(g) \rightarrow MgO(s)$

$2Al(s) + 1\frac{1}{2}O_2(g) \rightarrow Al_2O_3(s)$

$P_4(s) + 5O_2(g) \rightarrow P_4O_{10}(s)$

$S_8(s) + 8O_2(g) \rightarrow 8SO_2(g)$

3 $Na = +1, Mg = +2, Al = +3, P = +5, S = +4$

4 $Na(s) + \frac{1}{2}Cl_2(g) \rightarrow NaCl(s); +1$

$Mg(s) + Cl_2(g) \rightarrow MgCl_2(s); +2$

$2Al(s) + 3Cl_2(g) \rightarrow Al_2Cl_6(s); +3$

$Si(s) + 2Cl_2(g) \rightarrow SiCl_4(l); +4$

$P_4(s) + 10Cl_2(g) \rightarrow 4PCl_5(s); +5$

AQ. Should '9.02' be '9.01'

10.02

1 +3 in both

2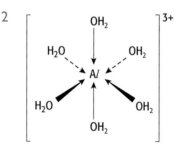

3 $AlCl_3$ dissolves in water to form $[Al(H_2O)_6]^{3+}$; the Al^{3+} is very small and highly charged and can attract the electrons in the water molecules strongly.

$[Al(H_2O)_6]^{3+} \rightarrow [Al(H_2O)_5OH]^{2+} + H^+$

The Na^+ ion is not highly charged and so does not have any effect on the water molecules.

4 $MgCl_2$ dissolves in water to produce $Mg^{2+}(aq)$; the Mg^{2+} attracts the electrons in the water molecules but is not as highly charged and is larger than Al^{3+} and cannot attract the electrons to the same extent.

Unit 11

11.01

1 Add $HCl(aq)$ with an acid–base indicator.

BaO dissolves to form a neutral solution. $BaCO_3$ effervesces.

$Ba(NO_3)_2$ dissolves to form an acidic solution.

2 Dissolve the MgO in $H_2SO_4(aq)$ with warming, filter, leave filtrate to cool and crystallise, pick out the crystals and dry them between sheets of filter paper.

3 Dissolve BaO in $HCl(aq)$ with warming, filter, add $Na_2SO_4(aq)$ to the filtrate, filter, dry the residue in a desiccator.

Unit 12

12.01

1 $Cl_2 + 2I^- \rightarrow 2Cl^- + I_2$

$Br_2 + 2I^- \rightarrow 2Br^- + I_2$

2 F is very electronegative and HF has a dipole: $^{\delta+}H–F^{\delta-}$.

F has three lone pairs of electrons and can form hydrogen bonds with neighbouring molecules: $^{\delta+}H–F^{\delta-}$ - - $^{\delta+}H–F^{\delta-}$ - - $^{\delta+}H–F^{\delta-}$.

More energy is needed to overcome the hydrogen bonding.

3 The intermolecular forces are dipole–dipole and van der Waals' forces. As the number of electrons increases down the group, so the van der Waals' forces increase and more energy is needed to overcome them.

4 Ionisation energy for Al is higher than for Na; too much energy is needed to remove three electrons from Al.

5 $3Cl_2 + 6OH^- \rightarrow 5Cl^- + ClO_3^- + 3H_2O$

12.02

1 Br: −1 to 0; S: +6 to +4

2 Purple vapour; fumes have a rotten egg smell

3 NaCl and NaI; $0.2 \, mol \, dm^{-3}$

Unit 13

13.01

1 A low temperature gives a higher equilibrium yield but a slow reaction; the catalyst is then used to speed up the reaction.

2 A high pressure gives both a higher equilibrium yield and a higher rate of reaction but the equipment to withstand a very high pressure is very expensive.

3 NaOH(aq) gives OH^- ions which form a green precipitate of $Ni(OH)_2$ with soluble nickel salts. $NH_3(aq)$ also gives OH^- ions which form a green precipitate of $Ni(OH)_2$; NH_3 can also act as a ligand and so dissolve the precipitate to form a soluble complex.

4 $NH_4^+(aq) + OH^-(aq) \xrightarrow{heat} NH_3(g) + H_2O(l)$

5 $3NH_3(aq) + H_3PO_4(aq) \rightarrow (NH_4)_3PO_4(aq)$

Unit 14

14.01

1 Any three of the following seven structures:

2 a i $C_7H_{11}O_2Cl$
   ii $C_7H_{11}O_2Cl$
  b i $C_6H_{12}O_2$
   ii C_3H_6O
  c i $C_9H_{12}O_3$
   ii C_3H_4O

14.02

1 a $CH_3CH_2CH(CH_3)CHO$
  b $(CH_3CH_2)_2C=CHCH_2OH$
  c $CH_3COC(CH_3)_2COCH_3$

2 a 3-methylpentan-2-one
  b 1,3-dibromobut-1-ene
  c 2-chloro-3-methylpentanoic acid

14.03

1 These are the five isomers: two are optical isomers of each other.

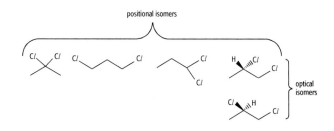

2 There are eight isomers, including three mirror-image pairs.

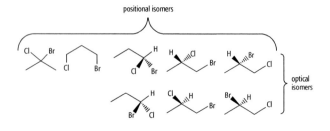

3 There are six isomers of C_5H_{10}, of which two are geometrical isomers of each other.

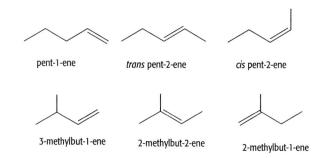

pent-1-ene   trans pent-2-ene   cis pent-2-ene

3-methylbut-1-ene   2-methylbut-2-ene   2-methylbut-1-ene

Unit 15

15.01

1 $CH_3CH_2CH_2CH_2CH_3$ $(CH_3)_2CHCH_2CH_3$ $(CH_3)_4C$

pentane   2-methylbutane   2,2-dimethylpropane

2 $CH_3CH_2CH(CH_3)CH_2CH_2CH_3$, 3-methylhexane

3 3,4-dimethylhexane,

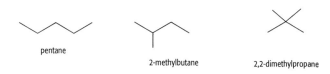

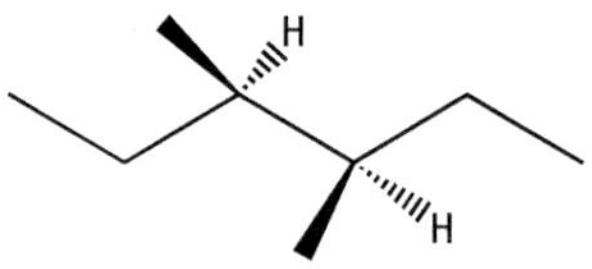

15.02

1 *Either* pass $CH_3CH_2CH(OH)CH_3$ vapour over heated Al_2O_3 or heat $CH_3CH_2CHBrCH_3$ with KOH in ethanol.

Note that either method is likely to produce a mixture of $CH_3CH=CHCH_3$ and $CH_3CH_2CH=CH_2$.

2 Heat *either* $(CH_3)_2CHCH_2CH(OH)CH_3$ or $(CH_3)_2CHCH(OH)CH_2CH_3$ with conc. H_2SO_4 at 180°C (or heat the corresponding bromides with KOH in ethanol). Note that either method is likely to produce a mixture of alkenes.

3

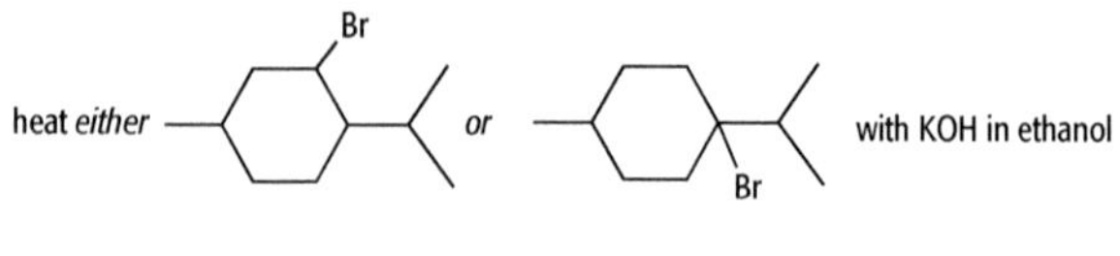

heat *either* [structure] or [structure] with KOH in ethanol

(*or* heat the corresponding alcohols with conc. H_2SO_4 at 180°C)

Note that the first bromo compound will give a mixture of

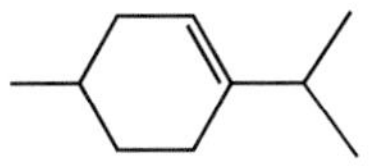

and

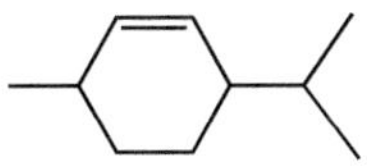

whilst the second bromo compounds will give a mixture of:

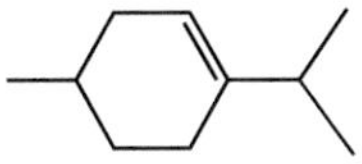

and

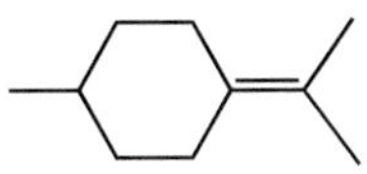

15.03

1.

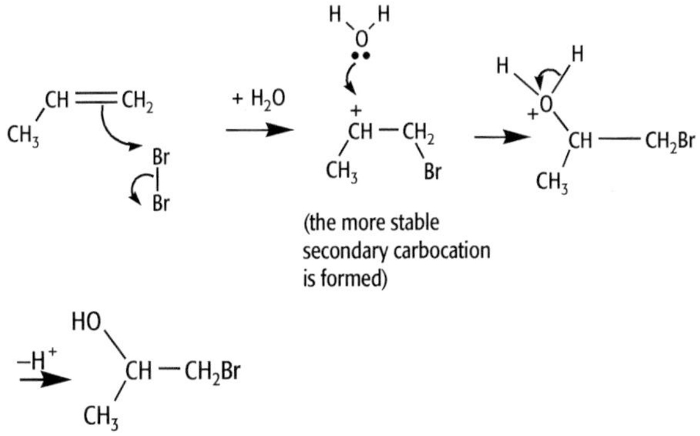

(the more stable secondary carbocation is formed)

2 a $CH_3–CHClCH_2CH_3$
   b $(CH_3)_3C–OH$
   c

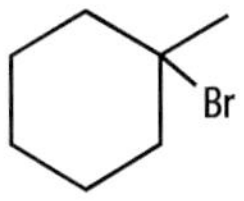

15.04

1 a $(CH_3)_2C=O$ and $CH_3COCH_2CH_3$
   b $CH_3CH_2CH(CH_3)COCH_3$ and CO_2
   c $(CH_3)_2CHCO_2H$ and $CH_3CH_2CO_2H$

2 a Y is   b Z is

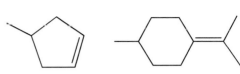

Unit 16

16.01

1 mixture of S_N1 and S_N2
2 mixture of S_N1 and S_N2
3 S_N2
4 S_N1

16.02

1 $CH_3CH_2CH_2Br + NaOH \rightarrow CH_3CH_2CH_2OH + NaBr$

2 $CH_3CH_2CHBrCH_3 + NaCN \rightarrow CH_3CH_2CH(CN)CH_3 + NaBr$

3 $CH_3CHClCH_3 + NH_3 \rightarrow CH_3CH(NH_2)CH_3 + HCl$

16.03

1 Heat 2-methyl-1-bromobutane, $CH_3CH_2CH(CH_3)CH_2Br$, and NaCN in ethanol.

2 Heat a mixture of 3-methyl-1-bromobutane, $CH_3CH(CH_3)CH_2CH_2Br$, and an excess of ammonia in ethanol in a sealed tube.

3 React 2-bromobutane with hot KOH in ethanol. Note: a mixture of but-2-ene and but-1-ene will be produced.

4 React 1-bromobutane with hot KOH in ethanol.

Unit 17

17.01

1 There are eight alcohols, four of which are primary; three are secondary and one is tertiary. Three of them contain chiral carbon atoms (*).

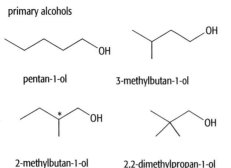

17.02

1. CH_3OH, methanol
2. $(CH_3)_3C-OH$, 2-methylpropan-2-ol
3. $CH_3CH(OH)C(CH_3)_2CH_2OH$, 2,2-dimethylbutan-1,3-diol

17.03

1. W is $(CH_3)_2CHCO_2H$.
2. X is $(CH_3)_2CHCH_2Br$.
3. Y is $(CH_3)_2CHCH_2CN$.
4. Z is $(CH_3)_2CHCH_2CO_2H$.

17.04

If all three react with sodium metal, all three must contain an –OH group. Since they each have only one oxygen atom, they must be alcohols.

Since **A** does not react with acidified potassium dichromate, it must be a tertiary alcohol. There is only one tertiary alcohol with this molecular formula:

A is therefore $(CH_3)_2C(OH)CH_2CH_3$, 2-methylbutan-2-ol.

Compound **B** gives an acidic product on oxidation, so it must be a primary alcohol, being oxidised to a carboxylic acid. Since it forms pent-1-ene on dehydration, **B** must be $CH_3CH_2CH_2CH_2CH_2OH$, pentan-1-ol.

Compound **C** is a secondary alcohol, being oxidised to a ketone, which is neutral. It gives the same alkene on dehydration as does compound **A**.

Compound **C** must be $(CH_3)_2CHCH(OH)CH_3$, 3-methylbutan-2-ol.

Compound **C** contains one chiral carbon atom.

Unit 18

18.01

1.

2

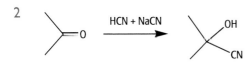

18.02

If **A** and **B** give orange precipitates with 2,4-DNPH, they must be carbonyl compounds. Compound **C** is likely to contain a C=C double bond, as it decolorises bromine water – it could therefore be an alkene-alcohol. Since **A** gives a silver mirror when warmed with Tollen's reagent, it must be an aldehyde.

All three contain a straight chain of four carbon atoms; in compound **A** the oxygen is attached to the end carbon atom, whereas in compounds **B** and **C** the oxygen is attached to the second carbon atom in the chain.

The structures are therefore as follows:

A is butanal, $CH_3CH_2CH_2CHO$

B is butanone, $CH_3CH_2COCH_3$

C is but-3-ene-2-ol, $CH_2=CHCH(OH)CH_3$

Note that compounds containing an –OH group on a C=C double bond are not usually stable, so **C** cannot be either $CH_3CH=C(OH)CH_3$ or $CH_3CH_2C(OH)=CH_2$.

Unit 19

19.01

1 The atomisation energy of $Na(s) = +107$ kJ mol^{-1}

The first ionisation energy of $Na(g) = +496$ kJ mol^{-1}

The atomisation energy of $Cl_2(g) = +121$ kJ mol^{-1}

The electron affinity of $Cl(g) = -349$ kJ mol^{-1}

The enthalpy of formation of $NaCl(s) = -411$ kJ mol^{-1}

2 a KBr, because Br$^-$ is smaller than I$^-$.

b Na_2O because O^{2-} is more highly charged than Br$^-$.

c MgO because O^{2-} is smaller than S^{2-}.

3 $\Delta H_f[CaO(s)] = \Delta H_{latt}[CaO(s)] + \Delta H_{at}[O(g)] + \Delta H_{ea}[O(g)] + \Delta H_{ea}[O^-(g)] + \Delta H_{at}[Ca(s)] + \Delta H_{ion}[Ca(g)] + \Delta H_{ion}[Ca^+(g)]$

$-635 = \Delta H_{latt}[CaO(s)] + 249 - 141 + 798 + 178 + 590 + 1145$

$\Delta H_{latt}[CaO(s)] = -635 - 249 + 141 - 798 - 178 - 590 - 1145 = -3454$ kJ mol^{-1}

19.02

1 Mixture is partially soluble; only the nitrate dissolves.

2 Brown gas is produced; nitrogen dioxide is produced. Glowing splint relights; oxygen is produced from thermal decomposition of a nitrate.

3 Effervescence; $BaCO_3$ does not decompose; produces $CO_2(g)$ with the acid.

4 Indicator colour is purple; BaO dissolves to become $Ba(OH)_2$.

19.03

1 $CoSO_4$ is soluble. The ionic radius of Co^{2+} is smaller than the ionic radius of Mg^{2+}. The lattice energy of $CoSO_4$ is more exothermic than the lattice energy of $MgSO_4$ and the hydration enthalpy of Co^{2+} is more exothermic than the hydration energy of Mg^{2+}. The ΔH_{sol} of $CoSO_4$ is more negative than the ΔH_{sol} of $MgSO_4$.

2 $Co(OH)_2$ is insoluble. The ionic radius of Co^{2+} is smaller than the ionic radius of Mg^{2+}. As OH^- is a smaller anion than SO_4^{2-}, the ΔH_{sol} of $Co(OH)_2$ is more positive than the ΔH_{sol} of $Mg(OH)_2$.

Unit 20

20.01

1 140 cm^3

2 Cl_2; some of this gas dissolves in the solution so the volume collected is smaller than expected.

3 33.6 kg

4 3.22 × 10²³

20.02

1 a $E^\ominus_{cell}$ = 0.46 V; Ag is the positive electrode; Cu is the negative electrode which is where oxidation takes place.

   b $E^\ominus_{cell}$ = 1.36 V; Cl_2 is the positive electrode; oxidation takes place at the hydrogen electrode.

2 No.

3 Yes.

20.03

1 1.1 V

2 1.01 V

3 1.49 V

Unit 21

21.01

1 a [H_3O^+(aq)] = 0.001 mol dm⁻³; pH = 3.0

   b [H_3O^+(aq)] = 0.05 mol dm⁻³; pH = 1.3

2 a [H_3O^+(aq)] = 1.26 × 10⁻³ mol dm⁻³; pH = 2.9

   b [H_3O^+(aq)] = 8.0 × 10⁻⁷ mol dm⁻³; pH = 6.1

3 a [H_3O^+(aq)] = 2.0 × 10⁻¹² mol dm⁻³; pH = 11.7

   b [H_3O^+(aq)] = 1.0 × 10⁻¹¹ mol dm⁻³; pH = 11.0

21.02

1 any

2 any with pKI_n less than 7

3 any

4 phenolphthalein

21.03

1

Acid	K_a / mol dm⁻³	Concentration of acid in solution / mol dm⁻³	Concentration of sodium salt in solution	[H_3O^+(aq)] / mol dm⁻³	pH
CH_3COOH	1.7 × 10⁻⁵	0.1	0.5 mol dm⁻³	**3.4 × 10⁻⁶**	**5.5**
HCOOH	1.6 × 10⁻⁴	1.0	**0.51 mol dm⁻³**	3.2 × 10⁻⁴	3.5
C_6H_5OH	**1.28 × 10⁻¹⁰**	0.1	0.2 mol dm⁻³	6.4 × 10⁻¹¹	**10.2**
CH_3COOH	1.7 × 10⁻⁵	0.5	2.5 g in 100 cm³	**2.8 × 10⁻⁵**	**4.6**

2 pH = $-\log_{10}$(3.2 × 10⁻⁴) = 3.49

3 [H_3O^+(aq)] = 1.58 × 10⁻⁵ mol dm⁻³; CH_3COOH seems most appropriate.

$$\frac{[CH_3COO^-(aq)]}{[CH_3COOH(aq)]} = 1.07:1$$

Prepare a solution of ethanoic acid of concentration 0.1 mol dm⁻³.

In 100 cm³ of the ethanoic acid solution, dissolve 0.877 g of sodium ethanoate.

21.04

1 $K_{sp}[NiCO_3]$ = [Ni^{2+}][CO_3^{2-}] mol² dm⁻⁶

   $K_{sp}[Cr(OH)_3]$ = [Cr^{3+}][OH^-]³ mol⁴ dm⁻¹²

   $K_{sp}[PbI_2]$ = [Pb^{2+}][I^-]² mol³ dm⁻⁹

   $K_{sp}[PbCrO_4]$ = [Pb^{2+}][CrO_4^{2-}] mol² dm⁻⁶

   $K_{sp}[Mg(OH)_2]$ = [Mg^{2+}][OH^-]² mol³ dm⁻⁹

   $K_{sp}[AgBr]$ = [Ag^+][Br^-] mol² dm⁻⁶

2 0.0214 mol dm⁻³

3 3.9 × 10⁻³ mol dm⁻³

Unit 22

22.01

1. zero order: $mol\,dm^{-3}\,s^{-1}$

   second order: $mol^{-1}\,dm^3\,s^{-1}$

2. first order

3. $m = 2$

22.02

1. first two half-lives = 125 s; $t_{1/2}$ is constant, so first order
2. $k = 1.25\,mol^{-2}\,dm^6\,s^{-1}$

22.03

1. $k = 8.82\,mol^{-2}\,dm^6\,s^{-1}$
2. rate = $k[NO(g)]^2[Br_2(g)]$

22.04

1. $O_3 + 2NO_2 \rightarrow O_2 + N_2O_5$
2. rate = $k[O_3][NO_2]$

Unit 23

23.01

1. a and e
2. $-200\,J\,mol^{-1}\,K^{-1}$

23.02

ΔS is positive as there are more gaseous products than reactants.

$\Delta G = \Delta H - T\Delta S$; both ΔH and $T\Delta S$ are positive.

- At low temperatures, $T\Delta S$ is less than ΔH, ΔG is positive; the reaction is not spontaneous.
- At high temperatures, ΔH is less than $T\Delta S$, ΔG is negative; the reaction is spontaneous.

4. a 31
   b 1.6 mg
   c 0.18 mg left in the water; 49.8 mg extracted; 1.4 mg

Unit 24

24.01

1. $Cu = +1$, $1s^2\,2s^2\,2p^6\,3s^2\,3p^6\,3d^{10}$
2. $Fe = +3$, $1s^2\,2s^2\,2p^6\,3s^2\,3p^6\,3d^5$
3. $Cr = +6$, $1s^2\,2s^2\,2p^6\,3s^2\,3p^6$
4. $V = +4$, $1s^2\,2s^2\,2p^6\,3s^2\,3p^6\,3d^1$
5. $V = +3$, $1s^2\,2s^2\,2p^6\,3s^2\,3p^6\,3d^2$
6. $Cu = +2$, $1s^2\,2s^2\,2p^6\,3s^2\,3p^6\,3d^9$

24.02

1. a $2V^{3+}(aq) + Zn(s) \rightarrow 2V^{2+}(aq) + Zn^{2+}(aq)$

   b The electrode potential for the reduction is less negative than the electrode potential of the oxidation so the reaction proceeds.
   $E^{\ominus}_{cell} = +0.5\,V$

2. $2CrO_4^{2-}(aq) + 2H^+(aq) \rightleftharpoons Cr_2O_7^{2-}(aq) + H_2O(l)$

24.03

1. octahedral
2. $Fe = +2$; $Cu = +1$; $V = +2$; $Cr = +3$; $Mn = +4$; $Co = +2$

3.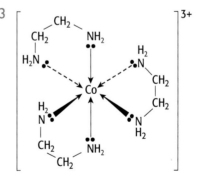

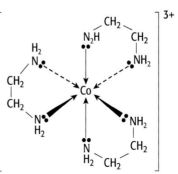

24.04

1. ox. no [Ti] is +4. Electron configuration = $1s^2 2s^2 2p^6 3s^2 3p^6$ so there are no d electrons and therefore no absorption in the visible region.

2. a $4I^- + 2Cu^{2+} \rightarrow 2CuI + I_2$

   b The blue colour of the $CuSO_4$(aq) disappears and a white precipitate of CuI is formed; this is coloured brown as iodine is also produced.

3. a ox. no. [Co] = +3

   b $1s^2 2s^2 2p^6 3s^2 3p^6 3d^6$

   c none

Unit 25

25.01

1. $CH_3CH_2Br + FeBr_3 \rightarrow CH_3CH_2^+ + FeBr_4^-$

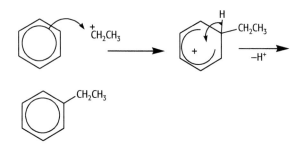

2. $CH_3COCl + AlCl_3 \rightarrow CH_3CO^+ + AlCl_4^-$

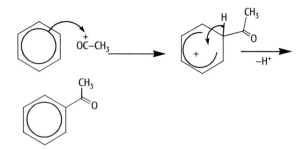

25.02

1. [structure: chlorobenzene with acetyl group]

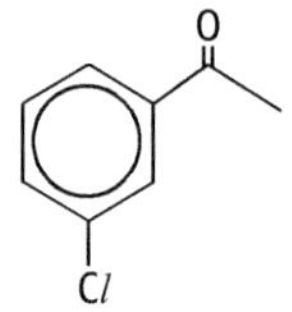

2. [structures: 4-nitrophenol + 2-nitrophenol]

   (if phenol is nitrated with a mixture of H_2SO_4 and HNO_3, trisubstitution usually occurs: 2,4,6-trinitrophenol)

3. [structures: 2-bromocumene and 2-bromocumene]
   CH(CH₃)₂ Br (and CH(CH₃)₂ Br)

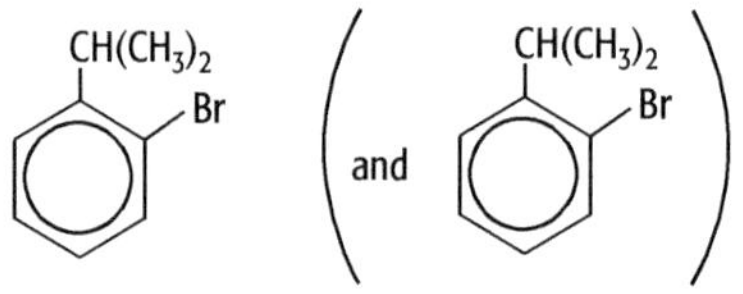

25.03

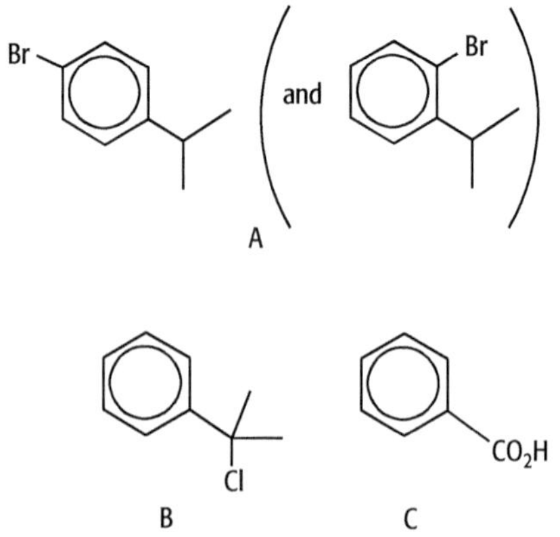

A [4-bromocumene and 2-bromocumene]

B [2-chloro-2-phenylpropane]

C [benzoic acid, CO_2H]

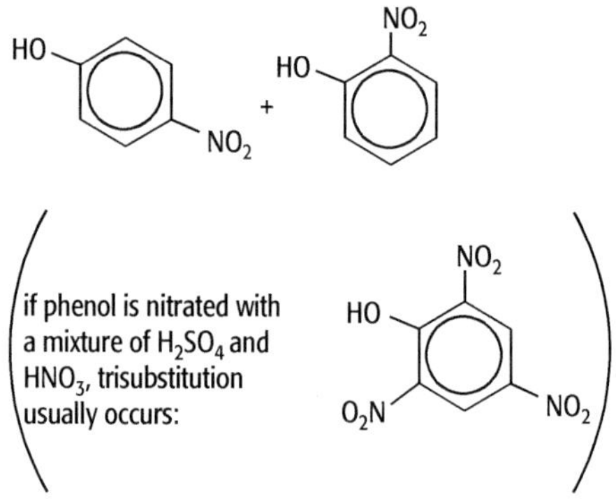

Unit 26

26.01

1. Iodine is less electronegative than chlorine, but more electronegative than hydrogen. There the pK_a of iodoethanoic acid should be between that of chloroethanoic acid (2.87) and ethanoic acid (4.76). Actual value = 3.16.

2. Fluorine is more electronegative than chlorine, so the pK_a of fluoroethanoic acid should be (slightly) lower than that for chloroethanoic acid. Actual value = 2.57.

3. With three electronegative fluorine atoms, the pK_a of trifluoroethanoic acid should be very low. Actual value = 0.59.

26.02

1. a React $(CH_3)_2CHCO_2H$ with $SOCl_2$ to give $(CH_3)_2CHCOCl$.
   b React this acyl chloride with $(CH_3)_2CHNH_2$ to give the target amide.

2. a React $CH_3CH_2CO_2H$ with $SOCl_2$ to give CH_3CH_2COCl.
   b React this acyl chloride with 4-nitrophenol to give the target ester.

Unit 27

27.01

Remember that diazo dyes are formed from a diazotised amine coupled with a phenol (in alkaline solution) or another aromatic amine. The phenol or amine group *stays* in the product, so the original diazotised amine is the half of the molecule on the *other* side of the –N=N– group to the phenol / amine.

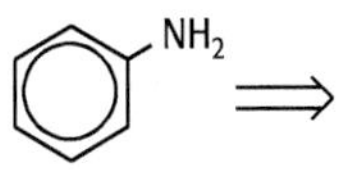

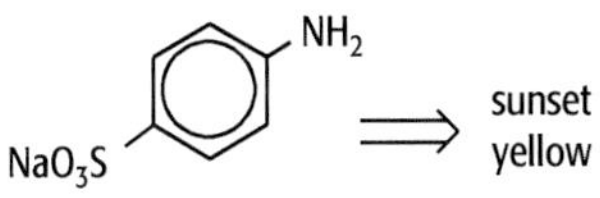

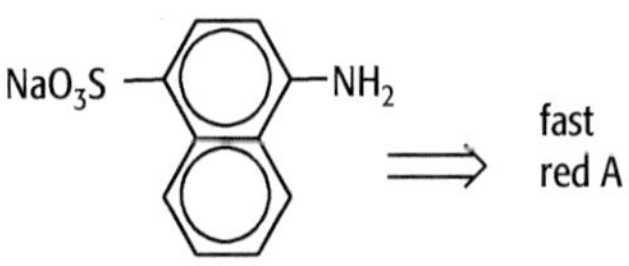

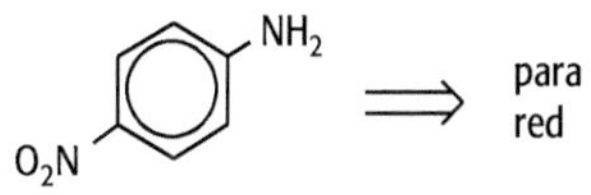

aniline ⇒ yellow

NaO₃S–C₆H₄–NH₂ ⇒ sunset yellow

NaO₃S–(naphthyl)–NH₂ ⇒ fast red A

O₂N–C₆H₄–NH₂ ⇒ para red

27.02

1. (Cys–Glu structure shown)

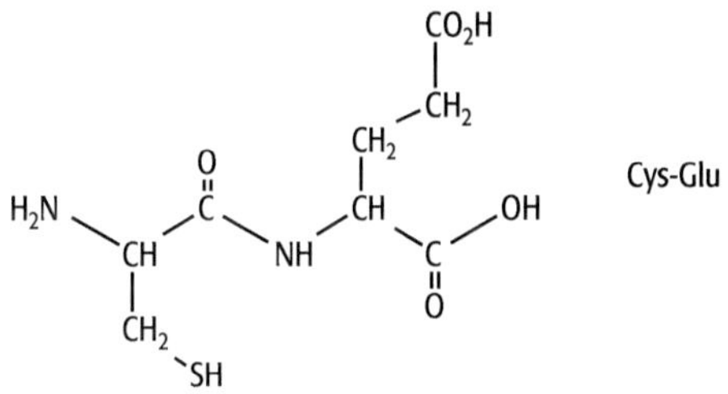

2. The peptide is Ser–Ala–Gly–Glu–Ala, so the amino acids obtained on hydrolysis will be serine (1 molecule), alanine (2 molecules), glycine (1 molecule) and glutamic acid (1 molecule).

Unit 28

28.01

1. a (polymer structure)

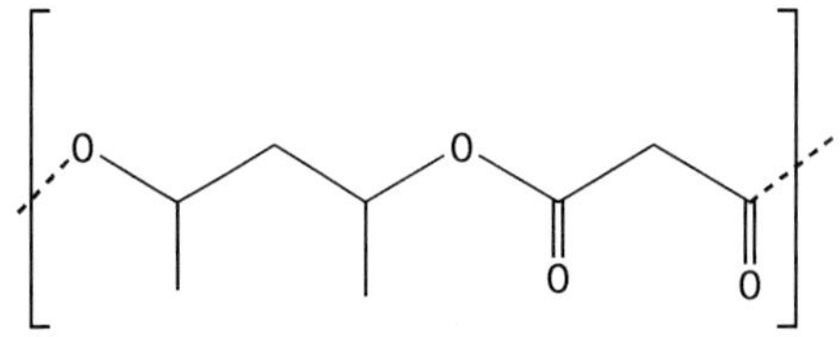

   b (polymer structure)

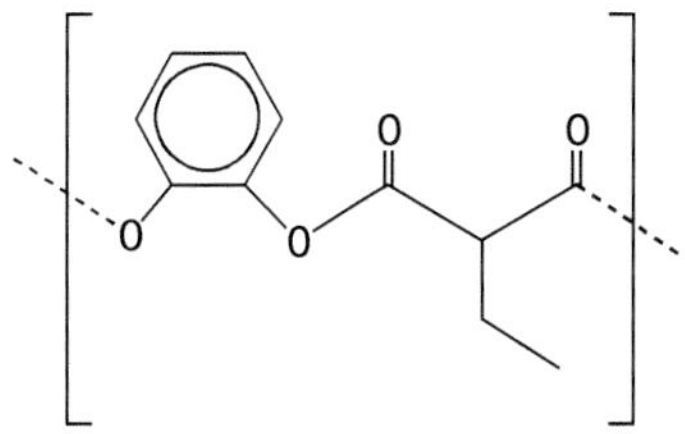

2

HO—C(CH₃)₂—OH and

HO—CH₂CH₂—C(=O)—OH (with another C=O)

(structures: 2,2-dimethylpropane-1,3-diol and succinic acid-like structure)

28.02

1 $CH_2=CH-CN$ addition polymerisation
2 $HOCH_2CH_2OH$ and $HO_2CCH_2CH_2CO_2H$
   condensation polymerisation
3 $NH_2CH_2CH_2CH_2CO_2H$
   condensation polymerisation

28.03

1 $HOCH_2CH_2CO_2H$ and $HOCH_2CO_2H$

2 $H_2N-C_6H_4-NH_2$ and $HO_2C-C_6H_4-CO_2H$ or
   $ClCO-C_6H_4-COCl$

3 a $HC\equiv C-COCH_3$
   b $CH_3CH(OH)CH_2CO_2H$

Unit 29

29.01

Compound **B** shows a peak at 2900 cm⁻¹ (C–H), a strong peak at 1750 cm⁻¹ (C=O) and a strong peak at 1250 cm⁻¹ (C–O). This fits with **B** being $CH_3CO_2CH_3$ (the spectrum might also fit with **B** being CH_3OCH_2CHO).

Compound **C** shows a very broad peak from 2500–3300 cm⁻¹ (O–H in CO_2H), a superimposed small peak at 2900 cm⁻¹ (C–H), a strong peak at 1710 cm⁻¹ (C=O) and a strong peak at 1230 cm⁻¹ (C–O). This fits with **C** being $CH_3CH_2CO_2H$.

29.02

1 $n = 6.005$, so 6 carbon atoms. Each carbon atom has a mass of 12 amu, so carbon contributes $6 \times 12 = 72$ to the M_r. The masses of oxygen + hydrogen have to add up to $132 - 72 = 60$. This is not large enough for 4 × oxygen atoms ($4 \times 16 = 64$). If there were just two oxygen atoms, that would leave $60 - (2 \times 16) = 28$ hydrogen atoms, which is too many for the number of carbon atoms. So we assume there are 3 × oxygen atoms. $3 \times 16 = 48$, leaving $60 - 48 = 12$ for the hydrogen atoms.

Suggested molecular formula is $C_6H_{12}O_3$.

2 m/e 74 is the molecular ion, so m/e 57 is due to the loss of 17 (–OH?).

m/e 45 could be the loss of CHO, leaving C_2H_5O, or the loss of C_2H_5 leaving CO_2H.

The three peaks at 27, 28 and 29 could be C_2H_5, C_2H_4, C_2H_3 (m/e 29 could also be CHO).

Conclusion: this compound could contain an ethyl group and an –OH group. Propanoic acid fits with the data:

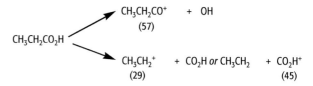

The data could also fit with $HOCH_2CH_2CHO$: $CH_3CH_2CHO = 57$; $HOCH_2CH_2 = 45$; $CHO = 29$

29.03

1 Pentan-2-one will have five peaks, as each carbon atom is different. Pentan-3-one will have just three peaks, as there is symmetry in the molecule.

2 The spectrum shows only four peaks, so two carbon atoms must be in the same environment. Compound is $(CH_3)_2CHCOCH_3$.

29.04

There are three peaks, so there are three types of hydrogen in the molecule.

The peak at δ 1.0 is a triplet, so is a –CH_3 group split by an adjacent –CH_2– group.

The peak at δ 2.2 is a quartet, so is a –CH$_2$– group split by an adjacent CH$_3$ group.

The peak at δ 11.8 is a singlet, so is the H in a –CO$_2$H group.

Structure of **D** is CH$_3$CH$_2$CO$_2$H.

The peak at δ 11.8 would disappear with D$_2$O.

Unit 30

30.01

1   CH$_3$CH$_2$OH $\xrightarrow[\text{distil immediately}]{Cr_2O_7^{2-} + H^+}$ CH$_3$CHO $\xrightarrow{HCN + NaCN}$ CH$_3$CH(OH)CN $\xrightarrow{\text{heat with H}^+\text{(aq)}}$ CH$_3$CH(OH)CO$_2$H

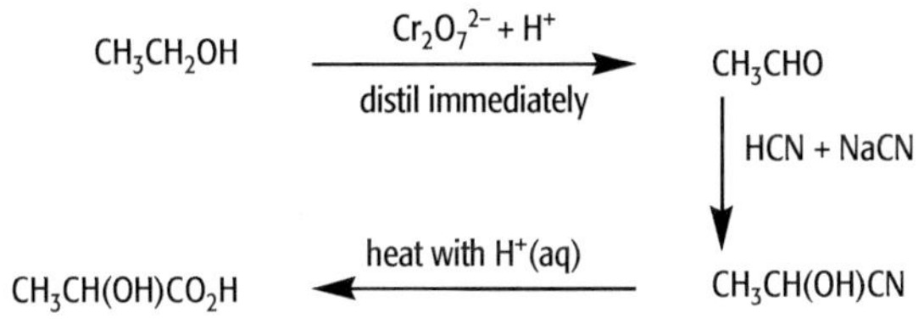

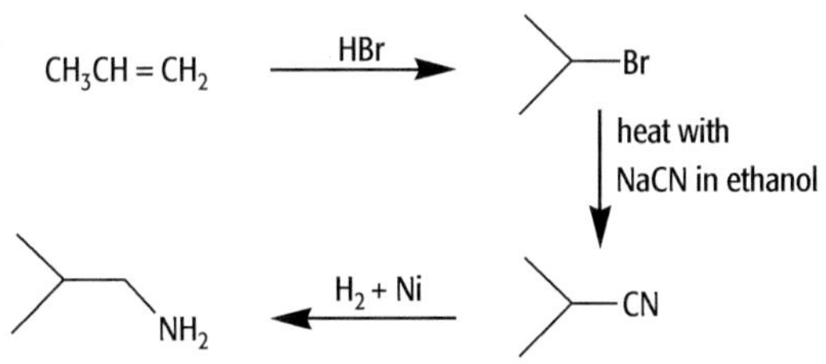

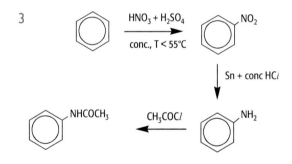

30.02

1   a one

    b two

2

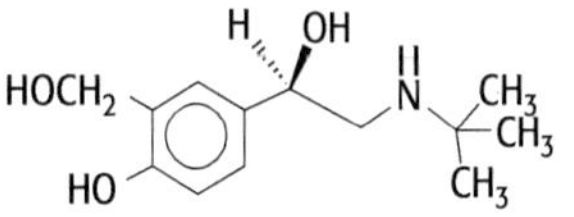

Isomer **A** is likely to be more physiologically active, since it has the same stereochemistry at the secondary alcohol group as adrenaline.

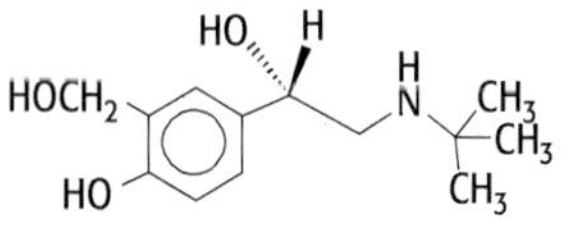

30.03

1   a **A** is CH$_3$CH(OH)CH$_2$CO$_2$H.

    b **B** is CH$_2$=CHCOCH$_3$.

2   a Either warm with Fehling's solution: first will form red ppt.; no reaction with second.

       Or add I$_2$ + OH$^-$(aq): no reaction with first; second will give pale yellow ppt.

    b Either add Br$_2$(aq): first will decolorise and give white ppt.; no reaction with second.

       Or add universal indicator solution: no (or very little) reaction with first; turns blue with second.

    c Either add water drops: fizzing occurs with first; no reaction with second.

       Or warm with Cr$_2$O$_7^{2-}$ + H$^+$(aq): no reaction with first; turns from orange to green with second.

30.04

I   electrophilic addition

II nucleophilic substitution

III oxidation

IV nucleophilic addition

V   nucleophilic substitution

VI reduction

Advanced practical skills

P.01

Decide on the following.

Volume of HCl solution:
- too much and it will overflow the beaker
- too little and the inaccuracy in measurement becomes too great
- measure it into the beaker using the graduated cylinder.

How much metal?
- it should all dissolve in the HCl
- work out the number of moles of acid taken: 50 cm^3 of 1.0 mol dm^{-3} HCl contains 0.05 moles
- the equation (M + 2HCl → MCl_2 + H$_2$) enables you to calculate the maximum number of moles of metal
- using the A_r of zinc and iron, you can work out the maximum mass of metal to be added
- these masses need to be weighed separately using the balance

What to do:
- the initial temperature of the acid needs to be measured
- add one of the metals and stir, noting the maximum temperature reached.

P.02

First identify the functional groups in the four compounds – there is an ether, a primary alcohol, a secondary alcohol and a tertiary alcohol.

The series of tests are as shown in the table.

	Add Na	Warm H$^+$ / Cr$_2$O$_7^{2-}$	NaOH / I$_2$ then warm
CH$_3$CH$_2$OCH$_2$CH$_3$ ether	no reaction		
CH$_3$CH$_2$CH$_2$CH$_2$OH primary alcohol	bubbles of hydrogen	orange to green	no reaction
CH$_3$CH$_2$CH(OH)CH$_3$ secondary alcohol	bubbles of hydrogen	orange to green	yellow precipitate
(CH$_3$)$_3$COH tertiary alcohol	bubbles of hydrogen	no change	
Conclusion	only the three alcohols react	tertiary alcohols are not oxidised	molecule containing CH(OH)CH$_3$ undergoes the triiodomethane reaction

Safety issues:
- Na and NaOH are corrosive
- $Cr_2O_7^{2-}$ is harmful
- wear safety spectacles
- use small amounts of chemicals
- dispose of $Cr_2O_7^{2-}$ as directed

Apparatus:
- suitable test tubes
- dropping pipettes for adding liquids

What to do:
- keep the temperature the same for the same test
- immerse the test tubes in a water bath

Answers to Exam-style questions

Unit 1

1 a i $Na_2CO_3 \cdot xH_2O + 2HCl \longrightarrow 2NaCl + CO_2 + (x+1)H_2O$ [2]

   ii 9.75×10^{-3} [1]

   iii 4.875×10^{-3} [1]

   iv 9.50 [3]

b 0.570 g [1]

c i $117 cm^3$ [1]

   ii Some gas will dissolve in the water [1]

   Total: 10

2 a $C_4H_3O_2$ [6]

b i 1.81×10^{-4} [1]

   ii 9.05×10^{-5} [1]

   iii 9.05×10^{-3} [1]

   iv 166 [1]

c $C_8H_6O_4$ [2]

Total: 12

Unit 2

1 a i proton:neutron:electron = +1:0:−1 [1]

   ii Isotopes are atoms which have the same number of protons but different numbers of neutrons in the nucleus. [2]

b i

Symbol	Number of neutrons	Number of protons	Number of electrons	
$^{23}_{11}A$	12	11	11	[1]
$^{22}_{10}B$	12	10	10	[1]
$^{27}_{13}C$	14	13	13	[1]
$^{25}_{12}D^+$	13	12	11	[1]
$^{26}_{12}E^{2+}$	14	12	10	[1]

   ii D and E [1]

   iii C [1]

Total: 10

2 a i two protons and two neutrons (an alpha particle) [2]

   ii

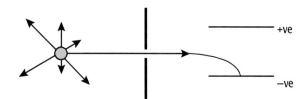

radioactive source   slit system   charged plates [2]

b i

Symbol	Number of protons	Number of neutrons
$^{234}_{90}Th$	90	144
$^{234}_{91}Pa$	91	143

[2]

   ii mass = 0; charge = −1 [2]

   iii an electron [1]

Total: 9

Unit 3

1 a i and ii

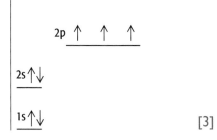

[3]

   iii

[2]

b i Cl $1s^2 2s^2 2p^6 3s^2 3p^5$ [1]

      Cl^- $1s^2 2s^2 2p^6 3s^2 3p^6$ [1]

   ii Ar [1]

Total: 8

2 a i Be = 900 kJ mol^{-1}

      B = 799 kJ mol^{-1} [1]

ii B less than Be as, although it has one more proton, the outer electron is in a 2p orbital which is slightly further from the nucleus than the 2s orbital. [2]

b $M^+(g) \longrightarrow M^{2+}(g) + e^-$ [1]

c Group 17, the halogens [1]

d i $F^-(g) \longrightarrow F(g) + e^-$
$F(g) \longrightarrow F^+(g) + e^-$ [2]

ii The ionisation energy of F is larger as the nucleus is attracting fewer electrons. [2]

Total: 9

Unit 4

1 a i CO_2 is linear as C is surrounded by two sets of bond pairs; SO_2 is bent as S is surrounded by two sets of bond pairs and a lone pair. [4]

ii NH_3 is pyramidal as N is surrounded by three bond pairs and a lone pair; BF_3 is a planar triangle as B is surrounded by three bond pairs only. [4]

iii CH_4 is tetrahedral as C is surrounded by four bond pairs; SF_4 is

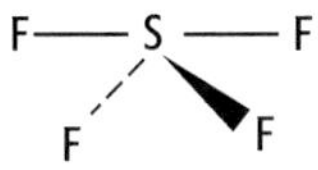

as S is surrounded by four bond pairs and a lone pair. [4]

b i F_2O [1]

ii BeF_2 SO_3 F_2O SF_6 [2]

Total: 15

2 a i [2]

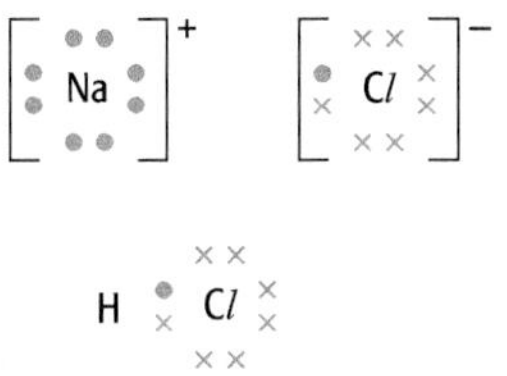

[diagram of Na+ with dots, Cl- with crosses and dots]

[diagram of H with dot, Cl with crosses and dots]

ii NaCl has strong electrostatic forces of attraction between the ions and a lot of energy is needed to overcome them; HCl exists as small covalent molecules and dipole–dipole forces attract them together which are much weaker than the ionic bonds. Less energy is needed to overcome them and so HCl has a much lower melting point. [3]

b C–C in ethane is a single bond; C=C in ethene is a double bond. The double bond, with two electron pairs, is stronger and shorter than the single bond. [2]

c i C–H is a σ bond; C=C is a σ and a π bond. The π bond has electron density above and below the plane of the molecule but the σ bond has electron density between the atoms. [4]

ii [diagrams of two isomers showing H₃C, H, CH₃ around C=C] [1]

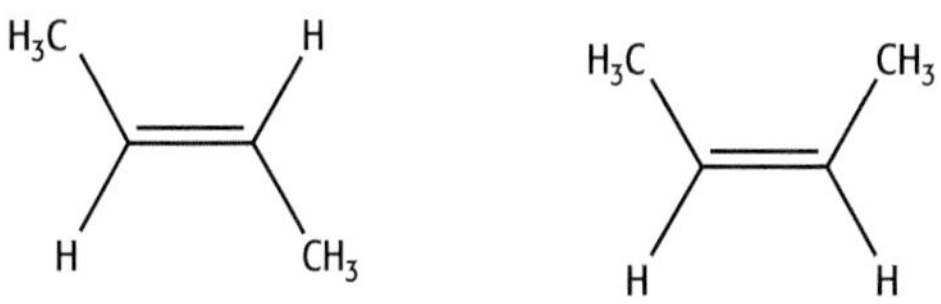

The π bond prevents the twisting of the double bond. [2]

iii No. Although the double bond cannot twist, having a CH_2 at one end of the double bond means only one compound can be formed. [1]

Total: 15

3 a It is a planar equilateral triangle. A pair of electrons on each O atom is in a p orbital and they overlap to form a molecular π orbital. [4]

b eight [1]

Total: 5

4 a NaI exists as a lattice of positive and negative ions held together by the electrostatic attraction between them. [2]

NaI does not conduct electricity as the ions are fixed in place and cannot move. [1]

Cu exists as a lattice of positive ions surrounded by delocalised electrons. It is held together by the attraction between positive ions and electrons. [2]

Cu conducts electricity as the delocalised electrons are free to move. [1]

b If the NaI was dissolved in water or if it was melted, it could conduct. [1]

Total: 7

Unit 5

1. a i No attractive forces between particles; negligible volume of particles; elastic collisions. [3]

   ii High temperature and low pressure. [2]

   iii NH_3 CO N_2 [1]

   NH_3 has hydrogen bonding; CO has a small dipole; N_2 has same number of electrons as CO (hence similar van der Waals' forces) but no dipole. [3]

   b 0.147 g [3]

   Total: 12

2. a NaI forms an ionic lattice with Na^+ surrounded by I^- ions and I^- ions surrounded by Na^+ ions. The lattice is held together by strong electrostatic attractions between the ions. [2]
   NaI has a high melting point, is brittle, has no electrical conductivity (unless at a high enough temperature to melt the salt or it is dissolved in water). [2]

   I_2 forms a lattice of I_2 molecules; although the atoms are held together by strong covalent bonds, the molecules are held together by weak van der Waals' forces. [2] The melting point / sublimation temperature is low and there is no electrical conductivity. [2]

   b i Both are small molecules and CH_3CH_3 only has weak van der Waals' forces between the molecules. [1] The boiling point of CH_3NH_2 is higher because there is a lone pair on N and also δ^+H so hydrogen bonds can also be formed which are stronger than van der Waals' forces. [2]

   ii Both are small tetrahedral molecules with van der Waals' forces between them. [1] $SnBr_4$ has more electrons and van der Waals' forces are stronger. [2]

   iii CO_2 is linear with no dipole and only van der Waals' attractions between molecules. [1] SO_2 is bent with a small dipole and also there are more electrons so van der Waals' forces will be stronger and there are permanent dipole forces as well. [2]

   c X = silicon, Y = graphene, Z = sulfur [2]

   Total: 19

Unit 6

1. a i $+53 \text{ kJ mol}^{-1}$ [2]

   ii

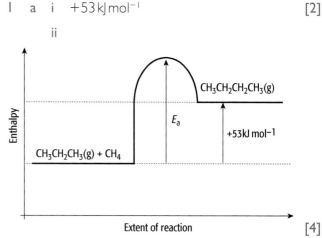

   [4]

   b i 43 kJ mol^{-1} [2]

   ii Bond energies are only average values. [1]

   c i -16 kJ mol^{-1} [2]

   ii The activation energy is very high. [1]

   Total: 12

2. a 0.675 g [2]

   b Some energy is lost to the surroundings / container. [1]

   Total: 3

Unit 7

1. a $2MnO_4^- + 5HXeO_4^- + 2H_2O \rightarrow 2Mn^{2+} + 9H^+ + 5XeO_6^{4-}$ [2]

   b i Cr, from +6 to +3 [2]

   ii N, from -1 to 0 [2]

   Total: 6

2. a $BrO_3^- + 2OH^- \rightarrow BrO_4^- + H_2O + 2e^-$
   $F_2 + 2e^- \rightarrow 2F^-$ [2]

   b i Sn, from 0 to +4 [2]

   ii S, from +6 to +4 [2]

   iii $Sn + 4H_2SO_4 \rightarrow Sn(SO_4)_2 + 4H_2O + 2SO_2$

   Total: 6

Unit 8

1. a $PV = nRT$

   $n = \dfrac{101 \times 10^3 \times 50 \times 10^{-6}}{8.31 \times 350} = 1.74 \times 10^{-3}$ [1]

b $n = 2.95 \times 10^{-3}$ [1]

c   Change in $n = 1.21 \times 10^{-3}$ [1]

d   i $n(N_2O_4) = 0.53 \times 10^{-3}$

    ii $n(NO_2) = 2.42 \times 10^{-3}$ [2]

e   i $p_{N_2O_4} = \dfrac{0.53 \times 101}{2.95}$ kPa = 18.15 kPa

    ii $p_{NO_2} = \dfrac{2.42 \times 101}{2.95}$ kPa = 82.85 kPa [2]

f   i $K_p = \dfrac{(p_{NO_2})^2}{p_{N_2O_4}}$ [1]

    ii $K_p = \dfrac{(82.85)^2}{18.15} = 378$ KPa [2]

Total: 10

2 a i   low temperature and high pressure [1]

      ii Exothermic reaction left to right and so low temperature moves equilibrium to right.

          High pressure moves the equilibrium to the side with fewer moles, which is the right. [2]

      iii High pressure, but cannot be too high for safety considerations.

          Too low a temperature means the reaction proceeds too slowly. [2]

   b $NH_3(aq) + H_2O(l) \rightleftharpoons NH_4^+(aq) + OH^-(aq)$ [1]

   c i $NH_4^+(aq) + OH^-(aq) \rightarrow NH_3(g) + H_2O(l)$ [2]

      ii Brønsted–Lowry acid = NH_4^+

          Brønsted–Lowry base = OH^- [2]

Total: 10

Unit 9

1 a by following the increase in the volume of gas [1]

   b For a reaction to take place the particles need to collide; oppositely charged ions collide and react when $AgCl$ forms.

      Molecules of N_2O_5 need to collide with energy greater or equal to E_a. The E_a is large and so few molecules have energy greater than E_a and can react. [5]

Total: 6

2 a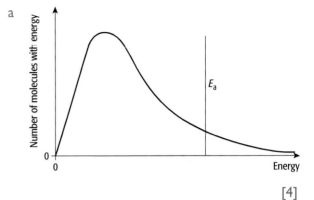
   [4]

   b Add a line to the right of E_a labelled C. [1]

   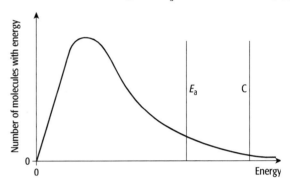

   c The maximum is higher and to the left of the original maximum. The two curves should only cross each other once. [2]

   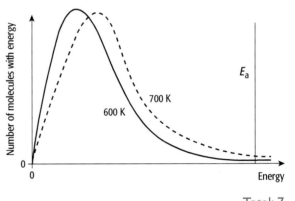

Total: 7

Unit 10

1 a i   sodium or magnesium [1]

      ii giant metallic [1]

      iii solid dissolves

          pH = 7 (or just below for $MgCl_2$)

          ions are hydrated [3]

      iv Either sodium oxide dissolves

          pH = 14

          $2Na + H_2O \longrightarrow 2NaOH$

Or magnesium oxide is insoluble or dissolves to only a small extent

pH = 9

$Mg + H_2O \longrightarrow Mg(OH)_2$ [3]

b i silicon / phosphorus / sulfur [1]

ii Silicon is a giant molecule; Phosphorus and sulfur are small molecules [1]

iii they hydrolyse or give HCl fumes
they form acidic solutions
pH = 1 to 3 [3]

iv One of the following: SiO_2 is insoluble; giant covalent structure has many strong covalent bonds to break

P_2O_5 dissolves; acidic solution / pH = 1

SO_2 dissolves; acidic solution / pH = 3–5 [2]

One of the following correctly balanced equations:

$SiCl_4 + 2H_2O \longrightarrow SiO_2 + 4HCl$
$PCl_5 + 4H_2O \longrightarrow H_3PO_4 + 5HCl$
$P_2O_5 + 3H_2O \longrightarrow 2H_3PO_4$
$SO_2 + H_2O \longrightarrow H_2SO_3$ [1]

Total: 16

2 a i It increases. [1]

ii At left-hand side, electronegativity difference between atom and O is large and this leads to ionic bonds.

At right-hand side, electronegativity difference is much smaller and so covalent bonds are formed. [2]

b i (Both are metals) Al has more delocalised electrons than Mg and so the electrostatic attractions between the electrons and positive ions are greater. [2]

ii Both are small molecules and only weak van der Waals' forces exist between molecules. S_8 has more electrons than P_4 and so has stronger van der Waals' forces. (Any mention of breaking the covalent bonds will negate the marks.) [2]

Total: 7

Unit 11

1 a i $X = Ba(NO_3)_2(aq)$
$Y = BaCl_2(aq)$

$Z = Ba(OH)_2(aq)$
$M = BaCO_3(s)$ [4]

ii $BaO(s) + 2HNO_3(aq) \longrightarrow Ba(NO_3)_2(aq) + H_2O(l)$

$BaO(s) + 2HCl(aq) \longrightarrow BaCl_2(aq) + H_2O(l)$

$BaO(s) + H_2O(l) \longrightarrow Ba(OH)_2(aq)$

$BaCl_2(aq) + Na_2CO_3(aq) \longrightarrow BaCO_3(s) + 2NaCl(aq)$ [4]

b M – no reaction

X – $2Ba(NO_3)_2 \longrightarrow 2BaO + 4NO_2 + O_2$ [3]

Total: 11

2 a i Radius of Mg^{2+} is smaller than the radius of Mg. A complete shell of electrons is lost on formation of the ion. [2]

ii The radius of M^{2+} increases going down Group 2 as another shell of electrons is added. [2]

b i gas / CO_2 evolved [1]

ii gas / CO_2 evolved, N dissolves completely in HCl(aq) [1]

iii On heating, $MgCO_3$ decomposes to form CO_2 but $CaCO_3$ does not at this temperature as it is further down Group 2. N is MgO and $CaCO_3$. On adding HCl(aq), $CaCO_3$ reacts to produce CO_2 gas and $CaCl_2$. Both $CaCl_2$ and MgO dissolve in HCl(aq) to form the chlorides. [4]

Total: 10

Unit 12

1 a The boiling point depends on the strength of the intermolecular forces [1]

For Br_2 and Cl_2, the intermolecular forces are weak van der Waals forces [1]

Br_2 has more electrons and so the van der Waals forces are larger [1]

Both HCl and HF have smaller van der Waals forces than Cl_2 and Br_2 [1]

HCl has stronger dipole – dipole forces [1]

HF has stronger hydrogen bonds between molecules [1]

b   H**I** pH approximately 1; KBr pH = 7   [1]

KBr in solution produces K$^+$(aq) and Br$^-$(aq) / aqueous ions   [1]

H**I** is a strong acid and completely ionises to produce H$_3$O$^+$(aq)   [1]

c   i   white AgC*l*   [1]

     ii   cream AgC*l* + Ag**I**   [2]

     iii   yellow Ag**I**   [1]

d   Purple fumes   [1]

H$_2$SO$_4$ + Ca**I**$_2$ ⟶ 2H**I** + CaSO$_4$

8H**I** + H$_2$SO$_4$ ⟶ H$_2$S + 4**I**$_2$ + 4H$_2$O   [2]

Total: 16

2   a   Br: 1s^2 2s^{2}2p^6 3s^{2}3p^{6}3d^{10} 4s^{2}4p^5

Br$^+$: 1s^2 2s^{2}2p^6 3s^{2}3p^{6}3d^{10} 4s^{2}4p^4 [1]

b   [Na]$^+$ [Br]$^-$ [1]

Br–P–Br with Br   [1]

c   i)   C*l*$_2$(aq) + H$_2$O(l) ⟶ HC*l*(aq) + HOC*l*(aq)   [1]

     ii)   BrC*l* is polar so the C*l* is slightly negative   [1]

BrC*l*(aq) + H$_2$O(l) ⟶ HC*l*(aq) + HOBr(aq)   [1]

d   Number of moles NaOH = 5 × 10^{-3} [1]

Number of moles H$^+$ in 25 cm^3 = 5 × 10^{-3} [1]

0.458 g gives 5 × 10^{-3} moles of acid so 91.6 g gives 1 mole of acid   [1]

Mr(C*l*$_2$O$_7$) = 183   [1]

so 1 mole C*l*$_2$O$_7$ gives 2 moles of acid   [1]

C*l*$_2$O$_7$ + H$_2$O ⟶ 2 HC*l*O$_4$ [1]

Ox no of C*l* = +7   [1]

Total: 13

Unit 13

1   a   i   H–N–H is about 107°; N–N–O is 180°.   [2]

     ii   In N$_2$H$_4$, each N has three bond pairs and a lone pair. There is sufficient bond polarisation for the δ− N to be able to donate the lone pair and accept a proton.

In N$_2$O, the O is more electronegative than N and so pulls the electrons and the N cannot donate the lone pair to a proton.   [4]

b   i   reducing agent   [1]

     ii   M; burn in air / O$_2$

N; bubble through cold water

P; BaC*l*$_2$(aq)

Q; bubble through H$^+$ / Cr$_2$O$_7^{2-}$(aq)

R; BaC*l*$_2$(aq)   [5]

Total: 12

2   a   Ammonia will be released.

Ca(OH)$_2$ + (NH$_4$)$_2$SO$_4$ ⟶ 2NH$_3$ + CaSO$_4$ + 2H$_2$O   [2]

b   i   air at high temperature in a car engine / air at high temperature in lightning

N$_2$ + O$_2$ ⟶ 2NO   [2]

     ii   acid rain

NO$_2$ + SO$_2$ ⟶ NO + SO$_3$

SO$_3$ dissolves in rain to form H$_2$SO$_4$ [3]

     iii   Lakes become acidic and fish die / limestone buildings are corroded / metal is corroded.   [2]

c   i   Nitrates enter lakes / algal bloom prevents sunlight getting through / water plants / fish die.   [2]

     ii   Use less fertiliser / do not use fertiliser before heavy rain / only use fertiliser at appropriate times of the year.   [1]

Total: 12

Unit 14

1   a   i   free radical substitution

     ii   *either* CH$_3$CH$_2$CH$_2$C*l* *or* CH$_3$CHC*l*CH$_3$

     iii   *either* 1-chloropropane *or* 2-chloropropane

b   i   oxidation

     ii   CH$_3$CH$_2$CO$_2$H

     iii   propanoic acid

c   i   elimination

     ii   (CH$_3$)$_2$C=CH$_2$

     iii   2-methylpropene

2 a i CH$_3$CH$_2$CH$_2$CH$_2$Cl + NaOH ⟶ CH$_3$CH$_2$CH$_2$CH$_2$OH + NaCl

ii 1-chlorobutane ⟶ butan-1-ol

other starting materials could be, e.g.
CH$_3$CH$_2$CHBrCH$_3$ + OH$^-$, CH$_3$(CH$_2$)$_3$OH + HCl

b i CH$_3$CH$_2$CH$_2$CHO + HCN ⟶ CH$_3$CH$_2$CH$_2$CH(OH)CN

ii butanal ⟶ (2-hydroxypentanonitrile)
other starting materials could be, e.g.
CH$_3$CH$_2$COCH$_3$ + HCN

c i CH$_3$CH=CHCH$_3$ + Br$_2$ ⟶ CH$_3$CHBrCHBrCH$_3$

ii but-2-ene ⟶ 2,3-dibromobutane
other starting materials could be, e.g.
CH$_3$CH$_2$CH=CH$_2$ + HCl

d i CH$_3$CH$_2$COCH$_3$ + 2[H] ⟶ CH$_3$CH$_2$CH(OH)CH$_3$

ii butanone ⟶ butan-2-ol
other starting materials could be, e.g.
CH$_3$CH$_2$CH=CH$_2$ + 2[H]

Unit 15

1 a *hydrocarbon*: a compound composed of the elements carbon and hydrogen only.

fractional distillation: the separation of a mixture of compounds by means of their different boiling points.

b gasoline (petrol for cars), about C$_5$–C$_{10}$

kerosene (aeroplane fuel), about C$_{10}$–C$_{16}$

gas oil (fuel for diesel engines), about C$_{12}$–C$_{20}$

c alkanes: C$_n$H$_{2n+2}$

alkenes: C$_n$H$_{2n}$

d i heat with steam at 800 °C

ii C$_{12}$H$_{26}$ ⟶ C$_6$H$_{14}$ + 3C$_2$H$_4$

e i Compounds with the same molecular formula but different structural formulae *or* having their atoms bonded in different ways.

ii five

iii Any three of the following:

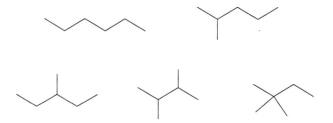

2 a An atom or group of atoms that has an unpaired electron.

b (UV) light

c C$_2$H$_6$ + Cl$_2$ ⟶ C$_2$H$_5$Cl + HCl

d initiation: Cl$_2$ + light ⟶ 2Cl$^•$

propagation: C$_2$H$_6$ + Cl$^•$ ⟶ C$_2$H$_5^•$ + HCl

and C$_2$H$_5^•$ + Cl$_2$ ⟶ C$_2$H$_5$Cl + Cl$^•$

termination: C$_2$H$_5^•$ + Cl$^•$ ⟶ C$_2$H$_5$Cl

or 2C$_2$H$_5^•$ → C$_4$H$_{10}$

or 2Cl$^•$ ⟶ Cl$_2$

e i (CH$_3$)$_2$CHCH$_2$Cl and (CH$_3$)$_3$CCl

ii Ratio will be 9:1; as there are 9 'CH$_3$' hydrogen atoms that could be replaced, compared to 1 'CH' hydrogen atom.

3 a When HX adds to a C=C, the H atom attaches itself to the least substituted carbon atom, to form the most stable carbocation.

b i (CH$_3$)$_2$CHCH(OH)CH$_3$

ii (CH$_3$)$_2$CBrCH$_2$CH$_3$

iii CH$_3$CBr(CH$_3$)CH$_2$CH$_3$

c None of these three alkenes have *cis–trans* isomers because for each alkene, one end of the C=C has two identical atoms or groups.

d (CH$_3$)$_2$CHCH=CH$_2$ gives (CH$_3$)$_2$CHCO$_2$H + CO$_2$

(CH$_3$)$_2$C=CHCH$_3$ gives (CH$_3$)$_2$CO + CH$_3$CO$_2$H

CH$_2$=C(CH$_3$)CH$_2$CH$_3$ gives CO$_2$ + CH$_3$CH$_2$COCH$_3$

Unit 16

1 a reaction 1: nucleophilic substitution

reaction 2: elimination

b

Reaction	Reagent	Conditions
1	NaOH	heat under reflux in water
2	KOH	heat under reflux in ethanol
3	NH_3	heat under pressure in ethanol

c **A** is $CH_3CH_2CH(CH_3)CN$.

B is $CH_3CH_2CH(CH_3)CO_2H$.

2 a i reactivity increases from R–Cl to R–I due to the decreasing strength of the C–Hal bond from R–Cl to R–I.

ii Add a few drops of R–Hal to $AgNO_3$(aq), shake well and allow to stand. R–I will form a pale yellow ppt. within a minute or so; R–Br will form a cream-coloured ppt. after 20 minutes or so; R–Cl will show a slight cloudiness after 1 hour.

b i CFCs are non-poisonous, non-flammable, non-corrosive and are volatile but easily liquefied under pressure.

ii UV light causes the homolytic breaking of a C–Cl bond.

e.g. $CF_2CCl_2 \rightarrow CF_2CCl^{\bullet} + Cl^{\bullet}$

The free $Cl^{\bullet}$ atoms can then react with ozone.

e.g. $Cl^{\bullet} + O_3 \rightarrow OCl^{\bullet} + O_2$

or $OCl^{\bullet} + O_3 \rightarrow Cl^{\bullet} + 2O_2$

Unit 17

1 a i **A** could be $BrCH_2CHO$.

B could be $NC–CH_2CHO$.

C could be $NC–CH_2CH(OH)CN$.

ii step 1: heat with HBr

step 2: heat with KCN in ethanol

step 3: add HCN + trace of NaCN

step 4: heat with H_3O^+ (e.g. H_2SO_4(aq))

b Yes, malic acid exhibits optical isomerism: it has a chiral carbon atom (the one with the OH group attached to it).

c **D** and **E** are the *cis* and *trans* isomers of $HO_2C–CH=CH–CO_2H$. This is geometrical isomerism, due to the C=C double bond not being able to rotate, and their being non-identical groups on each end of the C=C.

2 a i hydrolysis

ii $CH_3CH_2CO_2CH(CH_3)_2 + H_2O \rightarrow$

   G   H

   $CH_3CH_2CO_2H + CH_3CH(OH)CH_3$

b *either* add universal indicator solutions: **G** will turn it red; **H** will stay green

or warm with alkaline aqueous iodine: no reaction with **G**; a pale yellow ppt. with **H**

or warm with acidified $K_2Cr_2O_7$: no reaction with **G**; turns orange to green with **H**

c i **J** is $CH_3CO_2C(CH_3)_3$ (from $(CH_3)_3COH$) or $HCO_2C(CH_3)_2CH_2CH_3$ (from $CH_3CH_2C(CH_3)_2OH$)

ii **K** is $HCO_2CH_2CH(CH_3)CH_2CH_3$

iii **L** is $CH_3CH_2CH(CH_3)CO_2CH_3$

d The alcohol that makes up **F** is butan-2-ol (secondary).

The alcohol that makes up **J** is 2-methylpropan-2-ol or 2-methyl butan-2-ol (both are tertiary).

The alcohol that makes up **L** is methanol (primary).

test: warm with acidified $K_2Cr_2O_7$(aq):

no colour change $\Rightarrow$ tertiary alcohol

colour goes from orange to green $\Rightarrow$ primary or secondary

then test distillate with universal indicator paper:

no colour change (stays green): distillate is ketone from secondary alcohol

colour goes from green to red: distillate is carboxylic acid from primary alcohol

Unit 18

1. a i alcohol, alkene, aldehyde

   ii alcohol: add Na(s); effervescence (of $H_2(g)$) indicates –OH

   alkene: add $Br_2(aq)$ or dil $KMnO_4(aq)$; either reagent is decolorised

   aldehyde: warm with Fehling's solution or Tollens' reagent; gives a red ppt. or a silver mirror

   b $C_9H_{16}O_2$

   c Optical: the carbon that is joined to the –OH group is chiral; it has four different groups / atoms joined to it.

   Geometrical: the C=C in the hydroxynonenal molecule has non-identical groups at each end, so can exist as cis–trans isomers.

   d [reaction scheme: starting hydroxynonenal converted by $NaBH_4$ to diol (C_5H_{11}–CH(OH)–CH=CH–CH$_2$OH); by $Na_2Cr_2O_7 + H_2SO_4$ to ketone carboxylic acid (C_5H_{11}–CO–CH=CH–CO$_2$H with OH shown); by hot acidified $KMnO_4$ to C_5H_{11}–CO–CO$_2$H + (HO$_2$C–CO$_2$H) → ketone carboxylic acid, ethandioic acid or CO_2, CO_2]

2. a Reaction with 2,4-DNPH but not Fehling's suggests a ketone (not aldehyde).

   Reaction with Na but not Na_2CO_3 suggests alcohol (not carboxylic acid).

   Structure is $HOCH_2COCH_2OH$.

   b B is $HOCH_2CH(OH)CH_2OH$.

   C is $HO_2C–CO–CO_2H$.

   D is $(HOCH_2)_2C(OH)CN$.

   E is $(HOCH_2)_2C(OH)CO_2H$.

   c E is $C_4H_8O_5$.

   $M_r = 136$, so 1.00 g is 0.007353 mol

   1 mol gives 2 mol $H_2 = 48\,dm^3$

   so 0.007353 mol gives 0.353 dm³ (353 cm³)

   d [mechanism diagram showing nucleophilic addition of ^-CN to C=O: lone pair, charge and two curly arrows → correct intermediate with charge and lone pair → two curly arrows showing H^+ taken from HCN and ^-CN formed]

Unit 19

1. a The amount of energy released when gaseous ions are brought to their positions to make 1 mole of solid. [2]
   b i second ionisation energy Be(g); bond energy of O_2(g); electron affinity of O^-(g) [3]
   ii all less negative [3]
   c ΔH[BeO(s)] is more negative. [1]

   Total: 9

2. a $\Delta H_{sol} = {}^-\Delta H_{lat} + \Sigma \Delta H_{hyd}$

   Mg^{2+} is smaller than Ba^{2+}, lattice energy of $MgSO_4$ is greater than the lattice energy of $BaSO_4$, hydration energy of Mg^{2+} is greater than hydration energy of Ba^{2+}, hydration energy changes more than the lattice energy. ΔH_{sol} is more endothermic for $BaSO_4$. [6]

   b OH^- is much smaller than SO_4^{2-}, both lattice energy and hydration enthalpy are more exothermic than for the sulfates. Lattice energy changes more than hydration enthalpy / hydration enthalpy does not change so much. ΔH_{sol} of $MgSO_4$ is more endothermic. [4]

   Total: 10

Unit 20

1. a

   T = 298 K

   H_2(g) 1.0 atm

   salt bridge

   $[H^+]$ = 1.0 mol dm^{-3}

   platinum covered with platinum black

   bright platinum

   1.0 mol dm^{-3} Fe^{2+} and 1.0 mol dm^{-3} Fe^{3+} [5]

   b i 0.23 V
   ii $2Fe^{3+}(aq) + 2I^-(aq) \longrightarrow 2Fe^{2+}(aq) + I_2(aq)$ [3]
   c 0.77 V [1]

   Total: 9

2. $Q = I \times t = 0.5 \times 60 \times 60\,C = nF = 96500 \times n$

   no. of moles Cu^{2+} = ½ × no. of moles of electrons

   mass of copper = ½ × n × 63.5; cathode increases by 0.59 g [3]

   mass of Cu lost from anode = 0.84 × 0.59 g

   mass of Ni lost = 0.16 × 58.7 × ½ × n

   anode decreases by 0.585 g [3]

   Total: 6

Unit 21

1. a It is an aqueous mixture of a weak acid and its conjugate base which opposes changes to the pH when small amounts of an acid or base are added. [3]
   b 2.9 [2]
   c i 1.3
   ii 12.3
   iii 4.8 [7]

   Total: 12

2. a i It is the equilibrium constant for the concentration of solute in one solvent divided by the concentration of the solute in a second solvent. [2]

ii Use an immiscible solvent and shake the aqueous solution with the solvent in a separating funnel. Leave to allow the equilibrium to be established. Run off the lower layer and evaporate off the solvent to obtain the solute. [5]

b i 3.2 g [2]

ii 3.5 g [3]

iii Add in small portions; not too small a portion as a little is lost with each extraction. [2]

Total: 14

Unit 22

1 a Mix 100 cm³ ester + 100 cm³ H₂O. Put the flask in a water bath to keep the temperature constant. Add 1 cm³ of concentrated HCl. At known time intervals, withdraw 10 cm³ with a pipette and titrate with standard NaOH. Allow for the acid catalyst; find the number of moles of ethanoic acid formed. Plot a graph of the number of moles of ethanoic acid formed against time. [6]

b i Both are first order. [2]

ii rate = $k[HCl][CH_3CO_2CH_2CH_3]$ [2]

Total: 10

2 a i rate = $k[NH_3]^0 = k$

The rate is constant and does not depend on the ammonia concentration.

ii The catalysis occurs on the surface of the iron and so depends on the diffusion of the products off the surface rather than the concentration. [4]

b Axes should be labelled and include the units; y-axis shows [B] (in mol dm⁻³); x-axis shows time (in mins). Points need to be plotted accurately and a smooth curve of best fit drawn.
Two half-lives should be indicated with the values stated.
As consecutive half lives are the same, it is a first order reaction. [7]

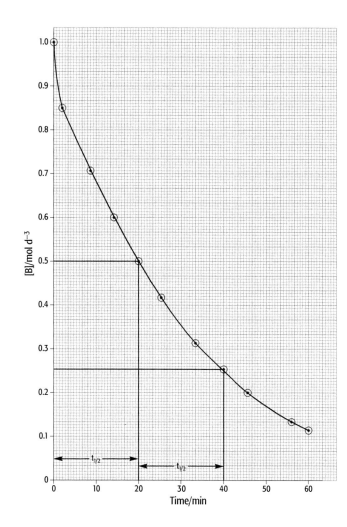

Total: 11

Unit 23

1 a reactions i and ii [2]

b $PCl_5(s) \longrightarrow PCl_3(l) + Cl_2(g)$ $\Delta G^\ominus$ = +42.3 kJ mol⁻¹ [3]

$2Mg(s) + O_2(g) \longrightarrow 2MgO(s)$ $\Delta G^\ominus$ = −1138.8 kJ mol⁻¹ [3]

Total: 8

2 a $\Delta S_{reaction}$ is positive as 1 mole of gas produces 2 moles of gas.

ΔH is positive as this is an endothermic reaction.

At low temperatures, $\Delta H > T\Delta S_{reaction}$ and so ΔG is positive and the reaction is not spontaneous.

At high temperatures, $\Delta H < T\Delta S_{reaction}$ and so ΔG is negative and the reaction is spontaneous. [4]

b i Both $\Delta H^\ominus_{reaction}$ and $\Delta S^\ominus$ are negative.

At high temperature, $T\Delta S^\ominus > \Delta H^\ominus_{reaction}$ so $\Delta G^\ominus$ is positive.

At low temperature, $\Delta H^\ominus_{reaction} > T\Delta S^\ominus$ so $\Delta G^\ominus$ is negative. [3]

ii 1135 K [2]

Total: 9

Unit 24

1 a i Fe is [Ar] $3d^6 4s^2$.

ii Ni^{2+} is [Ar] $3d^8$. [2]

b i octahedral [1]

ii In the octahedral complex, the 3d orbitals are split into three orbitals of lower energy and two orbitals of higher energy. The energy difference depends on the strength of the ligand bond. An electron can absorb light energy of just the right frequency / $\Delta E = hf$ to jump into the higher energy level. The absorbed frequency is in the visible region. Frequencies not absorbed are seen. NH_3 produces a larger / different splitting than does H_2O. [5]

c Both are Fe^{III} / ox. no. [Fe] = +3.

Electron configuration is [Ar] $3d^5$ for both.

CN^- gives a much larger / different splitting than H_2O. Splitting large; all electrons in the three lower energy orbitals – one unpaired electron. Splitting small; one electron in each orbital gives five unpaired electrons. [5]

Total: 13

2 a $Cu(H_2O)_6^{2+}(aq)$ is a pale blue solution.

In water, NH_3 produces some OH^-.

$NH_3(aq) + H_2O(l) \rightleftharpoons NH_4^+(aq) + OH^-(aq)$

Both $NaOH(aq)$ and $NH_3(aq)$ give a blue precipitate.

$Cu^{2+}(aq) + 2OH^-(aq) \longrightarrow Cu(OH)_2(s)$

Only $NH_3(aq)$ dissolves the blue precipitate to form a dark blue solution of $[Cu(NH_3)_4(H_2O)_2]^{2+}(aq)$.

$[Cu(H_2O)_6]^{2+}(aq) + 4NH_3(aq) \rightleftharpoons$
$[Cu(NH_3)_4(H_2O)_2]^{2+}(aq) + 4H_2O(l)$ [5]

b i $3Fe^{2+}(aq) \longrightarrow Fe(s) + 2Fe^{3+}(aq)$ [2]

ii $E^\ominus_{cell} = -1.21$ V. The reaction is not feasible as this is negative. [2]

Total: 9

Unit 25

1 a

A — 3-nitrobenzoic acid (structure with CO_2H and NO_2)

B — 2,6-dichloro-4-methylphenol (structure with OH, two Cl, and CH_3)

C — 4-ethylmethylbenzene or 4-methylethylbenzene

b i reaction 1: $Cl_2(g)$ + light (not (aq))

reaction 2: $Cl_2(g) + AlCl_3$ (not (aq))

ii Warm compounds **D** and **E** separately with $AgNO_3(aq)$. Compound **D** will give a white ppt. of $AgCl(s)$, but compound **E** will not react.

2 a i Phenol is more acidic than phenylmethanol, because the negative charge on the phenoxide anion can be delocalised over the ring, making it more stable.

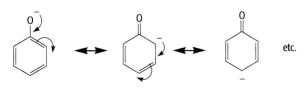 etc.

ii add $Br_2(aq)$; phenol will decolorise it, and form a white precipitate; no reaction with phenylmethanol

b Both **F** and **G** will be more acidic than phenol. The NO_2 group and Cl are both electron-withdrawing groups, so will stabilise the delocalised negative charge.

c i Reaction does not take place because the lone pair on oxygen in phenol is delocalised over the ring, making it not nucleophilic enough to react with the (protonated) ethanoic acid.

ii Reacting phenol with ethanoyl chloride (CH_3COCl) will give **H**.

Unit 26

1 a i Chloroethanoic acid is more acidic than ethanoic acid – the equilibrium

$ClCH_2CO_2H \rightleftharpoons ClCH_2CO_2^- + H^+$

is further over to the right because chlorine is more electronegative than hydrogen, it withdraws electron density from the $-CO_2^-$ anion, thus stabilising it.

ii CH_2FCO_2H is more acidic, as F is more electronegative than Cl.

CH_2BrCO_2H is less acidic, as Br is less electronegative than Cl.

$CHCl_2CO_2H$ is more acidic: two electronegative Cl atoms have greater electron-withdrawing effect.

$ClCH_2CH_2CO_2H$ is less acidic, as the Cl atom is further away from the CO_2^- group.

b **A** is $CH_3CH(OH)CO_2H$.

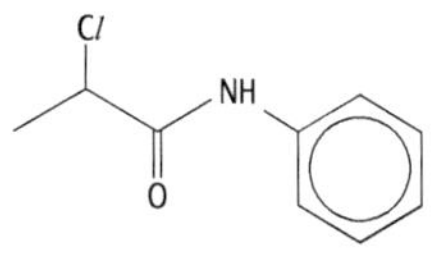

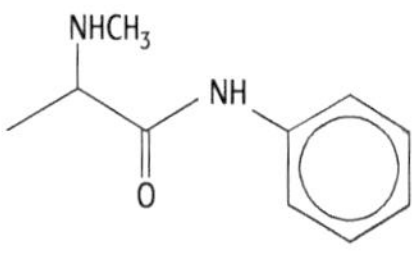

Acyl chlorides are more reactive than chloroalkanes, so the first amine (phenylamine) will react with $-COCl$ first.

2 a i heat with H_3O^+ (e.g. $H_2SO_4(aq)$)

ii **D** is $HCO_2CH_2CH_2OCHO$.

E is $HOCH_2CH_2OH$.

F is HCO_2H.

G is HO_2CCO_2H.

H is CO_2.

D to **E** + **F** is ester hydrolysis.

E to **G** is oxidation of a primary di-ol to a di-acid.

F + Fehling's solution is an oxidation (of –CHO).

G to **H** is also an oxidation.

b

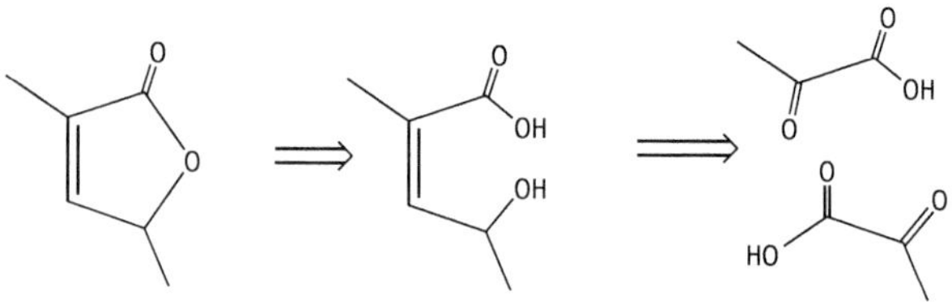

Unit 27

1 a reaction I: Sn + conc. HCl

reaction II: $NaNO_2 + HCl(aq)$ at $T < 5\,°C$

reaction III: heat in H_2O

reaction IV: phenol (C_6H_5OH) + NaOH(aq) at $5\,°C$

reaction V: dimethylphenylamine ($C_6H_5N(CH_3)_2$) at $5\,°C$

b **A** is an amine, **B** is a diazonium compound, **C** is an azo dye.

c

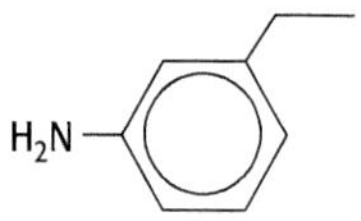

D

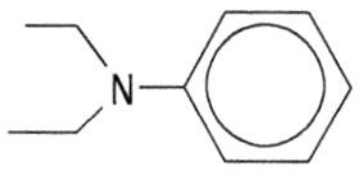

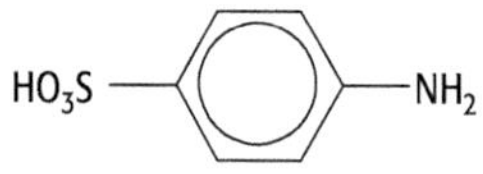

E

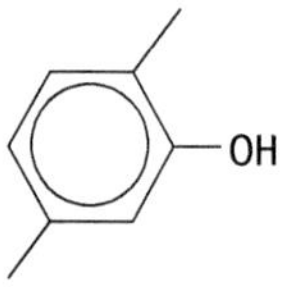

2 a i $(CH_3)_2CHCOCl$

ii reaction VI: $SOCl_2$ or PCl_5

reaction VII: $NH_3(aq)$

reaction VIII: $LiAlH_4$

iii $(CH_3)_2CHCONHCH_2CH(CH_3)_2$, amide, $C_8H_{17}NO$

b i

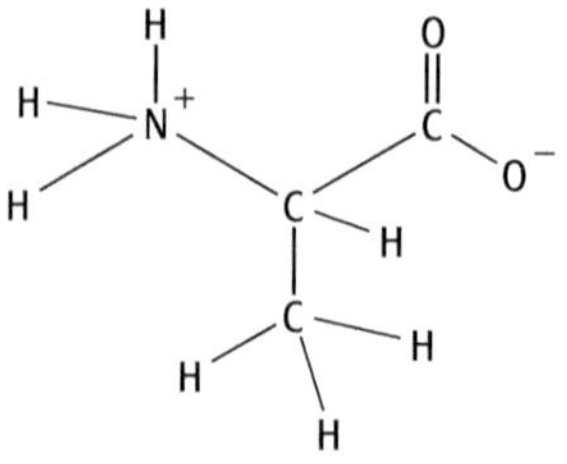

ii

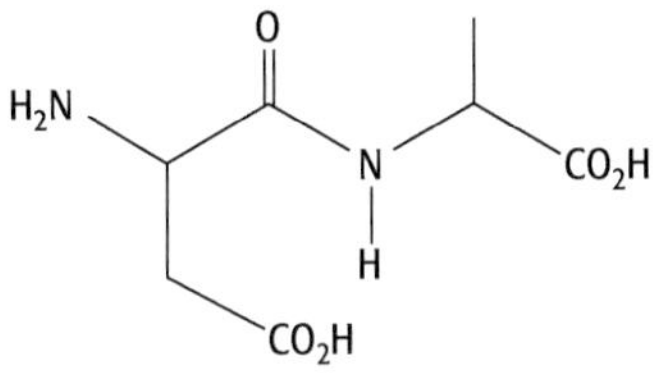

[structure: H₂N-CH(CH₂CO₂H)-C(=O)-NH-CH(CH₃)-CO₂H]

iii

[structure: H₃N⁺-CH(CO₂⁻)-CH₂-CO₂⁻]

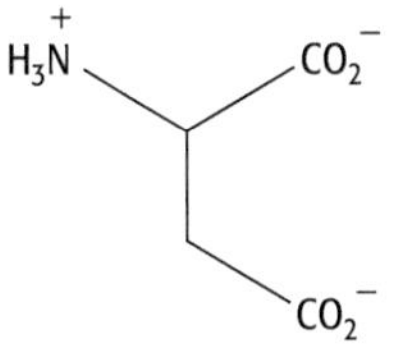

Unit 28

1 a A: addition, CH₂=CHCONH₂

B: condensation,
HO₂CCH(CH₃)CH(CH₃)CO₂H and
H₂NCH₂CH₂NH₂

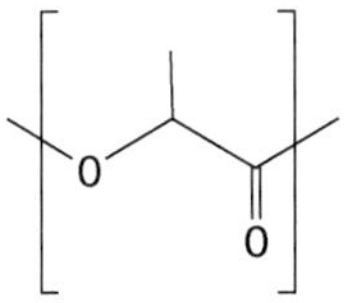

C: addition, CH₂=CHCO₂CH₃ and
CH₃CH=CHCN

b i [polymer repeat unit with C=C branching]

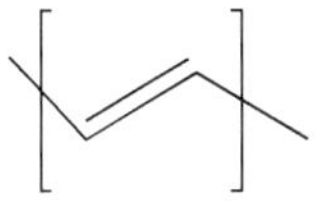

ii [polymer repeat unit with O and C=O]

iii [polymer repeat unit with NC and CO₂CH₃ substituents]
(other repeat units are possible)

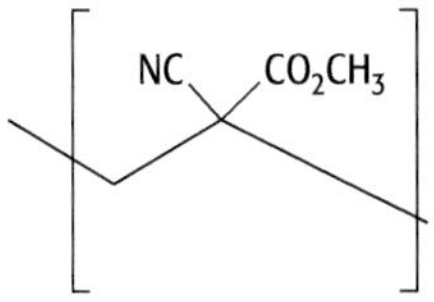

2 a i The order of the amino acids along the polypeptide chain, joined by (covalent) amide (peptide) bonds: –NH–CHR–CO–NH–CHR'–CO–.

ii The hydrogen bonding between the C=O and N–H groups of amide linkages in adjacent (antiparallel) polypeptide chains.

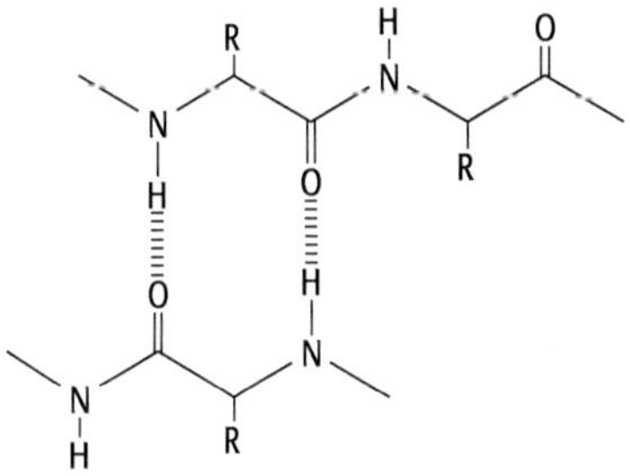

iii The covalent –S–S– bond formed by the oxidation of two cysteine residues, either in the same polypeptide chain or in different chains.

>CH–CH₂–SH + HS–CH₂CH< − 2[H]
⟶ >CH–CH₂–S–S–CH₂–CH<

b The bases on adjacent chains hydrogen bond to each other such that adenine (A) *only* bonds to thymine (T) and guanine (G) *only* bonds to cytosine (C). There are two hydrogen bonds between A and T, but three hydrogen bonds between G and C.

c similarity: in both polymers, the monomers are joined by amide bonds *or* both are polyamides.

differences: in proteins the monomers are usually different (each monomer can be any one of the 20 amino acids), whereas in nylon 6,6 the same pair of monomers alternate along the chain. Proteins have a specific length / number of monomers, whereas nylon 6,6 has a variable polymer chain length. (Other differences are possible.)

d i e.g. terylene

ii heat with aqueous acid or aqueous alkali

Unit 29

1 a $M_r = 88$

b 2500–3300: OH in –CO₂H group

1700: C=O (in –CO₂H)

1230: C–O (in –CO₂H)

so functional group is –CO₂H (carboxylic acid)

c three types of carbon atom, some in an alkyl group (δ 18), some next to C=O (δ 33), some in a –CO₂– group (δ 183) d The 1H at δ 11.8 is the H in $-CO_2H$, the 6H at δ 1.2 are two $-CH_3$ groups split by a single adjacent proton (⇒ doublet), the 1H at δ 2.5 is a single proton split by the six adjacent protons of the two CH_3 groups (⇒ multiplet).

Structure is $(CH_3)_2CH-CO_2H$.

e $m/e = 43$ is $(CH_3)_2CH^+$

2 a **B** is $CH_3CH_2CO_2H$.

C is $CH_3CO_2CH_3$.

b for B: δ 160–185 (**C** in $-CO_2H$), δ 25–50 (**C** next to CO_2H), δ 0–50 (alkyl **C**)

for **C**: δ 160–185 (**C** in $-CO_2-$), δ 50–70 (**C** next to –O–), δ 25–50 (**C** next to $-CO_2-$)

c for B: δ 9–13, singlet (CO_2H), δ 2.2–3.0, quartet (CH_2 next to C=O and CH_3), δ 0.9–1.7, triplet (CH_3 next to CH_2)

for C: δ 3.2–4.0, CH_3 next to –O–, δ 2.2–3.0, CH_3 next to C=O

d B would have –O–H absorption at 2500–3300 cm^{-1}, but **C** will not.

e **C** will have the shorter retention time, as it is more volatile.

Unit 30

1 a $KMnO_4(aq)$ tests for C=C (alkene); Tollen's reagent tests for –CHO (aldehyde); $I_2(aq)$ + $OH^-(aq)$ tests for $-COCH_3$ (methyl ketone); 2,4-DNPH tests for >C=O (carbonyl)

b

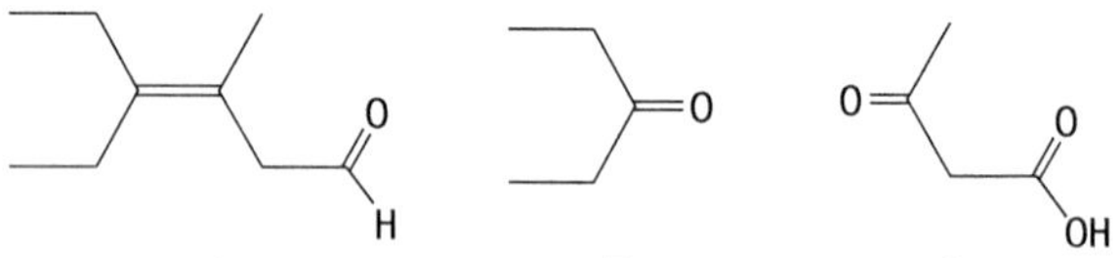

A B C

c

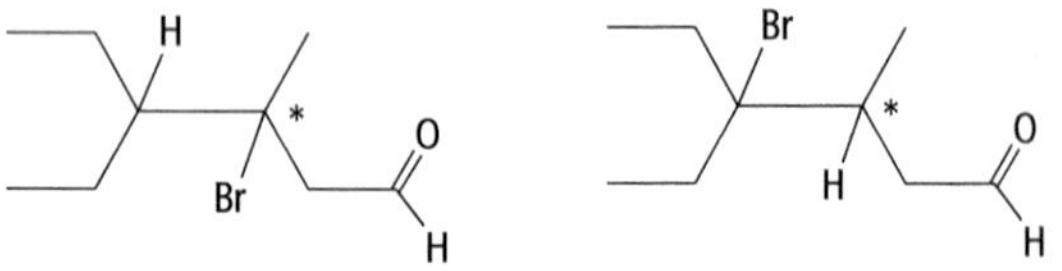

2 a

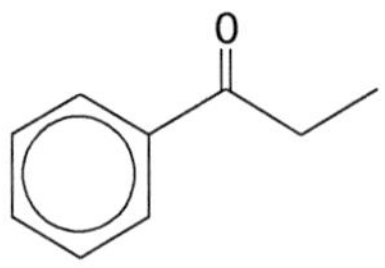

D

b step 2: $NaBH_4$ *or* $LiAlH_4$

step 3: heat with conc. H_2SO_4 *or* H_3PO_4

step 4: add $Br_2(aq)$

step 5: CH_3NH_2, heat in ethanol under pressure

c step 1: electrophilic substitution

step 2: reduction

step 3: elimination

step 4: electrophilic addition

step 5: nucleophilic substitution

d There are two chiral centres: each can be (+) or (−), so there are four optical isomers.

e i insoluble: the alcohol –OH is not acidic enough to react with NaOH

ii soluble: the basic (secondary) amine will form a salt: $R_2NH_2^+Cl^-$

Advanced practical skills

a i $Mg(NO_3)_2 \longrightarrow MgO + 2NO_2 + \frac{1}{2}O_2$

ii nitrogen (IV) oxide, 4.8 dm^3

oxygen, 1.2 dm^3

b

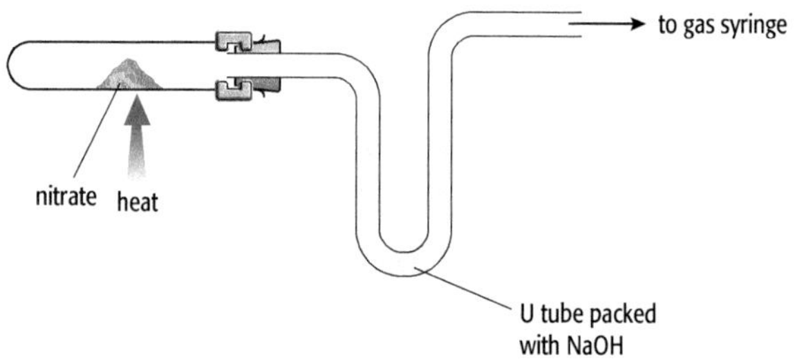

The U tube packed with solid NaOH could be replaced by apparatus to allow the gases to be bubbled through NaOH(aq).

c 80 cm^3; $Mg(NO_3)_2$ = 0.99 g; $Ca(NO_3)_2$ = 1.09 g

d Need to keep the Bunsen with the same flame / same distance from the tube.

Read the gas syringe at time = 0.

Read the gas syringe at measured time intervals whilst heating.

e Plot two graphs of volume of oxygen (on *y*-axis) against the time (*x*-axis) for the two nitrates.

Draw tangents at t = 0 and find the gradients.

The gradients are proportional to the rate of oxygen production at t = 0.

f Carry out the experiment in a fume hood.

Glossary

¹³C NMR NMR spectroscopy that investigates the environments of the ¹³C isotope of carbon in molecules.

¹H NMR NMR spectroscopy that investigates the environments of protons (¹H) in molecules.

acid dissociation constant, K_a the equilibrium constant for a weak acid: $K_a = [H^+][A^-] / [HA]$.

acid a proton (hydrogen ion) donor.

acid–base indicator a substance that changes colour over a narrow range of pH values.

activation energy the minimum energy that colliding particles must possess for a successful collision that results in a reaction taking place.

active site the 'pocket' on an enzyme surface where the substrate binds and undergoes catalytic reaction.

acyl chloride a compound whose functional group is –COCl.

addition polymer a large molecule formed by the addition of many monomers containing C=C bonds.

addition reaction a reaction in which an organic molecule (usually containing a double bond) reacts with another molecule to give only one product.

adsorption (in catalysis) the first stage in heterogeneous catalysis – molecules of reactants (usually gases) form bonds with atoms on the surface of the catalyst.

alcohol a compound whose functional group is just an –OH group bonded to a carbon atom.

aldehyde a carbonyl compound whose functional group is –CH=O, at the end of a carbon chain.

alicyclic an alicyclic compound contains at least one non-aromatic ring, e.g. cyclohexane.

aliphatic the carbon atoms in an aliphatic compound form a chain (straight or branched), and are not in a ring.

alkali a base that is soluble in water.

alkaline earth metals the elements in Group 2 of the Periodic Table.

alkane a saturated hydrocarbon containing only single bonds, with the general formula C_nH_{2n+2}.

alkene an unsaturated hydrocarbon containing at least one C=C double bond, with the general formula C_nH_{2n}.

allotrope different crystalline or molecular forms of the same element. Graphite and diamond are allotropes of carbon.

alloy a mixture of two or more metals or a metal with a non-metal.

alpha-helix

amide a compound whose functional group is –CONH$_2$.

amine a compound formed by replacing one or more hydrogens in the ammonia (NH$_3$) molecule by organic groups, e.g. methylamine, CH$_3$NH$_2$.

amino acid a compound containing an amino and a carboxylic acid group: H$_2$N–CHR–CO$_2$H.

amphoteric able to behave as both an acid and a base. Aluminium oxide is amphoteric.

anion a negatively charged ion.

anode the positive electrode.

arene a hydrocarbon that contains a benzene ring.

aromatic in chemistry, an aromatic compound is one which contains a benzene ring.

atomic orbitals regions of space outside the nucleus that can be occupied by one or, at most, two electrons. Orbitals are named s, p, d and f. They have different shapes.

atomic radius half the distance between two neighbouring nuclei in the solid element.

average bond energy a general bond energy value used for a particular bond, e.g. a C–H, when the exact bond energy is not required. Average bond energies are often used because the strength of a bond between two particular types of atom is slightly different in different compounds.

Avogadro constant the number of atoms (or ions, molecules or electrons) in a mole of atoms (or ions, molecules or electrons): its numerical value is 6.02×10^{23}.

azo (diazo) compound a compound containing the functional group –N=N–. Many dyes are azo compounds.

base a proton (hydrogen ion) acceptor.

base-pairing the specific pairing of adenine (A) with thymine (T), and of cytosine (C) with guanine (G), formed by hydrogen bonds between the bases on opposite chains of DNA.

beta-pleated sheet

bidentate ligands that can form two coordinate bonds from each ion or molecule to the central transition metal ion.

biofuels renewable fuels, sourced from plant or animal materials.

boiling point the temperature at which the vapour pressure is equal to the atmospheric pressure.

Boltzmann distribution a graph showing the distribution of energies of the particles in a sample at a given temperature.

bond energy (bond enthalpy) the energy needed to break 1 mole of a particular bond in 1 mole of gaseous molecules.

bond polarisation the way a bond pair is closer to one of the atoms, which is then more negative than the other atom.

Born–Haber cycle a type of enthalpy cycle used to calculate lattice energy.

Brønsted–Lowry theory of acids acids are proton donors and bases are proton acceptors.

buffer solution a solution that minimises changes in pH when moderate amounts of acid or base are added. Common forms of buffer consist of either a weak acid and its conjugate base or a weak base and its conjugate acid.

carbocation a positive ion that contains a carbon atom surrounded by only six electrons.

carbonyl compound a compound whose functional group is C=O. Carbonyl compounds are either aldehydes or ketones.

carboxylic acid a compound whose functional group is $-CO_2H$.

catalyst a substance that increases the rate of a reaction but remains chemically unchanged itself at the end of the reaction.

cathode the negative electrode.

cation a positively charged ion.

ceramic an inorganic non-metallic solid that is prepared by heating a substance or mixture of substances to a high temperature.

CFCs chlorofluorocarbons are compounds that were once used as aerosol propellants and refrigerants, but have now been withdrawn due to their detrimental effect on the ozone layer.

chemical shift (δ) the difference between the frequency of absorption of a proton (or ^{13}C atom) in a particular chemical environment and the frequency at which TMS absorbs.

chiral centre this is an atom (usually carbon) that has four different groups or atoms bonded to it. The four groups can be arranged in two different ways, which are non-superimposable mirror images of each other (optical isomers).

cis–trans isomers see isomerism: geometrical.

closed system a system in which matter or energy is not lost or gained, e.g. gases in a closed vessel.

common-ion effect the reduction in the solubility of a dissolved salt by adding a compound that has an ion in common with the dissolved salt. This often results in precipitation of the salt.

complex a central transition metal ion surrounded by ligands.

compound a substance made up of two or more elements bonded (chemically joined) together.

condensation polymer a large molecule formed by condensation reactions between many monomers, usually forming either a polyester or a polyamide.

condensation reaction a reaction between two molecules to form a larger molecule, with the removal of a molecule of water (or another small molecule).

condensation the change in state when a vapour changes to a liquid.

conjugate pair (acid/base) an acid and base on each side of an equilibrium equation that are related to each other by the difference of a proton; e.g. the acid in the forward reaction and the base in the reverse reaction or the base in the forward reaction and the acid in the reverse reaction.

coordinate bond a covalent bond in which both electrons in the bond come from the same atom.

coordination number the number of coordinate (dative) bonds formed by ligands to the central transition metal ion in a complex.

covalent bond a bond formed by the sharing of pairs of electrons between two atoms.

cracking the breaking of higher alkane molecules into smaller alkanes and alkenes by means of heat and/or a catalyst.

cross-linking the joining together of two polymer chains by forming covalent bonds between them. The simplest example is the disulfide bridge.

curly arrow used in describing the mechanisms of organic reactions. It represents the movement of a pair of electrons, from the position at the 'tail' of the arrow to the position at the 'head'.

cyanoacrylates see superglue.

dative covalent bond another name for a coordinate bond.

degenerate orbitals atomic orbitals at the same energy level.

delocalised electrons electrons that occupy a delocalised orbital, spread over three or more atoms, rather than the usual localised electrons that occupy an orbital spread over or between just two adjacent atoms.

delta (δ) see chemical shift.

deoxyribose the 5-carbon sugar ($C_5H_{10}O_4$) containing a 5-membered ring which forms part of the backbone chain of DNA.

desorption the last stage in heterogeneous catalysis. The bonds holding the molecule(s) of product(s) to the surface of the catalyst are broken and the product molecules diffuse away from the surface of the catalyst.

deuterium exchange the use of heavy water, D_2O, to exchange protons in OH and NH groups with deuterium atoms (ROH + $D_2O \longrightarrow$ ROD + HDO), thus removing those proton absorptions from the NMR spectrum.

dimer a species (often a molecule) formed by the reacting together of two identical atoms or molecules – see also polymer.

dipeptide a dimer formed of two amino acids joined by one peptide (amide) bond.

dipole a separation of charge in a molecule. One end of the molecule is permanently positively charged and the other is negatively charged.

dipole moment a measure of the charge separation within a molecule.

discharge the conversion of ions to atoms or molecules at electrodes during electrolysis, e.g. during the electrolysis of concentrated sodium chloride solution, chlorine is discharged at the anode by the conversion of Cl^- ions to Cl atoms, which then combine to form Cl_2 molecules.

displayed formula

disproportionation the simultaneous reduction and oxidation of the same species in a chemical reaction.

dissociation the break-up of a molecule into ions, e.g. when HCl molecules dissolve in aqueous solution, they dissociate completely into H^+ and Cl^- ions.

disulfide bridge the –S–S– bond formed between two cysteine residues, bringing together either two separate polypeptide chains, or two parts of the same chain.

DNA deoxyribonucleic acid is a condensation polymer in the form of a double helix, containing two chains of alternating deoxyribose-phosphate units. To each unit is attached one of four different nitrogenous bases. The two strands of the helix are joined by hydrogen bonds between the bases. The order of bases along the polymer chains is the genetic code 'blueprint' for making a specific protein.

dot-and-cross diagram a diagram showing the arrangement of the outer-shell electrons in an ionic or covalent element or compound. The electrons are shown as dots or crosses to show their origin.

double covalent bond two shared pairs of electrons bonding two atoms together.

dynamic (equilibrium) in an equilibrium mixture, molecules of reactants are being converted to products at the same rate as products are being converted to reactants.

electrochemical cell two half-cells in separate compartments joined by a salt bridge. When the poles of the half-cells are joined by a wire, electrons travel in the external circuit from the half-cell with the more negative $E^\ominus$ value to the half-cell with the more positive $E^\ominus$ value.

electrode a rod of metal or carbon (graphite) which conducts electricity to or from an electrolyte.

electrode potential the voltage measured for a half-cell compared with another half-cell.

electrolysis the decomposition of a compound into its elements by an electric current.

electrolyte a molten ionic compound or an aqueous solution of ions that is decomposed during electrolysis.

electron tiny subatomic particles found in orbitals around the nucleus. They have a negative charge and have negligible mass.

electron affinity (first electron affinity), $\Delta H_{ea1}^\ominus$ the enthalpy change when 1 mole of electrons is added to 1 mole of gaseous atoms to form 1 mole of gaseous 1− ions under standard conditions.

electron affinity (second electron affinity), $\Delta H_{ea2}^\ominus$ the enthalpy change when 1 mole of electrons is added to 1 mole of gaseous 1− ions to form 1 mole of gaseous 2− ions under standard conditions.

electron pair repulsion theory a way of predicting the geometry of molecules from the number of electron pairs surrounding the central atom.

electronegativity the ability of an atom to attract the bonding electrons in a covalent bond.

electronic configuration a way of representing the arrangement of the electrons in atoms showing the principal quantum shells, the subshells and the number of electrons present, e.g. $1s^2 2s^2 2p^3$. The electrons may also be shown in boxes.

electrophile an atom or group of atoms that reacts with electron-rich centres in other molecules. Electrophiles usually have an empty orbital in their valence shell. Many are positively charged.

electrophoresis the separation of charged molecules (ions) by their different rates of movement in an electric field through a buffer.

electrovalent bond another name for an **ionic bond**.

element a substance made of only one type of atom.

elimination reaction a reaction where a small molecule such as H_2O or HBr is removed from an organic molecule, usually forming a C=C double bond.

empirical formula see formula: empirical.

end point the exact point in a titration at which an indicator changes colour.

endothermic the term used to describe a reaction in which energy is absorbed from the surroundings: the enthalpy change is positive.

energy levels (of electrons) the regions at various distances from the nucleus in which electrons have a particular amount of energy. Electrons further from the nucleus have more energy. See **principal quantum number**.

enhanced global warming the increase in average temperatures around the world as a consequence of the huge increase in the amounts of CO_2 and other greenhouse gases produced by human activity.

enthalpy change the energy transferred in a chemical reaction (symbol ΔH).

enthalpy change of atomisation, $\Delta H_{at}^\ominus$ the enthalpy change when 1 mole of gaseous atoms is formed from its element under standard conditions.

enthalpy change of combustion, $\Delta H_c^\ominus$ the enthalpy change when 1 mole of element or compound is completely burnt in oxygen.

enthalpy change of formation, $\Delta H_f^\ominus$ the enthalpy change when 1 mole of compound is formed from its elements.

enthalpy change of hydration, $\Delta H_{hyd}^\ominus$ the enthalpy change when 1 mole of a specified gaseous ion dissolves in sufficient water to form a very dilute solution.

enthalpy change of neutralisation, $\Delta H_{neut}^\ominus$ the amount of energy released when an acid reacts with a base to form a mole of water.

enthalpy change of solution, $\Delta H_{sol}^\ominus$ the energy absorbed or released when 1 mole of an ionic solid dissolves in sufficient water to form a very dilute solution.

enthalpy cycle a diagram showing alternative routes between reactants and products that allows the determination of one enthalpy change from other known enthalpy changes by using Hess's law.

enthalpy profile diagram a diagram showing the enthalpy change from reactants to products along the reaction pathway.

entropy a measure of the disorder of a system. The system becomes energetically more stable when disordered.

enzyme a protein molecule that is a biological catalyst. Most act on a specific substrate.

epoxide a compound containing a 3-membered ring of two carbon atoms and one oxygen atom.

equilibrium constant a constant calculated from the equilibrium expression for a reaction.

equilibrium expression a simple relationship that links K_c to the equilibrium concentrations of reactants and products and the stoichiometric equation.

equilibrium reaction a reaction that does not go to completion and in which reactants and products are present in fixed concentration ratios.

equivalence point the exact point in a titration at which the exact amounts of reagents have been added together for complete reaction.

ester a compound containing the $-CO_2-R$ group.

esterification the reaction between an alcohol and a carboxylic acid (or acyl chloride) to form an ester and water (or HCl).

eutrophication an environmental problem caused by fertilisers leached from fields into rivers and lakes. The fertiliser then promotes the growth of algae on the surface of water. When the algae die, bacteria thrive and use up the dissolved oxygen in the water, killing aquatic life.

exothermic the term used to describe a reaction in which energy is released to the surroundings: the enthalpy change is negative.

Faraday constant the charge carried by 1 mole of electrons (or 1 mole of singly charged ions). It has a value of 96 500 coulombs per mol (C mol^{-1}).

feasibility (of reaction) is the likelihood or not of a reaction occurring when reactants are mixed. We can use $E^\ominus$ values to assess the feasibility of a reaction.

Fehling's solution an alkaline solution of Cu^{II} which is used to test for the presence of a mild reducing agent such as an aldehyde. A positive result is the formation of a red precipitate.

formula: displayed shows all the bonds and atoms within a molecule of the compound.

formula: empirical the simplest ratio of elements the compound contains.

formula: general a formula that represent a homologous series of compounds. For example, alcohols have the general formula $C_nH_{2n+1}OH$, where n is a positive integer.

formula: molecular this tells us the number of atoms of each element that are in one molecule of the compound. The molecular formula of a compound is always a multiple of its empirical formula (e.g. ×1, ×2, ×3).

formula: skeletal this is a simplified version of the displayed formula: it shows all the bonds in a molecule, except those to hydrogen atoms, and does not label the carbon atoms.

formula: stereochemical a displayed formula that shows the three-dimensional shape of the molecule, using 'wedges' and 'hatched' bonds.

formula: structural this indicates which atoms are joined to which in the molecule of the compound.

fractional distillation the separation of the components of a liquid mixture by making use of their different boiling points.

fragmentation the breaking-up of a molecular ion in a mass spectrometer, to form ionic fragments of the original molecule.

free energy see Gibbs free energy.

free radical very reactive atom or molecule that has a single unpaired electron.

Friedel–Crafts reactions Electrophilic substitution reactions in which an alkyl or acyl group is attached to a benzene ring.

fuel cell a source of electrical energy that comes directly from the energy stored in the chemicals in the cell, one of which is oxygen (which may come from the air).

functional group an atom or group of atoms in an organic molecule that is responsible for its characteristic reactions.

general gas equation $pV = nRT$, an equation relating the volume of a gas to the temperature, pressure and number of moles of gas. Also called the ideal gas equation.

giant molecular structure/giant covalent structure structures having a three-dimensional network of covalent bonds throughout the whole structure.

Gibbs free energy the energy change that takes into account both the entropy change of a reaction and the enthalpy change. Reactions are likely to be feasible if the value of the Gibbs free energy change of reaction is negative. The Gibbs free energy change of reaction is given by the relationship $\Delta G^\ominus = \Delta H^\ominus - T\Delta S^\ominus$

GLC or GC gas–liquid chromatography.

half-cell half of an electrochemical cell. The half-cell with the more negative $E^\ominus$ value supplies electrons. The half-cell with the more positive $E^\ominus$ value receives electrons.

half-equation in a redox reaction, an equation showing either an oxidation or a reduction.

half-life the time taken for the amount (or concentration) of the limiting reactant in a reaction to decrease to half its value.

halogenoalkane a compound whose functional group is just a halogen atom (F, Cl, Br or I) bonded to a carbon atom.

halogens Group 17 elements.

HDPE high-density polyethene, which contains long chains with no side branches. More rigid and harder than LDPE.

Hess's law the total enthalpy change for a chemical reaction is independent of the route by which the reaction takes place.

heterogeneous catalysis the type of catalysis in which the catalyst is in a different phase from the reactants. For example, iron in the Haber process.

homogeneous catalysis a reaction in which all the reactants and the catalyst are in the same phase (usually either gaseous or liquid). For example, sulfuric acid catalysing the formation of an ester from an alcohol and carboxylic acid.

homologous series a series of organic molecules whose molecular formulae differ by CH_2. Members of a homologous series have the same general formula, and the same (or very similar) chemical reactions.

hybridisation (of atomic orbitals) the process of mixing atomic orbitals so that each has some character of each of the orbitals mixed.

hydrocarbon a compound whose molecules contain carbon and hydrogen only.

hydrogen bond the strongest type of intermolecular force. It is formed between molecules having a hydrogen atom bonded to one of the most electronegative elements (F, O or N).

hydrogenation the addition of hydrogen to an unsaturated compound. See also **reduction**.

hydrolysis reaction a hydrolysis reaction is one where water reacts with a molecule and splits it into two smaller molecules.

ideal gas a gas whose volume varies in proportion to the temperature and in inverse proportion to the pressure. Noble gases such as helium and neon approach ideal behaviour because of their low intermolecular forces.

infrared (IR) spectroscopy the identification of the functional groups present in a molecule by measuring the frequency of absorption of infrared radiation.

intermolecular forces the weak forces between molecules.

iodoform see **triiodomethane**.

ion polarisation the distortion of the electron cloud on an anion by a neighbouring cation. The distortion is greatest when the cation is small and highly charged.

ionic bond the electrostatic attraction between oppositely charged ions.

ionic product of water, K_w the equilibrium constant for the ionisation of water: $K_w = [H^+][OH^-]$.

ionisation energy, ΔH_i the energy needed to remove 1 mole of electrons from 1 mole of atoms of an element in the gaseous state to form 1 mole of gaseous ions.

isomer compounds that have the same molecular formula, but have different arrangements of atoms.

isomerism: chain chain isomers have the same number of carbon atoms as each other, but their carbon backbones are different.

isomerism: functional group compounds having the same molecular formula, but different functional groups.

isomerism: geometrical (also known as *cis–trans* isomerism) isomers that contain a C=C double bond that does not have two identical groups on either end of it.

isomerism: optical molecules with the same structural formula, which are non-superimposable mirror images of each other. They usually contain at least one chiral centre.

isomerism: positional positional isomers have the same carbon backbone, but the positions of their functional group(s) differ.

isomerism: structural structural isomers have different structural formulae from each other, but the same molecular formula.

isotopes atoms of an element with the same number of protons but different numbers of neutrons.

isotopic symbol the symbol of an element showing the proton number and the mass number.

ketone a carbonyl compound whose functional group is >C=O, in the middle of a carbon chain.

Kevlar a strong, hard-wearing polyamide formed from benzene-1,4-dicarboxylic acid and 1,4-diaminobenzene, used in bulletproof vests, high performance tyres and bridge cables.

kinetic theory the theory that particles in gases and liquids are in constant movement. The kinetic theory can be used to explain the effect of temperature and pressure on the volume of a gas as well as rates of chemical reactions.

lattice a regularly repeating arrangement of ions, atoms or molecules in three dimensions.

lattice energy the enthalpy change when 1 mole of an ionic compound is formed from its gaseous ions under standard conditions.

LDPE low-density polyethene, which contains chains with side-branches. More flexible and softer than HDPE.

le Chatelier's principle when any of the conditions affecting the position of equilibrium are changed, the position of that equilibrium shifts to minimise the change.

LED light-emitting diode.

ligand a molecule or ion with one or more lone pairs of electrons available to donate to a transition metal ion.

lock-and-key mechanism a model used to explain why enzymes are so specific in their activity. It is suggested that the active site of the enzyme has a shape into which the substrate fits exactly – rather like a particular key fits a particular lock.

lone pairs (of electrons) are pairs of electrons in the outer shell of an atom that are not bonded.

M+1 peak the peak in a mass spectrum having a mass one greater than the molecular mass. The abundance ratio of the $M + 1$ peak and the M peak allows the number of carbon atoms in a molecule to be calculated.

M+2 peak the peak in a mass spectrum having a mass two greater than the molecular mass. The presence of this peak, and the abundance ratio of the $M + 2$ peak and the M peak, allows the identification of Cl or Br in the molecule.

mass number see nucleon number.

mass spectrometer an instrument for finding the relevant isotopic abundance of elements and to help identify unknown organic compounds.

mass spectrometry (MS) the measuring of the relative masses of ions and ionic fragments formed from a compound, to determine its M_r and structure.

mechanism the series of steps a reaction goes through from reactants to products.

mesomerism the representing of a molecule by two or more 'classical' structures, one of which can be converted to the other by the movement of a pair of electrons. The actual bonding arrangement is somewhere in between those represented by the classical structures.

molar mass the mass of a mole of substance in grams.

mole the unit of amount of substance. It is the amount of substance that has the same number of particles (atoms, ions, molecules or electrons) as there are atoms in exactly 12g of the carbon-12 isotope.

mole fraction the number of moles of a particular component divided by the total number of moles.

molecular formula the formula that tells us the actual numbers of each type of atom in a molecule.

molecular orbital the region of space formed by combining two or more atomic orbitals.

monodentate ligands, such as water and ammonia, that can form only one coordinate bond from each ion or molecule to the central transition metal ion.

monomer a small molecule that can be joined to many similar molecules to form a polymer.

Nernst equation an equation used to predict quantitatively how the value of an electrode potential varies with the concentration of the aqueous ion.

neutron a subatomic particle found in the nucleus of an atom. It has no charge and has the same mass as a proton.

nitrile a compound whose functional group is $-C\equiv N$.

non-degenerate orbitals atomic orbitals that have been split to occupy slightly different energy levels.

non-polar (molecule) a molecule with no separation of charge; it will not be attracted to a positive or negative charge.

nuclear magnetic resonance (NMR) spectroscopy the measuring of the frequencies of absorption of radio waves by certain nuclei when placed in a strong magnetic field, allowing the molecular environments of those nuclei to be determined, and hence the structure of the molecule.

nucleon number the total number of protons and neutrons in the nucleus of an atom.

nucleophile this is an atom or group of atoms that reacts with electron-deficient centres in molecules (such as the $C^{\delta+}$ in $C^{\delta+}-Br^{\delta-}$ compounds). Nucleophiles always contain a lone pair of electrons. Can act as a donor of a pair of electrons.

nucleus the small dense core at the centre of every atom, containing protons (positively charged) and neutrons (no charge). Nuclei are therefore always positively charged.

nylon see polyamide.

OPLED organic polymer light-emitting diode, in which light is produced when a current is passed through a conducting polymer.

optical isomers stereoisomers that exist as two non-superimposable mirror images.

orbital a region of space in an atom which is outside the nucleus and which can be occupied by one or two electrons.

order of reaction the power to which the concentration of a reactant is raised in the rate equation. If the concentration does not affect the rate, the reaction is zero order. If the rate is directly proportional to the reactant concentration, the reaction is first order. If the rate is directly proportional to the square of the reactant concentration, the reaction is second order.

oxidation the addition of oxygen, removal of electrons or increase in oxidation number of a substance; in organic chemistry refers to a reaction in which oxygen atoms are added to a molecule and/or hydrogen atoms are removed from a molecule.

oxidation number (oxidation state) a number given to an atom in a compound that describes how oxidised or reduced it is.

oxidising agent a reactant that increases the oxidation number of (or removes electrons from) another reactant.

partial pressure the pressure that an individual gas contributes to the overall pressure in a mixture of gases.

partition coefficient the ratio of the concentrations of a solute in two different immiscible solvents when an equilibrium has been established.

peptide bond an amide bond formed between two amino acids.

periodicity the repeating patterns in the physical and chemical properties of the elements across the periods of the Periodic Table.

permanent dipole–dipole forces a type of intermolecular force between molecules that have permanent dipoles.

pH the hydrogen ion concentration expressed as a logarithm to base 10: $pH = -\log_{10}[H^+]$.

phenol a class of compound in which the hydroxy (–OH) group is directly attached to a benzene ring.

pi (π) bonds multiple covalent bonds involving the sideways overlap of p atomic orbitals.

pK_a values of K_a expressed as a logarithm to base 10: $pK_a = -\log_{10}[H^+]$.

PLA (polylactic acid) a biodegradable polyester whose repeat unit is $[-OCH(CH_3)CO-]$.

polar (covalent bond) a covalent bond in which the two bonding electrons are not shared equally by the atoms in the bond. The atom with the greater share of the electrons has a partial negative charge, $\delta-$ and the other has a partial positive charge, $\delta+$.

polarising power (of a cation) is the ability of a cation to attract electrons and distort an anion.

polyalkene a long-chain molecule formed by the joining of many hundred alkene monomers. Polyethene (polythene) is the simplest example.

polyamide a polymer containing amide groups between its monomers. The trade name for some polyamides is 'nylon'.

polyester a polymer containing ester groups between its monomers.

polymer a long-chain molecule formed by the joining of many hundred of small molecules, called monomers.

polymerisation the formation of a long-chain molecule by the joining of many hundred of small molecules, called monomers.

polypeptide a condensation polymer formed of many amino acids joined by peptide bonds.

primary a primary amine contains just one organic group attached to a nitrogen atom, e.g. $CH_3CH_2NH_2$.

primary structure the sequence of amino acids in a polypeptide chain, joined by covalent peptide (amide) bonds.

principal quantum number, *n* regions at various distances from the nucleus that may contain up to a certain number of electrons. The first quantum shell contains up to 2 electrons, the second up to 8 and the third up to 18.

proton a positively charged subatomic particle in the nucleus.

proton number the number of protons in the nucleus of an atom; it places the atom in the Periodic Table.

PTFE poly(tetrafluoroethene), $[-CF_2-CF_2-]_n$; a hard polymer with only small intermolecular forces, used for non-stick surfaces and bridge bearings.

radical an atom or group of atoms that has an unpaired electron.

rate constant the proportionality constant in the rate equation (see **rate equation**).

rate equation an equation showing the relationship between the rate constant and the concentrations of those reactants that affect the rate of reaction. The general form of the rate equation is: rate = $k[A]^m[B]^n$ where k is the rate constant, [A] and [B] are the concentrations of those reactants that affect the rate of reaction, m is the order of the reaction with respect to A and n is the order of reaction with respect to B.

rate-determining step the slowest step in a reaction mechanism.

rate of reaction a measure of the rate at which reactants are used up or the rate at which products are formed. The units of rate are $mol\,dm^{-3}\,s^{-1}$.

real gases gases that do not obey the ideal gas law, especially at low temperatures and high pressures.

redox reaction a reaction in which oxidation and reduction take place at the same time.

reducing agent a reactant that decreases the oxidation number of (or adds electrons to) another reactant.

reduction the removal of oxygen, addition of electrons or decrease in oxidation number of a substance; in organic chemistry, a reduction reaction involves the addition of hydrogen to, or the removal of oxygen from, a compound. See also **hydrogenation**.

relative atomic mass the weighted average mass of the atoms of an element, taking into account the proportions of naturally occurring isotopes, measured on a scale on which an atom of the carbon-12 isotope has a mass of exactly 12 units.

relative formula mass the mass of one formula unit of a compound measured on a scale on which an atom of the carbon-12 isotope has a mass of exactly 12 units.

relative isotopic mass the mass of a particular isotope of an element on a scale in which an atom of the carbon-12 isotope has a mass of exactly 12 units.

relative molecular mass the mass of a molecule measured on a scale in which an atom of the carbon-12 isotope has a mass of exactly 12 units.

repeat unit the part of a polymer that repeats along the chain, e.g. the repeat unit of polypropene is $[-CH_2CH(CH_3)-]$.

retardation factor (R_f) the ratio of the distance travelled by a compound to the distance the solvent front has travelled on a TLC plate.

retention time the time that has elapsed between the injection of a sample into a GC column and the detection of one of its components by the analyser at the end of the column.

reversible reaction a reaction in which products can be changed back to reactants by reversing the conditions.

R_f see **retardation factor**.

salt bridge a piece of filter paper soaked in potassium nitrate solution used to make electrical contact between the half-cells in an electrochemical cell.

saturated hydrocarbon a hydrocarbon containing only C–C single bonds.

secondary a secondary alcohol or halogenalkane has its functional group joined to a carbon atom to which only one hydrogen is also joined. Examples are propan-2-ol ($CH_3CH(OH)CH_3$) and 2-chlorobutane ($CH_3CH_2CHClCH_3$). A secondary amine has two organic groups attached to a nitrogen atom, e.g. $(CH_3)_2NH$.

secondary structure the hydrogen bonds formed between the N–H and C=O groups of a polypeptide chain, forming more rigid structures such as the α-helix or the β-sheet.

shielding the ability of inner shells of electrons to reduce the effective nuclear charge on electrons in the outer shell.

sigma (σ) bonds single covalent bonds, formed by the 'end-on' overlap of atomic orbitals.

single covalent bond a shared pair of electrons bonding two atoms together.

S_N1 mechanism a nucleophilic substitution reaction in which only one molecule is involved in the slow step.

S_N2 mechanism a nucleophilic substitution reaction in which two molecules (or ions) are involved in the slow step.

solubility product, K_{sp} the equilibrium expression showing the product of the concentrations of each ion in a saturated solution of a sparingly soluble salt at 298 K, raised to the power of the relative concentrations: $K_{sp} = [C^{y+}(aq)]^a[A^{x-}(aq)]^b$ where a is the number of C^{y+} ions in one formula unit of the compound and b is the number of A^{x-} ions in one formula unit of the compound.

solute a substance that is dissolved in a solution.

specific most enzymes are described as specific because they will only catalyse one reaction involving one particular molecule or pair of molecules.

specific heat capacity the amount of energy required to raise the temperature of 1.0 g of a material by 1.0 °C.

spectator ions ions present in a reaction mixture that do not take part in the reaction.

spin-pair repulsion electrons repel each other as they have the same charge. Electrons arrange themselves so that they first singly occupy different orbitals in the same sub-level. After that they pair up with their spins opposed to each other.

spin pairing the property of an electron which indicates a spinning charged particle; if two electrons spin in opposite directions they can become a spin pair and occupy an orbital together.

spontaneous change a change which proceeds by itself without an external source of energy to drive it.

stability constant, K_{stab} the equilibrium constant for the formation of the complex ion in a solvent from its constituent ions or molecules.

standard cell potential the difference in standard electrode potential between two half-cells.

standard conditions conditions of temperature and pressure that must be the same in order to compare moles of gases or enthalpy changes accurately. Standard conditions are a pressure of 10^5 pascals (100 kPa) and a temperature of 298 K (25 °C).

standard electrode potential the electrode potential of a half-cell when measured with a standard hydrogen electrode as the other half-cell.

standard enthalpy change an enthalpy change that takes place under the standard conditions of pressure (10^5 Pa) and temperature (298 K).

standard hydrogen electrode a half-cell in which hydrogen gas at a pressure of 1 atmosphere (101 kPa) bubbles into a solution of 1.00 mol dm^{-3} H$^+$ ions. This electrode is given a standard electrode potential of 0.00 V. All other standard electrode potentials are measured relative to this value.

state symbol a symbol used in a chemical equation that describes the state of each reactant and product: (s) for solid, (l) for liquid, (g) for gas and (aq) for substances in aqueous solution.

stereoisomers compounds whose molecules have the same atoms bonded to each other but with different arrangements of the atoms in space.

stoichiometry the mole ratio of the reactants and products in the balanced equation for a reaction.

strong acid/base an acid or base that is (almost) completely ionised in water.

structural formula the formula that tells us about the atoms bonded to each carbon atom in an organic molecule, e.g. $CH_3CH=CH_2$.

subshells regions within the principal quantum shells where electrons have more or less energy depending on their distance from the nucleus. Subshells are given the letters s, p, d and f.

substitution reaction a reaction in which one atom or group replaces another atom or group in a molecule.

substrate a molecule that fits into the active site of an enzyme and reacts.

successive ionisation energy ΔH_{i1}, ΔH_{i2}, etc.: the energy required to remove the first, then the second, then the third electrons and so on from a gaseous atom or ion, producing an ion with one more positive charge each time. Measured in kJ per mole of ions produced.

superglue poly(methyl cyanoacrylate), $[-CH_2C(CN)CO_2CH_3-]_n$. The monomer easily polymerises in the presence of water vapour.

surroundings in enthalpy changes, anything other than the chemical reactants and products, e.g. the solvent, the test tube in which the reaction takes place, the air around the test tube.

tertiary a tertiary alcohol or halogenalkane has its functional group joined to a carbon atom to which no hydrogens are also joined. An example is 2-methylpropan-2-ol, $(CH_3)_3COH$. A tertiary amine has three organic groups attached to a nitrogen atom, e.g. $(CH_3)_3N$.

tertiary structure the interactions between the side-chains of amino acids in a polypeptide chain, allowing the protein to form a specific three-dimensional shape. Interactions include hydrogen bonding, ionic attractions, induced dipole (van der Waals') attractions and disulfide bridges.

titre in a titration, the final burette reading minus the initial burette reading.

TLC thin-layer chromatography.

TMS tetramethylsilane, $(CH_3)_4Si$; an inert volatile liquid added to an NMR sample to 'zero' the δ scale as $\delta(TMS) = 0.0$.

Tollens' reagent an alkaline solution of Ag^I which is used to test for the presence of a mild reducing agent such as an aldehyde. A positive result is the formation of a silver mirror on the side of the test tube.

triiodomethane reaction a test for the groupings $-CH(OH)CH_3$ or $-COCH_3$ by adding an alkaline aqueous solution of iodine. A positive result is the formation of a pale yellow precipitate of triiodomethane, CHI_3.

unsaturated hydrocarbon a hydrocarbon containing at least one C=C double bond.

van der Waals' forces the weak forces of attraction between molecules caused by the formation of temporary dipoles.

vaporisation the change in state when a liquid changes to vapour.

vapour pressure the pressure exerted by a vapour in equilibrium with a liquid.

weak acid/base an acid or base that is only slightly ionised in water.

zwitterion a compound which contains both a positive ionic group and a negative ionic group. Amino acids form zwitterions: $H_3N^+-CHR-CO_2^-$.

Index

A

Acid dissociation constant, K_a, 137
Acid–base indicator, 138
Activation energy, 40, 53
    endothermic reaction, 40
    exothermic reaction, 40
Acyl chlorides
    formation of, 175
    reactions, 175
        with alcohols, 175
        with amines, 175
        with phenols, 175
        with water, 175
addition polymers. see Polyalkenes
Alcohols
    formulae and properties, 108–109
        boiling points, 108
        carboxylic acid, 108–109
        nucleophilic substitution, 110, 111
    preparation, methods of
        electrophilic addition, 110
        reducing agents, 110
    reactions of, 111–112
        elimination, 112
        of –O–H bond, 112
        oxidation, 111
        triiodomethane, 112
    tests for, 114
    types of, 110, 114
    uses of, 115
    vs. aldehydes, boiling points, 118
Aldehydes, 117, 120
    and ketones, reactions
        condensation reactions, 120
        mechanism of, 119
    oxidation reactions
        acidified dichromate(VI), 120
        Fehling's solution, 120
        Tollens' reagent, 120
    vs. alcohols, boiling points, 118
Aliphatic amines, 179
Alkanes
    preparation, 89
    reactions
        combustion of, 91
        with halogens, 91–92
Alkenes
    oxidation of, 95
    polymerisation of, 96
    preparation, 89
    reactions of
        catalytic hydrogenation, 94
        electrophilic addition, 93
        Markovnikov's rule, 93
Alkylation, acylation, 167
Amides
    properties and reactions, 182
        hydrolysis, 182
        reduction, 182
Amines
    reactions
        as bases, 180
        of diazonium salts, 181
        as nucleophiles, 180
    synthesis of
        aliphatic amines, 179
        aromatic amines, 180
Amino acids, 182–183
Ammonia
    and ammonium ion, 74–75
    Brønsted–Lowry base, 74
    eutrophication, 75
    manufacture of, 74
    nitrate fertilisers, 75
        uses of, 75
    as weak base, 75
Amphoteric oxides, 62
Arenes, 166
    side-chains, reactions
        halogenation of, 170
        oxidation of, 169
Aromatic amines, 180
Atomic radius, 59
Atoms
    electrons in, 8, 12–16
    energy levels, 12
    high ionisation energies, 19
    isotopic symbol, 9
    low ionisation energies, 18
    neutrons, 8
    and nucleons, 9
    particles in, 8
    protons, 8
Avogadro constant, 2
    determination of, 131
Azo, dyes, 181

B

Bacterial fermentation, 98
Bases, amines, reactions of, 180
Batteries, 136
Benzene, structure of, 166
Benzene ring
    reactions of
        alkylation and acylation, 167
        halogenation, 167
        hydrogenation, 167
        nitration, 167
BF_3
    bonding in, 20
    molecule of, 20
Bleach, 72
Boltzmann distribution, 54
    for gas, 55
    temperature, 54
Bond
    breaking and making, 24
    energy, 23
    length, 23
    polarity, 23
Bond energy, 39
Bonding
    in BF_3, 20
    in methane, CH_4, 19
    Period 3 elements, 59
    properties, 24
    types of, 24
Born–Haber cycles, 123–124
    electron affinity, 124
    for $NaCl$, 124
Bromination, 168
Brønsted–Lowry theory, 50
Buffer solutions
    definition, 140
    pH, in blood, 141
    pH of, 140
    uses of, 141
    working, 140

C

^{13}C NMR spectra, 203
Carbocations, 102
Carbonyl compound
    2,4-dinitrophenylhydrazones, 120
    aldehydes, 117
    ketones, 117
    mesomeric, 118
    preparation, methods of, 117
    reduction of, 119
Carboxylic acids, 108–111
    alcohols, oxidation of, 110
    chlorine substituted, 174
    hydrolysis, 110
        of acid, 111
        of nitriles, 110
    molecules, 108
    reaction with
        alcohols, 112
        carbonates, 112
        sodium hydroxide, 112
    reduction to, alcohols, 113
    tests for, 114
    uses of, 114
Catalyst
    heterogeneous, 56, 151–152
    homogeneous, 56, 119, 151
    transition elements, 158
Catalytic cracker, 90
Cell potentials
    calculating, 131–132, 133
Ceramics, 31
CFCs, 106
Chain isomerism, 109
Chain isomers, 83
chemical shift (δ), 203
Chemical tests, functional groups, 211–212
Chiral centre, 83
Chlorine, bromine isotopes, 201
Chromate(VI) ions, 159
Chromatography, types of, 197–198
Complex formation
    ligand, 160
    oxidation number, 160
    shapes of, 160–161
Concentration/time graphs, 146
Condensation polymers
    polyamides, 188
    polyesters, 187
    polymer vs. monomer(s), 189
Conducting polymers, 194
Constructing equations, 4
Covalent bonding, 19–20
    dative, 19
    properties, 24
Crude oil, 89
    fractions from, 90
Crystalline solids
    ionic solids, 28
    metallic solids, 28
Cyanoacrylates. see Glue

D

d orbitals
    and colour, 163–164
    degenerate orbitals, 163
    octahedral complexes, 163
    splitting of, 163
    tetrahedral complexes, 164
Dative covalent bond, 19
Degradable polymers, 194
Deoxyribose, 192
Depolymerisation, 98
Deuterium exchange, 204
Dipeptide, 183
DNA, 192
Dot-and-cross diagram
    for covalent bond, 19
    for NaCl, 18
Drugs, design of, 211

E

Electrode potentials, 131–132
    and halogens, reactivity, 134
    measuring, 132
    oxidation states, 133
    variation of, 135
Electrolysis
    calculations, 130
    substance, liberated
        aqueous solution, 130
        concentration effects, 130
        electrodes, 130
        ions, 130
        molten lead bromide, 129
Electrolyte, 129
Electron affinity, 124
Electron configurations
    3d orbitals, 157
    of Period 4, 157
    transition elements, 157
        properties of, 158
Electron flow
    reaction, feasibility of, 133
Electron pair repulsion theory, 20
Electronegativity, 22
    bond polarisation, 22
    dipole moment, 22
Electronic configuration, 16
Electrons
    behaviour of, 8
Electrophiles, 86, 168
Electrophilic addition, 110
    to alkenes, 101
    mechanism of, 93
    organic reactions, 85
Electrophilic substitution, reactions
    mechanism of, 167–168
Electrophoresis
    of amino acids, 183
    of peptides, 183
Elements
    orientation in, 169
    oxidation states, 159
    physical properties, 158
    redox reactions, 159
Empirical formula, 3
Endothermic reactions, 33–34
Energy levels
    in hydrogen atom, 12–13
    and orbitals, 12–14
Energy profile diagrams, 40, 53
Enthalpy changes, 33
    of atomisation, 37
    of combustion, 33–34
    of formation, 35
    of hydration, 37
    of neutralisation, 37–38
    specific heat capacity, 35
    standard conditions, 34
Entropy, 153
    calculation of, 154
    changes, 153
Enzymes, 56–57
Epoxyresins. see Glue
Esters
    formation of, 112
    hydrolysis of, 114
    uses of, 114
Ethanedoic acid, 175
Ethanol, formation of, 110
Eutrophication, 75
Exothermic reactions, 33–34

F

$F = Le$, 129
Faraday constant, 129, 135
Fehling's solution, 120
Free energy. see Gibbs free energy
Free radicals, 86, 91, 101
Friedel–Crafts reactions, 167
Fuel cells, 135
Functional group isomerism, 109
Functional group isomers, 83

G

Gas–liquid chromatography (GLC), 198
Geometrical isomerism, 83
Giant molecular structures, 29
    diamond and silica, 29
Gibbs free energy
    entropy *vs.* enthalpy changes, 154
    surroundings, 154
    total entropy change, 154
Glue, 193–194
Graphs, concentration/time, 146
Group 2
    carbonates, 65, 125
        decomposition of, 65
    electronic configuration, 64
    hydroxides, 65
    nitrates
        decomposition of, 65
    oxides, 65
    reaction
        with aqueous acid, 64
        with oxygen, 64
        with water, 64
    uses of, 65
Group 2 nitrates, 125
Group 2 sulfates, 126–127

H

H^+ concentration, calculating
    strong bases, 138
    strong monobasic acids, 137
    weak acids, 138
^{1}H NMR, 204
Halogenated hydrocarbons, 72
Halogenation, 167
Halogenoalkanes, 101
    elimination reactions, 105
    environmental impact, 105–106
    preparation methods, 101–102
    properties, 101
    silver nitrate in, 104
Halogens
    as oxidising agents, 68
        chlorine, reaction of, 69
        Fe^{2+}, 69
        and halide ions, 68
    production of, 69
    uses of, 72
        bleach, 72
        halogenated hydrocarbons, 72
        PVC, 72
        water purification, 72
Hess's law, 36
Homologous series, 79
Hydrides
    formation, 69
    physical properties, 69
    reactions of
        halide ions, tests, 71
        nonoxidising acid, 70
    thermal stability, 69
    water, 69
Hydrocarbons
    classes of, 88
    physical properties, 89
Hydrogen bonds, 22
    importance of, 30
    in proteins and DNA, 30
    and solubility, 30
    in water, 30
Hydrogenation, 167
Hydroxides, solubility of, 126–127
Hydroxypropanoic acid, 194

I

Incineration, 98
Indicators, choices of, 140
Infrared spectroscopy, 198–199
Initial rates, calculating, 149
Intermolecular forces
    boiling points, 23
Iodoform. see triiodomethane
Ionic bonding
    dot-and-cross diagram, 18
    properties, 24
Ionic solids, formation of, 126
Ionisation energy, 13–14
    consecutive, 15–16
IR. see infrared spectroscopy
Isomerism
    chain, 89, 109
    complex, 161
    functional group, 109
    of hydrocarbon, 89
    and nomenclature
        alcohols, 109
        halogenoalkanes, 101
    optical, 89, 109
    positional, 109, 166
    primary, 109
    secondary, 109

stereoisomers
- geometrical isomerism, 83
- optical isomerism, 83

structural isomers
- chain isomers, 83
- functional group isomers, 83
- positional isomers, 83

tertiary, 109

Isotopes, 1

Isotopic symbol, 9

K

K_a, 137

Ketones, 117
- and aldehydes, reactions
  - mechanism of, 119
- condensation reactions, 120

Kevlar, 193

Kinetic theory
- of gases, 26
- ideal behaviour, 26
- and ideal gases, 26

K_w, 137

L

Lattice energy, 123
- and decomposition, 126

LDPE, 96

Ligand, 160
- ammonia, 75
- exchange reactions, 162

Liquid state, 27–28
- water temperature, 27–28

M

$[M + 2]^+$ and $[M + 4]^+$ peaks, 201

$[M + 1]^+$ peak, 201

Markovnikov's rule, 93

Mass spectra
- and relative atomic mass calculations, 2–3

Mass spectrometry, 2, 200–202
- molecular fragments, 202
- molecular ion, 201

Mesomeric pair, 118

Metallic bonding, 23
- properties of, 24

Metals, 30–31

Methanoic acids, 174

Mole, 1

Mole and Avogadro constant, 1–2

Mole calculations, 4–5
- combustion data calculations, 5
- reacting gas volumes, 5
- reacting masses, 4

Molecular formula, 3

Molecular orbital, 21

Molecular shape
- σ and π bonds, 21

Molecules
- and bond angles, 20–21
- shapes of, 20–21

Molten lead bromide, 129

N

Nernst equation, 135
- combination of, 135
- complex electrodes, 135
- electrode system, 135
- metal dipping, 135

Neutrons, behaviour of, 8

Nitration
- reactions of
  - benzene ring, 167
  - electrophilic substitution, 168

Nitrile, 104

Nitrogen, unreactivity of, 74

Nitrogen oxides
- internal combustion engine, 76
- nitrogen cycle, 76

Nomenclature
- and isomerism, 166–167

Nuclear magnetic resonance (NMR) spectroscopy, 203–204
- basis of, 203
- organic molecules, 203

Nucleons
- mass number, 9
- proton number, 9

Nucleophiles
- amines, reactions of, 180
- organic reaction, 86, 180

Nucleophilic addition
- of HCN, 119

Nucleophilic substitution
- alcohol, preparation, 110
- halogenoalkanes, 102
- reactions
  - alcohol, 111
  - halogenoalkanes, 102–103

Nucleus, 8

O

Octahedral complexes, 163
Optical isomerism, 83–84, 109
Orbital, 12
Order of reaction, 145, 147
    half-life, 147
    rate method, 148
Organic compounds
    classes of, 79–80
    formulae, 80
        displayed formula, 80
        empirical formula, 80
        molecular formula, 80
        skeletal formula, 80
        stereochemical formula, 80
        structural formula, 80
    methylpropanoic acid, formulae for, 80
    names of, 81
        stem, 81
        stem-end, 81
Organic reactions
    types of
        addition reaction, 86
        condensation reaction, 86
        electrophile, 86
        electrophilic reaction, 85
        elimination reaction, 86
        hydrolysis reaction, 86
        substitution reaction, 86
Organochlorine compounds, reactivities, 175–176
Oxidation
    alcohols, reaction, 110, 111
    aldehydes reaction, 120
    of alkenes, 95
    of ethanedioic acids, 175
    with manganate(VII), 159
    organic reactions, 86
    redox processes, 42
    of side-chain, 169
Oxides, and chlorides, Period 3, 61

P

π bonds, 21
Partition coefficient, 142–143
Peptide bond, electrophoresis, 183
Period 3
    chlorides, 61
    elements
        oxygen, reaction, 60
        water, reaction, 60
    periodicity of, 59
Period 3, reactions of, 61
    chlorides with water, 62
    oxides
        aqueous acid, 62
        aqueous sodium hydroxide, 62
        sodium hydroxide, 62
Periodicity
    atomic radius, 59
    of Period 3 elements, 59
pH, 137
    calculating
        strong bases, 138
        strong monobasic acids, 137
        weak acids, 138
    titrations, changes of, 138
        strong acid *vs.* weak base, 139
        weak acid *vs.* strong base, 139
Phenol
    acidity, 170
    electrophilic substitution, 171
    esterification, 171
    properties and reactions of, 170–172
pK_a, 137
PLA. *see* Polylactic acid
Polyalkenes, 96, 98, 187, 193
    polymers
        by ethenes, 96–97
Polyalkenes, PTFE, 193
Polyamides, 188
Polydentate ligands
    bidentate, 161–162
    exchange reactions, 162
Polyesters, 187
poly(ethene), properties of, 96
Polylactic acid, 194
Polymers
    disposal of, 98
    by ethenes, 96–97
    properties of, 193–194
Polypeptide, 183
Positional isomerism, 109
Positional isomers, 83
Primary alcohols, 109
Primary alcohols, oxidising agent, 111
Primary structure, 190
Principal quantum number
    s, p and d orbitals, 12–13
    spin paired, 13
Proteins, 190–192
    primary structure, 190
    secondary structure, 190
    tertiary structure, 191

Protons
    behaviour of, 8
PTFE, 193
$pV = nRT$, 27
PVC, 72

R

Radical, 86
Rate
    activation energy, 53
    concentration, effects of, 54
    of reaction, 53
    temperature, effect of, 55
Rate constants, 145
    calculating, 149
    temperature, effect of, 151
Rate-determining step, 150
Rate equations, 145
Rate of reaction, 53
    techniques, for studying, 150
        initial rate, 151
        progress of, 151
Reaction mechanisms, 150
Reaction types, recognising, 212–214
Recycling, 98
Redox equations, balancing, 134
Relative masses
    atomic mass, 1
    isotopic mass, 1
    molecular mass, 1
Resources, materials as
    ceramics, 31
    metals, 30
Retardation factors, 197–198
R_f, see Retardation factors

S

σ bonds, 21
Secondary alcohols
    oxidising agent, 111
Secondary structure, 190
Simple molecular structures, of iodine, 29
S_N1 mechanism, 102
S_N2 mechanism, 102
Solubility product, 142
    and common-ion effect, 142
    concentration, calculating, 142
Solutions, formation of, 126
sp^3 hybridisation, 21
Spin pairing, 13
Spontaneous change, 153
Stability constant, K_{stab}, 162
Standard electrode potential, 132
Standard hydrogen electrode, 132
Structure, bonding, 59
Sulfur dioxide
    from human activity, 77
    from natural, 77
    uses of, 77
Synthetic routes, 209–211

T

Tertiary alcohols, 109
Tertiary structure, 191
Thin-layer chromatography (TLC), 197
Tollens' reagent, 120
Transition elements
    electron configurations, 154
    properties of, 158
Triiodomethane, 112
    reaction, 121

V

Van der Waals' forces, 23, 30
Vaporisation, 28

W

Water purification, 72

Z

Zwitterion, 183